ILLINOIS

A GEOGRAPHICAL SURVEY

Illinois Geographical Society

Ronald E. Nelson, *Editor*

George Huseman, Assistant Editor

KENDALL/HUNT PUBLISHING COMPANY
4050 Westmark Drive   Dubuque, Iowa 52002

CONTENTS

PREFACE

Two decades ago a team of Illinois geographers was assembled under the auspices of the Illinois Geographical Society to prepare a book on Illinois that might be used as a text and a reference work by teachers in the state. That volume, *Illinois: Land and Life in The Prairie State,* published in 1978 by Kendall/Hunt Publishing Company, was the predecessor of the current book. All of the original chapter authors, with the notable exception of the late Martin W. Reinemann of Northern Illinois University, have again contributed their work to the new volume. In the preparation of two chapters, however, new authors have been added to the writing team: Edwin N. Thomas (Chapter 5) and Donald W. Clements (Chapter 7). Furthermore, a new appendix has been prepared for the current volume by Paul S. Anderson, Albert D. Hyers, and Michael D. Sublett.

For *Illinois: A Geographical Survey*, we have retained the organization and the approach of its predecessor. A series of topical chapters, each covering the state as a whole, is followed by regional chapters on the two most populated segments of Illinois—the Chicago Metropolitan Area and Metro East (the Illinois portion of the St. Louis Metropolitan Area). Although the chapters complement one another and collectively provide a thorough survey of the state, each is capable of being used independently by a reader interested in only the topic or region it covers.

Our goal for this book, like that for its predecessor, is to further the reader's understanding of Illinois in three ways:

1. By comprehending the state's landscape variety through examination of geographical patterns of environmental and human phenomena
2. By gaining an awareness of the intimate relationships between human populations and their habitats in the state
3. By tracing the development of these landscapes and relationships through time in order to determine how the state's present land and life came to be

To support the written descriptions and analyses, well over a hundred maps are included in the book.

As before, we have adopted a broad scope, and our search of the past for origins of present conditions and features encompasses both the human history and the geologic history of Illinois. We have endeavored to support our extensive coverage with adequate detail for the reader to acquire a reasonably thorough understanding of the state. To cover an area as large and complex as Illinois in a single volume requires abbreviated treatment of some topics and the omission of others, however. To guide the reader who is interested in obtaining additional or more specialized information, we have provided an extensive bibliography at the end of the book.

In addition to the authors (who are identified with their contributions on later pages), many people in other roles assisted with the completion of this book and deserve to have their efforts acknowledged. George Huseman,

a teacher at Brown County High School, provided editorial assistance with some of the
chapters. Word processing of the entire manu-
script was accomplished by Cindy Hare, the
Geography Department secretary at Western
Illinois University. Most of the original maps
were prepared in the Western Illinois University Cartographic Laboratory by Scott Miner.
Assistance was provided by graduate student
Sara Wood in locating data and information
sources and in generating computer maps for
one chapter. Also, a number of colleagues
around the state donated photographs and
slides to be considered for inclusion in the
book. To all involved with this project, many
thanks for your valuable input.

Ronald E. Nelson

AN INTRODUCTION
TO THE PRAIRIE STATE

Ronald E. Nelson
Western Illinois University

The person who is only casually acquainted with Illinois may tend to associate the state with the legend of Abraham Lincoln, whose restored home village of New Salem and the state capital of Springfield where he practiced law and became a national political figure have become favored attractions among tourists. Others may be inclined to link the state with the seemingly endless fields of corn and other crops now occupying the vast

FIGURE 1-1. New Salem, where young Abraham Lincoln lived and worked, is now a popular attraction for tourists. The village has been meticulously restored to reflect the nature of pioneer life in central Illinois during the mid-nineteenth century. (Courtesy Illinois Office of Tourism.)

grasslands that posed such a difficult habitat for nineteenth century pioneer settlers and gave Illinois its nickname—the Prairie State. In the minds of many, Illinois is closely linked with Chicago and that city's political and social reputation. The inadequacies of such tendencies to stereotype the state should become apparent to readers as they progress through this book; Illinois is a complex entity that defies easy characterization.

As perceived by most Illinoisans, the state is divided into two segments—the Chicago Metropolitan Area and "Downstate"—and the inhabitants of each tend to be suspicious of the other area. The Chicago Metropolitan Area, of course, contains the state's greatest industrial complex and urban agglomeration; its inhabitants represent nearly two-thirds of Illinois' total population. Although Downstate Illinois encompasses numerous smaller cities, it is fundamentally an agricultural region of enormous productivity. The massive output of its farmers enables Illinois to be one of the country's leading states in agricultural income. Nevertheless, agriculture is far less important than other sectors of the state's economy, and the wealth of Illinois is concentrated in the Chicago Metropolitan Area.

In terms of area, Illinois is not distinctive; with about 56,400 square miles, it ranks only 24th among the 50 states. At its maximum breadth from the Mississippi River near Quincy to the boundary with Indiana, the distance is barely over 200 miles. On the other hand, the north-south axis from Cairo at the confluence of the Ohio and Mississippi rivers to the Wisconsin boundary is considerably greater—about 380 miles. As a consequence of this pronounced elongation (involving over 5° of latitude), there are significant climatic differences between southern and northern Illinois. Winters are distinctly milder and the growing season begins about a month earlier in the southern part of the state than in the north.

In addition to such climatic variations, other environmental variety contributes to the diversity of the Illinois landscape. Fairly rugged and scenic terrain, with a local relief of several hundred feet, exists in the unglaciated ("driftless") areas of the state: the Shawnee Hills south of Carbondale; the Galena area in the far northwest; and part of the western margin, especially Calhoun County. Although the remainder of the state is a glaciated plain with low relief, it contains several distinctive features including the lake plain on which Chicago was built; floodplains along the major rivers; and glacial moraines, particularly in northern and central Illinois. The extensive prairie grasses and occasional marshes and swamps that once were distinctive elements of the northern and central Illinois environment are now gone, however. In converting these areas into productive croplands, the state's early farmers plowed under virtually all of the prairie grasses and dug drainage ditches and installed tile networks to carry away excess water. Despite being cut over extensively, the forests of southern Illinois have survived to a greater degree. A large part of the timberland in that part of the state is now contained in the Shawnee National Forest.

The Illinois environment also includes an assortment of mineral resources—bituminous coal, oil and natural gas, stone, sand and gravel, and fluorspar—that have been important economic assets for the state. Coal is the dominant mineral, presently accounting for more than two-thirds of the value of all minerals produced in Illinois. The vast majority of coal mining operations are in the southern one-third of the state. In recent years, Illinois has ranked fifth among the nation's coal producing states. The relatively high sulfur content of Illinois coal, representing potential air pollution when burned, has diminished its attractiveness as a modern fuel. Oil is the second leading mineral in Illinois, providing nearly 15 percent of the state's income from mineral production. It, too, is mostly concentrated in southern Illinois.

Much of the landscape variety in the state is vividly portrayed on James A. Bier's physiographic map of Illinois (Figure 1-2). Recognizing the educational potential of this map, Professors Paul S. Anderson, Albert D. Hyers, and Michael D. Sublett have written a series of exercises and questions focused on it that appears as Appendix A in this book. Their work incorporates an inquiry and discovery approach, and it is complete with answers to the questions posed. Teachers of geography, social studies, and environmental science should find this appendix to be a practical addition to the informational chapters in this book.

THE SIGNIFICANCE OF LOCATION

The character of Illinois is to a great extent a reflection of its location. Positioned in the midsection of the country, the state is adjacent not only to Lake Michigan but also to the Mississippi and Ohio rivers (Figure 1-3). In earlier times, these waterways served as routes for explorers and pioneer settlers to reach Illinois and they provided a means for early farmers to ship grain and livestock products to market. This excellent access to water transportation, now involving huge barge tows on the rivers and even ocean vessels on the Great Lakes, has enabled Illinois to become a leading exporter of a variety of agricultural and industrial products. In addition, relatively inexpensive water transportation has facilitated the assembly of such bulky industrial raw materials as iron ore for Chicago area steel mills and petroleum for southwestern Illinois oil refineries. Chicago, now the leading port city on the Great Lakes, has had its accessibility by ocean vessels enhanced with completion of the St. Lawrence Seaway in 1959.

Partly because of marketing advantages associated with its central location within the country, Illinois has developed into a leading

FIGURE 1-3. The development of Illinois has been aided by the state's central location within the United States and its direct access to major rivers and the Great Lakes.

FIGURE 1-4. A view of Chicago from the air. The wealth of Illinois is markedly concentrated in the city and its suburbs. (Illinois Geographical Society)

industrial and commercial state. Numerous firms involved in manufacturing and marketing a variety of products such as steel, construction equipment, agricultural machinery, automobiles, electrical appliances, and foodstuffs have chosen to locate in Illinois because of the state's situational advantages. Central location was also a major factor in the establishment of the nation's leading mail-order firms, notably Sears, Roebuck and Company, Montgomery Ward, and J.C. Penney, in Chicago.

Illinois is positioned in a part of the North American continent that was covered by shallow seas during the geologic period known as the Paleozoic. Formed on the bottom of these ancient water bodies were the layers of sedimentary rock that now exist beneath the surface of most of Illinois and are exposed at the surface in a few areas, particularly in the southern and northern extremities of the state. Contained within these sedimentary strata are the major mineral resources of Illinois, notably bituminous coal and petroleum. During a much more recent geologic period—the Pleistocene, which began about two million years ago—four enormous continental glaciers formed in Canada around the margins of Hudson Bay. These ice sheets, each separated in time by a warmer and drier interglacial period, advanced outward from their source regions and eventually covered parts of northern United States. At least two of the Pleistocene glaciers covered large segments of Illinois, the margin of one reaching as far south as Giant City State Park near Carbondale. As a result of being located within the glaciated area of North America, most of Illinois was covered by a blanket of glacial debris as the ice melted and had its surface form and drainage pattern altered by the advance and subsequent melting of the ice sheets. Rough terrain in Illinois today is mostly restricted to small areas in the northwestern corner and the far southern parts of the state that escaped glaciation.

Because Illinois occupies a middle latitude position in the interior of a large landmass, its climate involves four distinct seasons, a large annual temperature range, and moderate amounts of precipitation. Maximum summer temperatures are consistently high and occasionally exceed 100°F, while in winter the thermometer may plunge to subzero readings. The moderately long growing season (ranging from about 200 days in the far south to 160 days in the north) and the moderately abundant precipitation (varying from an average of 45 inches in the south to 34 inches near the Wisconsin border) are nearly ideal for most middle latitude crops. As throughout eastern United States, tropical air masses from the Gulf of Mexico provide most of the moisture that falls as precipitation in Illinois. Dry and relatively cool conditions occur when polar air masses from Canada migrate across Illinois. Its position between the Gulf and Canada therefore allows Illinois to experience the alternating passage of tropical and polar air masses and the consequence of frequent weather changes.

SETTLEMENT AND EARLY DEVELOPMENT

More than 300 years have passed since the land of Illinois was first observed by European explorers (Louis Jolliet and Jacques Marquette in 1673) and, of course, several centuries earlier Indian civilizations

FIGURE 1-5. A Norwegian freighter at Chicago to be loaded with grain. (Illinois Geographical Society)

flourished in portions of the Illinois and middle Mississippi River valleys. The first permanent European settlements in what is now Illinois were French forts and small villages (Cahokia, Kaskaskia, Fort de Chartres, Prairie du Rocher) founded at the turn of the eighteenth century along the Mississippi River in the southwest. Although these French communities were the most significant European settlements in Illinois before 1800, they never attained a population of more than a few hundred. Not until several years after Illinois became a state in 1818 did the number of settlers and degree of land development reach a substantial level. In 1810, the year of the first census enumeration in Illinois Territory, the number of inhabitants was only 12,282 (Table 1-1).

TABLE 1-1. Population of Illinois, 1810–1990.

Year	Population	Percent Increase
1810*	12,282	349.2
1820	55,162	185.2
1830	157,445	202.4
1840	476,183	78.8
1850	851,470	101.1
1860	1,711,951	48.4
1970	2,539,891	21.1
1890	3,077,871	24.3
1900	4,821,550	26.0
1910	5,638,591	16.9
1920	6,485,280	15.0
1930	7,630,654	17.7
1940	7,897,241	3.5
1950	8,712,176	10.3
1960	10,081,158	15.7
1970	11,110,285	10.2
1980	11,427,409	2.8
1990	11,430,602	0.0

*Illinois Territory

Following the War of 1812 there occurred expansion of settlement in wooded southern Illinois, and by 1820 the settled area containing at least two people per square mile extended completely across the southern part of the state from the Mississippi to the Wabash River (Figure 1-6). The northward push of settlers by this time also resulted in a population density of at least two people per square mile in the Wabash Valley, the Mississippi Valley to a point about 50 miles north of the mouth of the Illinois River, the Big Muddy Valley, and the lower valley of the Kaskaskia.

The frontier was largely in central Illinois by 1820 when the population of the state totaled 55,162. The Sangamon country and the southern part of the Military Tract in western Illinois were settled rapidly during the 1820s. In addition, a detached nucleus of settlement, the Galena Lead District in the northwestern corner of the state, had become established by this time. All of these areas as well as the valley of the Illinois River to a point about midway between its mouth and Lake Michigan had attained at least two people per square mile by 1830 (Figure 1-7). Illinois numbered 157,445 inhabitants in 1830, an increase over the 1820 figure of nearly three fold.

Although a complex of factors influenced the areal pattern of settlement in Illinois up to the 1830s, most authorities are in agreement that accessibility was of greatest importance. The Ohio River, the paramount route to the West during these early decades of settlement, led to southern Illinois; the northern part of the state, in contrast, was distinctly isolated prior to the 1830s. In addition to accessibility, the early settlement pattern was influenced by the opening of lead deposits in the Galena area and the tendency of pioneers to locate in or close to timber. The settlers who reached Illinois prior to the 1830s were predominantly from the Upland South and tended to have a greater aversion for the prairies of the central and northern parts of the

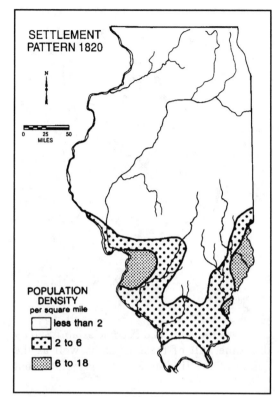

FIGURE 1-6. Settlement pattern, 1820. (Adapted from a map in Harlan H. Barrows, *Geography of the Middle Illinois Valley,* 1910.)

state than did their successors. The need for timber to provide material for such pioneering necessities as shelter, fuel, and fencing was critical, and the prairies were notorious for their poor drainage, seasonal fires, and isolation.[1] It is clear that all but a few of the early settlers located in or near timbered areas.

The delay in the settlement of northern Illinois was not entirely because of its inaccessibility and domination by prairie, however. Not until after the infamous Black Hawk War in 1832 did white settlers feel that the Indian threat was eliminated. (In reality, the Indians did not pose a significant threat to nineteenth century pioneers in Illinois because they were few in number and their civilizations had seriously deteriorated by that

time.) Also, early pioneers found the acquisition of land titles difficult because the government was tardy in establishing land offices in northern Illinois and often the choicest tracts had been obtained relatively early by land speculators. Many early pioneers, by necessity or choice, became squatters and improved land that they did not legally own.

The vast expansion of the settled area and the great increase in population in Illinois during the 1830s and 1840s was facilitated by the opening up of transportation routes to the northern part of the state, particularly the establishment of steamboat service on the Great Lakes; accelerated sales of public lands; and the development of a steel plow to ease the task of breaking the tough prairie sod. In 1840 the population reached 476,183 and by 1850

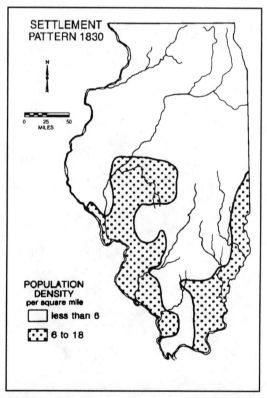

FIGURE 1-7. Settlement pattern, 1830. (Adapted from a map in Harlan H. Barrows, *Geography of the Middle Illinois Valley,* 1910.)

Illinois could claim 851,470 inhabitants. Between 1830 and 1850 only two areas failed to surpass a density of six people per square mile, one northwest and the other southeast of the Illinois Valley (Figs. 1-8 and 1-9). The largest of these two areas, the Grand Prairie of eastern Illinois, was the most nearly woodless, poorly drained, and inaccessible part of the state. Frontier conditions persisted here into the decade of the 1850s, longer than in any other part of the state.

The settlement of northern Illinois was conducted largely by pioneers from northeastern United States and foreign immigrants. Most of them travelled westward by way of the Great Lakes, landing at Chicago and subsequently fanning out over the northern prairies. Douglas McManis has found that they

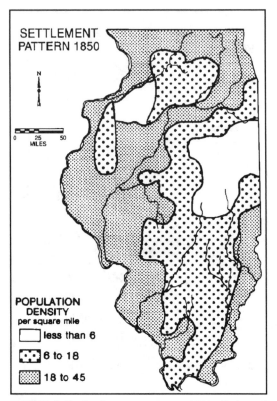

FIGURE 1-9. Settlement pattern, 1850. (Adapted from a map in Harlan H. Barrows, *Geography of the Middle Illinois Valley*, 1910.)

were inclined to place a higher evaluation on the prairie and were more willing to venture out onto the grasslands than their earlier counterparts from the Upland South.[2] Numerous placenames and many cultural institutions in Chicago and other parts of northern Illinois today clearly reflect the "Yankee" and foreign origin of the area's earliest settlers.

The development of Illinois during the 1850s was highlighted by transportation improvements, particularly the construction of railroads; the development of Chicago as the state's premier urban center; final organization of the state's 102 countries (Figure 1-10); and the passing of frontier conditions. Chicago was incorporated in 1833 and by 1850 its population had grown to about

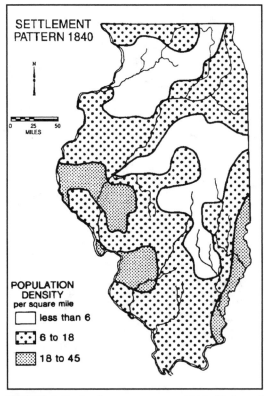

FIGURE 1-8. Settlement pattern, 1840. (Adapted from a map in Harlan H. Barrows, *Geography of the Middle Illinois Valley*, 1910.)

ILLINOIS COUNTIES AND (COUNTY SEATS)

1. ADAMS (QUINCY)
2. ALEXANDER (CAIRO)
3. BOND (GREENVILLE)
4. BOONE (BELVIDERE)
5. BROWN (MT. STERLING)
6. BUREAU (PRINCETON)
7. CALHOUN (HARDIN)
8. CARROLL (MT. CARROLL)
9. CASS (VIRGINIA)
10. CHAMPAIGN (URBANA)
11. CHRISTIAN (TAYLORVILLE)
12. CLARK (MARSHALL)
13. CLAY (LOUISVILLE)
14. CLINTON (CARLYLE)
15. COLES (CHARLESTON)
16. COOK (CHICAGO)
17. CRAWFORD (ROBINSON)
18. CUMBERLAND (TOLEDO)
19. DEKALB (SYCAMORE)
20. DEWITT (CLINTON)
21. DOUGLAS (TUSCOLA)
22. DUPAGE (WHEATON)
23. EDGAR (PARIS)
24. EDWARDS (ALBION)
25. EFFINGHAM (EFFINGHAM)
26. FAYETTE (VANDALIA)
27. FORD (PAXTON)
28. FRANKLIN (BENTON)
29. FULTON (LEWISTOWN)
30. GALLATIN (SHAWNEETOWN)
31. GREENE (CARROLLTON)
32. GRUNDY (MORRIS)
33. HAMILTON (MCLEANSBORO)
34. HANCOCK (CARTHAGE)
35. HARDIN (ELIZABETHTOWN)
36. HENDERSON (OQUAWKA)
37. HENRY (CAMBRIDGE)
38. IROQUOIS (WATSEKA)
39. JACKSON (MURPHYSBORO)
40. JASPER (NEWTON)
41. JEFFERSON (MT. VERNON)
42. JERSEY (JERSEYVILLE)
43. JO DAVIESS (GALENA)
44. JOHNSON (VIENNA)
45. KANE (GENEVA)
46. KANKAKEE (KANKAKEE)
47. KENDALL (YORKVILLE)
48. KNOX (GALESBURG)
49. LAKE (WAUKEGAN)
50. LASALLE (OTTOWA)
51. LAWRENCE (LAWRENCEVILLE)
52. LEE (DIXON)
53. LIVINGSTON (PONTIAC)
54. LOGAN (LINCOLN)
55. MCDONOUGH (MACOMB)
56. MCHENRY (WOODSTOCK)
57. MCLEAN (BLOOMINGTON)
58. MACON (DECATUR)
59. MACOUPIN (CARLINVILLE)
60. MADISON (EDWARDSVILLE)
61. MARION (SALEM)
62. MARSHALL (LACON)
63. MASON (HAVANA)
64. MASSAC (METROPOLIS)
65. MENARD (PETERSBURG)
66. MERCER (ALEDO)
67. MONROE (WATERLOO)
68. MONTGOMERY (HILLSBORO)
69. MORGAN (JACKSONVILLE)
70. MOULTRIE (SULLIVAN)
71. OGLE (OREGON)
72. PEORIA (PEORIA)
73. PERRY (PINCKNEYVILLE)
74. PIATT (MONTICELLO)
75. PIKE (PITTSFIELD)
76. POPE (GOLCONDA)
77. PULASKI (MOUND CITY)
78. PUTNAM (HENNEPIN)
79. RANDOLPH (CHESTER)
80. RICHLAND (OLNEY)
81. ROCK ISLAND (ROCK ISLAND)
82. ST. CLAIR (BELLEVILLE)
83. SALINE (HARRISBURG)
84. SANGAMON (SPRINGFIELD)
85. SCHUYLER (RUSHVILLE)
86. SCOTT (WINCHESTER)
87. SHELBY (SHELBYVILLE)
88. STARK (TOULON)
89. STEPHENSON (FREEPORT)
90. TAZEWELL (PEKIN)
91. UNION (JONESBORO)
92. VERMILLION (DANVILLE)
93. WABASH (MT. CARMEL)
94. WARREN (MONMOUTH)
95. WASHINGTON (NASHVILLE)
96. WAYNE (FAIRFIELD)
97. WHITE (CARMI)
98. WHITESIDE (MORRISON)
99. WILL (JOLIET)
100. WILLIAMSON (MARION)
101. WINNEBAGO (ROCKFORD)
102. WOODFORD (EUREKA)

• Dot Represents County Seat Location

FIGURE 1-10. Illinois counties and county seats. The division of the state into 102 counties was completed in 1859. Only 19 counties had been organized 40 years earlier.

30,000. Founded on the shore of Lake Michigan, the young and bustling city was joined with the Illinois River by the Illinois and Michigan Canal in 1848. Its role as the dominant transportation center of the midcontinent was firmly established by 1855 when it became the focus of 10 railroad trunk lines. The railroads linked Chicago with the rapidly developing farm lands of northern Illinois and the markets of the East, resulting in the city's role as a major agricultural processing and distribution center. The building of the enormous Chicago stockyards reflects that function. The railroads also gave access to previously isolated prairies and provided a means of importing wood products in areas where timber was scarce. The last remaining major unoccupied areas of Illinois were thereby settled and frontier conditions were superseded on the eve of the Civil War. In 1860 the federal census recorded 1,711,951 inhabitants of Illinois.

The disposal of the public domain in Illinois was carried out primarily by two means:

1. Cash sales of land at government land offices in various parts of the state
2. Land bounties awarded to the enlisted noncommissioned veterans (or their heirs) of the War of 1812 in the Military Tract located between the Mississippi and Illinois rivers (Table 1-2)

Cash sales, usually for $1.25 per acre, was the dominant means, accounting for over half the land converted to private ownership in the state. The other methods of disposal of land, of which railroad grants and swamp and saline land grants were the most important, involved relatively small acreages. Homesteading was insignificant in Illinois, as nearly all the land was entered prior to the passage of the first homestead law in 1862. The chronology of

land sales in Illinois (Table 1-3) closely paralleled that in the country as a whole during the period between the War of 1812 and the Civil War. Sales reached their highest levels immediately preceding the economic crises of 1819, 1837, and 1857. It has been pointed out that the upward trends in land sales were closely associated with speculation and increases in business activity.[3]

TABLE 1–2. Disposal of Federal Land in Illinois.

Method of Disposal	Percent of Total Area
Cash Sales	56.1
Military Bounties	26.9
Homesteads	0.1
Miscellaneous	0.7
State Grants:	
Swamp and Saline Lands	4.5
Educational Grants	2.9
Internal and River Improvements and Public Buildings	1.5
Railroad Construction Grants	7.3
	100.0

Source: Adapted from Allan G. Bogue, *From Prairie to Corn Belt: Farming on the Illinois and Iowa Prairies in the Ninteenth Century* (Chicago: University of Chicago Press, 1963), p. 30. Bogue acknowledges that these statistics are approximations but is of the opinion that they are reasonably accurate.

Maximum annual land sales in both the state and the nation occurred in 1836, a year of widespread speculation and prosperity. Following the depression of 1837, on the other hand, the decade of the 1840s was a time of tight money and modest land sales in Illinois. Improvements in economic conditions were accompanied by increases in sales in the early 1850s, but by the middle of that decade relatively little unclaimed land remained for purchase in Illinois.

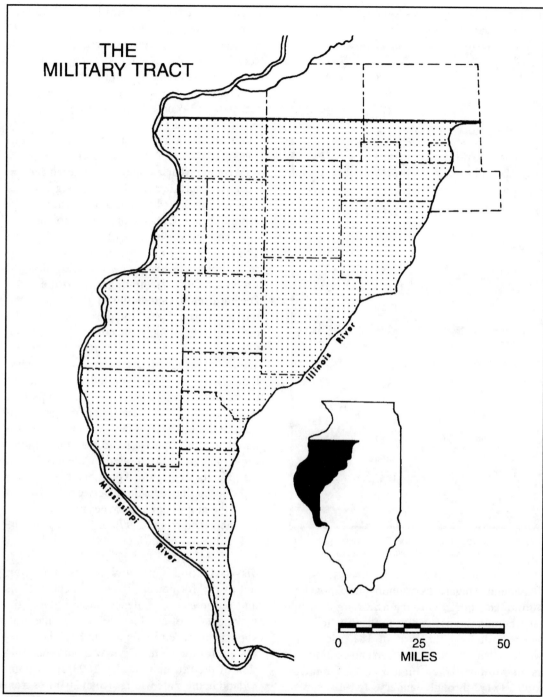

THE
MILITARY TRACT

FIGURE 1-11. The Illinois Military Tract. The federal government granted land bounties totalling nearly 3 million acres between the Illinois and Mississippi rivers to vererans of the War of 1812.

TABLE 1–3. Public Land Sales in Illinois
(in thousands of dollars).

Year	Amount	Year	Amount
1814	168	1838	983
1815	53	1839	1,421
1816	207	1840	492
1817	572	1841	440
1818	1,491	1842	554
1819	611	1843	520
1820	87	1844	616
1821	64	1845	611
1822	335	1846	600
1923	76	1847	615
1824	58	1848	374
1825	82	1849	319
1826	110	1850	313
1827	81	1851	421
1828	121	1852	492
1829	282	1853	1,218
1830	402	1854	1,562
1831	420	1855	897
1832	261	1856	473
1833	381	1857	155
1834	440	1858	12
1835	2,688	1859	12
1836	4,003	1860	.....
1837	1,271		

Source: Adapted from Vernon Carstensen (ed.) *The
Public Lands* (Madison: University of Wisconsin Press,
1963), pp. 234–5.

Among the various methods of disposal of
public land in Illinois, the awarding of military bounty lands was second only to cash
sales. Following the War of 1812, lands in
western Illinois were surveyed and designated
as the Military Tract (Figure 1-11). From October, 1817, through January, 1819, the War
Department issued patents to veterans for
over 2.8 million acres in the Military Tract,
according to calculations by Theodore
Carlson.[4] An individual holding a patent was
allowed to draw by lot a quarter section (160
acres) in the Military Tract. The great majority of the veterans holding rights to bounty
lands decided to sell or trade those rights
rather than become actual settlers, however.
In this way, large segments of the Military
Tract were acquired for a low price by eastern
speculators.

With the prairies finally conquered and the
Civil War terminated, the following decades
brought rapid growth and development interrupted by temporary economic problems and
outbreaks of worker dissatisfaction in Illinois.
The availability of jobs, particularly in railroad construction and new factories, brought
waves of European immigrants to the state.
During each decade between 1870 and 1900,
the population grew by more than 20 percent—a rate of increase not exceeded since
(Table 1-1). As farmers improved the land and
acquired pieces of horse-drawn machinery,
their increased production led to a surplus of
agricultural commodities. As a consequence,
the prices of farm products declined and agriculture experienced a depression that culmi-
nated in the Populist Revolt as the century
drew to a close. Stimulated by the Civil War,
manufacturing grew at a phenomenal rate during succeeding decades. Between 1870 and
1890 the number of employees in manufacturing increased from 82,979 to 312,198 and the
net value of manufactured products grew
from $78,020,595 to $379,621,191.[5] Worker
dissatisfaction with low wages, however,
led to the formation of labor organizations
and outbreaks of violence, most notably the
Haymaker Riot in Chicago. Perhaps the most
significant accomplishment in Illinois immediately following the Civil War was construction of the railroad net. By 1893 every point
of land in the state was less than 20 miles from
a railroad and 85 percent of the land was
within four miles of a track.[6]

From the end of the Civil War to the beginning of the twentieth century, Illinois made the transition from a predominantly rural and agricultural state to one in which a majority of the people lived in urban places and engaged in nonfarm occupations. With transportation improvement brought on by the boom in railroad construction, with rapid industrial growth, and with the surge of immigration to swell population numbers, small towns and villages became bustling cities in a matter of only a few years. Chicago recovered from its most famous disaster, the fire of 1871 that left nearly 100,000 people homeless, and attained a population of over one and a half million by the end of the century. Despite a high rate of population growth for the state as a whole, predominantly rural counties began losing population in the 1870s; by 1900 only 45.7 percent of the state's 4.8 million inhabitants were classified as rural. Although Illinois would continue to be one of the country's leading agricultural states, an urban-industrial way of life had moved to the forefront by the beginning of the new century.

ILLINOIS IN THE TWENTIETH CENTURY

Illinois has amassed a record of impressive demographic growth and economic development during the twentieth century. An assortment of events—most notably two world wars, the Great Depression of the 1930s, conflicts in Korea and Vietnam, social restructuring, and economic competition from new sources—have caused irregularities in these trends, but the state's overall direction has been marked by expansion and improvement.

Population Growth and Urbanization

Although the high rates of population growth of the nineteenth century have not been maintained, the number of Illinois inhabitants has increased with each decade of the twentieth century (Table 1-1). Natural population increase was heavily supplemented by the arrival of immigrants primarily from Europe during the early decades of the century. The depression years between 1930 and 1940 recorded a sharp reduction in the rate of demographic growth, but recovery followed World War II in the form of a "baby boom" and the resumption of large-scale immigration. Asians and Hispanics have formed the dominant foreign immigrant groups since the 1950s. African-Americans, migrating to Illinois from the South, also have been an important supplement to the state's population growth. This group was first attracted to Illinois on a significant scale by World War I labor shortages in the industrial cities, particularly East St. Louis and Chicago. Between 1910 and 1920 the African-American population of Chicago more than doubled, and now African-Americans account for about 40 percent of the city's total inhabitants. The 1990 census recorded 1.7 million blacks in Illinois—about 15 percent of the state's population.

Since 1970 the population of Illinois has increased at the slowest rate in the state's history (Table 1-1). In fact, the census in 1990 found a mere 3,193 more people than in 1980—an increase of less than 0.1 percent over the decade. Undoubtedly the state's sluggish economy is part of the explanation, but changes in personal attitudes and goals also are important. Delayed marriages, an increase in the proportion of women undertaking careers outside the home, and greater family planning have significantly reduced birth rates and lowered the rate of population growth. A return to the burgeoning population increases of earlier decades in Illinois seems highly unlikely.

Throughout the twentieth century the urban population of Illinois has exceeded the number of people in rural areas. Mechanization of agriculture has greatly reduced the need for farm workers, while a growing number of jobs in industry and services has attracted people to urban centers. Consequently, predominantly rural counties have suffered a persistent population decline. Population loss was experienced by 56 counties between 1900 and 1910, 63 counties between 1920 and 1930, 49 counties between 1960 and 1970, and 50 counties between 1970 and 1990. On the other hand, urban population has increased from 54.3 percent of the state's total in 1900 to 84.6 percent in 1990. There were 57 Illinois cities with 30,000 or more inhabitants in 1990, most

TABLE 1-4. Cities in Illinois With More Than 30,000 People in 1990.

Rank	City	Population	Rank	City	Population
1.	Chicago	2,783,726	29.	E. St. Louis	40,944
2.	Rockford	139,426	30.	Bolingbrook	40,843
3.	Peoria	113,504	31.	Rock Island	40,552
4.	Springfield	105,227	32.	Normal	40,023
5.	Aurora	99,581	33.	Quincy	39,681
6.	Naperville	85,351	34.	Lombard	39,408
7.	Decatur	83,885	35.	Palatine	39,523
8.	Elgin	77,010	36.	Calumet City	37,840
9.	Joliet	76,836	37.	Tinley Park	37,121
10.	Arlington Heights	75,460	38.	Glenview	37,093
11.	Evanston	73,233	39.	Buffalo Grove	36,427
12.	Waukegan	69,392	40.	Urbana	36,344
13.	Schaumburg	68,586	41.	Park Ridge	36,175
14.	Cicero	67,436	42.	Orland Park	35,720
15.	Champaign	63,502	43.	North Chicago	34,978
16.	Skokie	59,432	44.	DeKalb	34,925
17.	Oak Lawn	56,182	45.	Danville	33,828
18.	Oak Park	53,648	46.	Galesburg	33,530
19.	Des Plaines	53,223	47.	Elk Grove Village	33,492
20.	Mount Prospect	53,170	48.	Chicago Heights	33,072
21.	Bloomington	51,972	49.	Alton	32,905
22.	Wheaton	51,464	50.	Hanover Park	32,895
23.	Downers Grove	46,858	51.	Granite City	32,862
24.	Hoffman Estates	46,561	52.	Northbrook	32,308
25.	Berwyn	45,426	53.	Pekin	32,254
26.	Moline	43,202	54.	Addison	32,058
27.	Belleville	42,785	55.	Carol Stream	31,716
28.	Elmhurst	42,029	56.	Streamwood	30,987
			57.	Highland Park	30,575

of them in the counties of the Chicago Metro-politan Area (Table 1-4). This concentration is the result of burgeoning developing of suburbs around Chicago.

Economic Trends and Patterns

After entering the century on an economic upswing, Illinois responded to demands created by World War I with an enormous increase in production on its farms and in its factories. Between 1917 and 1918 Illinois farmers increased their wheat production by 100 percent, and that critical food grain temporarily displaced corn as the state's leading crop.[7] Heavy industries were quickly converted to produce munitions and weapons for the war; in 1918 one-third of the state's industrial output was to fulfill direct war contracts.[8] Upon successful completion of "the war to end all wars," optimism and prosperity pre-

vailed until they were replaced by the stock market crash, bank closures, and widespread unemployment of the depression. In 1933, 1.5 million Illinoisans were without gainful employment. As the state and nation struggled through this period of economic collapse, the jobless and poor formed lengthy food lines in the cities and some farmers found greater reward in burning their grain as fuel than selling it on the market.

The depression persisted until World War II created demands again for food and equipment to aid the Allies and support the nation's soldiers. Illinois farmers responded with new production records, and the state's factories accounted for about one-tenth of the national war production between 1940 and 1943.[9] In addition, scientists at the University of Chicago were successful in achieving the first controlled nuclear reaction to make possible

FIGURE 1-12. Machinery assembled to harvest a field of Illinois corn. (Photograph by Scott D. Miner.)

FIGURE 1-13. A view from the air of Illinois Cereal Mills, a corn-processing industry in Paris, Illinois. (Photograph by Allen Englebright.)

the atomic bombs that ended the war with Japan in 1945. Several years of postwar prosperity was followed by major economic and social changes that dramatically influenced life in Illinois during subsequent decades.

Although farming no longer provides employment for a significant part of the state's labor force, Illinois remains a major producer of agricultural products. In the early 1990s Illinois ranked fifth among the states in value of agricultural production. Over 8 billion dollars worth of agricultural commodities were sold by Illinois farmers in 1993, with crops accounting for $6.1 billion and livestock about $2.2 billion. To attain such high levels of production, farmers have invested not only their labor but also vast amounts of capital in land, buildings, machinery, fuel, seed, and

chemical fertilizers and pesticides. Choice farmland in the state commonly sells for more than $3,000 per acre, and the cost of machinery to operate a successful grain farm can total hundreds of thousands of dollars. The staggering costs of farming have caused operators of small holdings to find second jobs or sell their land to neighbors who are acquiring additional acreage to improve the viability of their farms. Illinois farms now average nearly 370 acres in size, and grain farms often are two to three times larger.

With extensive mechanization, technological developments, and changing economic conditions, Illinois farming has drastically changed in character. Farming now is the primary source of employment for only two percent of the state's labor force. The remaining farmers increasingly have specialized in their operations. Many have chosen to abandon the rearing of livestock and concentrate on the production of cash grain crops, primarily corn and soybeans. In areas of cash grain farming, therefore, fences and barns have been demolished and animal manure has been replaced as a fertilizer by such chemical products as nitrogen and anhydrous ammonia. Virtually all grain crops are now raised from commercially produced hybrid seeds, first developed in the 1930s. Even the traditional practice of crop rotation to maintain soil fertility has been largely abandoned as hybrid seeds and chemical fertilizers are employed to maintain high yields.

Initially raised primarily to feed the livestock on most farms, the Illinois grain crop over the past half century has been increasingly destined for export to foreign markets. For barge lines on the Mississippi River and its eastern tributaries, Illinois grain is one of the major commodities transported to the Gulf of Mexico for export. Among the states, only Iowa exceeds Illinois in the export of corn and soybeans.

The symbiotic relationship established between the farms and factories in Illinois during the nineteenth century has continued to the present. Most early industries in Illinois were engaged in the processing of grain and livestock and the production of tools, fencing, wagons, machinery, and implements for sale to farmers. Moline, Kewanee, Peoria, and Chicago became important centers of the farm implement and machinery industry, and East St. Louis and Chicago acquired a national reputation for their livestock slaughtering and

FIGURE 1-14. Along the shore of Calumet Harbor in the Chicago metropolitan area is one of the greatest concentrations of industries and shipping facilities in the Middle West. (Illinois Geographical Society)

meat packing functions. The famous Chicago stockyards were closed in 1971, but livestock processing continues to be an important industry in several of the state's smaller cities and towns that are able to cope with the objectionable odors and noise on a more limited scale. Decatur also is a noted industrial center with a specialty related to agriculture—the processing of soybeans into a growing variety of consumer goods, foodstuffs, and livestock feeds. The conversion of corn into ethanol for motor fuel is the basis for a relatively new agriculturally-related manufacturing specialty in Illinois.

The most important group of products manufactured in Illinois factories is durable goods, especially primary metals, fabricated metal products, machinery, and electrical equipment and supplies. Chicago and its industrial satellite cities in the Chicago Metropolitan Area contain the greatest concentration of these factories. However, major production facilities also are operated in Rockford, Peoria, Rock Island-Moline, Danville, Bloomington-Normal, and some of the Metro East cities. Among the newer developments are the emergence of a complex of producers of electronic goods in the suburbs of Chicago and the establishment of a huge automobile factory in Bloomington-Normal.

Significant changes in the geographic pattern of manufacturing have occurred in both the state and the nation during recent decades. The North American Manufacturing Belt, of which Illinois is a major component, has diminished in importance since the 1950s. Many of the region's industries have suffered from their failure to modernize operations and reduce production costs. A relocation of factories to different parts of the country—usually the South or Southwest—as a means of lowering production costs was accelerated by growing competition from foreign producers. Soaring energy costs during the 1970s, combined with foreign competition, resulted in

the closure of many mills and factories. For Illinois and other states in the Manufacturing Belt, the result has been a sharp decrease in manufacturing employment. In 1967 Illinois had nearly 1.4 million people employed in manufacturing; by the mid-1980s the number had fallen to less than 1 million. Manufacturing now accounts for only about one-fifth of the labor force in Illinois.

The greatest concentration of Illinois manufacturing traditionally has been in Chicago, but the city's dominance is declining. Between 1973 and 1993 the number of manufacturing plants within the city limits of Chicago fell from 7,330 to 4,720. Factories, like people, have been migrating from the city to the suburbs. When the suburban counties of DuPage, Kane, Lake, McHenry, and Will are combined with Cook County, this area—the Chicago Metropolitan Area—accounts for almost 70 percent of the total employment in Illinois manufacturing, however.

In Illinois, as in the nation, the decline of employment in agriculture and manufacturing has been offset by an increase in people working in the service sector of the economy. Over 60 percent of the Illinois workforce is now engaged in the services. This includes people in the medical, legal, and teaching professions; wholesale and retail sales; insurance and finance; the maintenance and repair of machinery; governmental services; and many others. The dominance of services as a source of employment suggests the progression of Illinois to a postindustrial economy.

ACCOMPLISHMENTS AND PROSPECTS

As a result of its development over nearly two centuries, Illinois has become the nation's sixth most populated state and a leader in many areas of economic and social endeavor. It is the wealthiest state in the Middle

West, ranking among the country's leaders in agriculture, manufacturing, and financial activities. Its dominant urban center, Chicago, is the nation's third largest city. Long noted for its importance as a lake port and railroad center, Chicago's role as a leading transportation center is now enhanced by O'Hare airport—the nation's busiest air transportation facility. These and other major achievements and events in the development of Illinois are recorded in the pages and chapters that follow.

With the approach of the twenty-first century, a number of challenges as well as opportunities await for Illinoisans. Environmental deterioration in such forms as air and water pollution, soil erosion, and disturbance of the land's natural surfaces has been a detrimental result of the state's industrial and agricultural development. To reverse these trends and undertake necessary measures to improve the quality of the environment can be expected to require new technology, the expenditure of vast sums of money, and at least slower rates of economic growth in the future. Another necessity for the years ahead is increased effort to eliminate racial and ethnic frictions, a particularly critical problem in Chicago, East St. Louis, and other cities where diversified populations must live and work in close contact with one another. Legislation will need to be supplemented by personal commitments to understanding, compassion, compromise, and even sacrifice in order for people of all races and ethnic backgrounds to enjoy a peaceful and fulfilling future.

NOTES

1. Carol O. Sauer, *Geography of the Upper Illinois Valley and History of Development,* Bulletin No. 27 (Urbana: Illinois State Geological Survey, 1916), p. 155.

2. Douglas R. McManis, *The Initial Evaluation and Utilization of the Illinois Prairies 1813-1840,* Department of Geography Research Paper No. 94 (Chicago: University of Chicago, 1964), p. 92.

3. Vernon Carstensen, ed. *The Public Lands* (Madison: University of Wisconsin Press, 1963), p. 238.

4. Theodore L. Carlson, *The Illinois Military Tract: A Study in Land Occupation, Utilization and Tenure,* Vol. XXXII of Illinois Studies in the Social Sciences (Urbana: University of Illinois Press, 1951), p. 7.

5. Theodore Calvin Pease and Marguerite Jenison Pease, *The Story of Illinois* 3rd ed. (Chicago: University of Chicago Press, 1965), p. 188.

6. *Ibid.*, p. 193.

7. *Ibid.*, pp. 231-232.

8. Robert P. Howard, *Illinois: A History of the Prairie State* (Grand Rapids, Mich.: William B. Eerdmans Publishing Co., 1972), p. 444.

9. Pease and Pease, *Story of Illinois*, p. 251.

2

THE PHYSICAL ENVIRONMENT: LANDFORMS

Arlin D. Fentem
Western Illinois University

The advent of humans into the middle Mississippi region happened only a few thousand years ago; since that time, the physical landscape has both shaped the nature of people's activities and has itself been altered by their presence. Geographers describe and try to understand the differences from place to place in human institutions, economics, and works and their interrelations with the physical environment. The subjects of this chapter and the one that follows are the physical world that was here before humans arrived, the changes that resulted from their coming, and the importance of physical environments for present-day activities.

Physical environment is a general term that includes the shapes and forms of the earth's surface (landforms) and the earth materials which compose it; conditions and patterns of drainage; climate and its changes during human occupation; natural vegetation and its evolution during the same period; soils; and finally, the minerals that have been useful to people. All of these components are interrelated; none can be entirely separated from the others. For example, each combination of surface form, earth materials, drainage

conditions, and climate produces a different natural vegetation, which in turn has been continuously altered by human action.

We might simply describe and map all these elements, but true understanding requires that we look for answers to the questions *Why?* and *How?* as well as *What?* The key to geographical interpretation is the map, which shows the variation from place to place in an element such as landforms or vegetation. We will also rely a great deal on another perspective, that of changes in mapped distributions through time, in order to understand the evolution of the present landscape. Finally the chain of events that produced the present landscapes has been very complicated and we are often unsure of our interpretations. Therefore, some distributions will be presented as problems for which solutions are being sought.

THE SHAPE AND COMPOSITION OF THE LAND SURFACE[1]

To a visitor from Colorado or New England, Illinois seems monotonously flat and

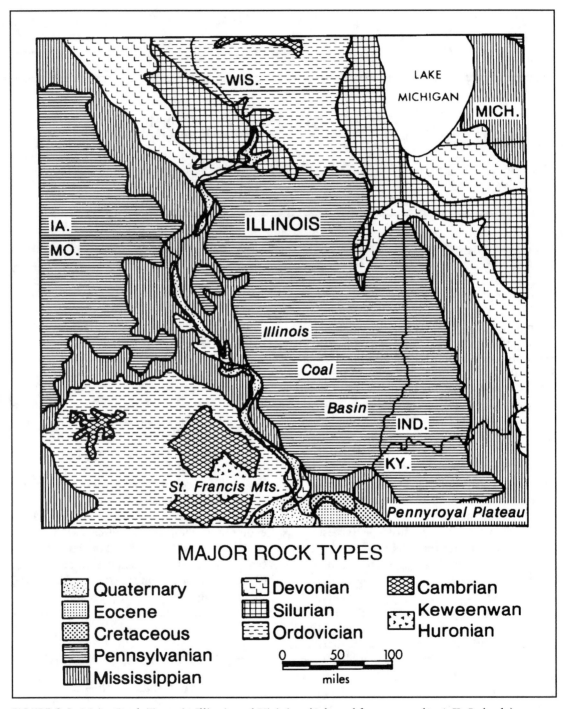

MAJOR ROCK TYPES

Quaternary   Devonian   Cambrian
Eocene   Silurian   Keweenwan
Cretaceous   Ordovician   Huronian
Pennsylvanian
Mississippian

0   50   100

miles

FIGURE 2-1. Major Rock Types in Illinois and Vicinity. (Adapted from a map by A.K. Lobeck.)

lacking in interest; nevertheless, the state has areas of unusual scenic beauty, and its subtle differences in form and materials have important consequences in the uses of the land and in the prosperity of its people.

Bedrock

One of the most important controls bringing about these differences is the bedrock that lies just below a thin "skin" of unconsolidated (loosely compacted) materials. Figure 2-1 shows how the major rock types are arranged in and near the state. In the pages to follow, each of the rock types will be described, with particular reference to qualities important for shaping the surface and for utility to humans.

Almost all the rocks that appear at the surface are made of materials deposited in or along the shores of successive oceans during the Paleozoic geological epoch, which began about 600 million years ago and ended about 230 million years before the present. The exceptions are rocks of Cretaceous age (about 100 million years old), which occupy small areas in west central Illinois and in the extreme south (Figure 2-2). These rocks are very similar to the older Pennsylvanian sedimentary rocks that surround them and which range in age from 300 to 280 million years. In both cases, soft and easily broken (friable) shales predominate, although there are occasional layers and lenslike bodies of sandstone. Shales were originally laid down at the bottoms of small seas and shallow lakes as mud or clay. The sandstone bodies and layers mark the courses of ancient streams or the sites of beaches and deltas at the margins of lakes and seas which expanded, shrank, and sometimes disappeared as the multimillion year prehistory of the Pennsylvanian unfolded. The shales and weak sandstones of both Cretaceous and Pennsylvanian age weather and erode very easily when exposed to the atmosphere and running water. Areas floored by

such rocks often become lowland plains with a thick mantle of weathered material over the bedrock.

The Pennsylvanian rocks also contain two major economic resources—bituminous coal and petroleum. Coal is organic in origin; it was formed from remains of the profuse tropical vegetation that grew and were subsequently preserved in Pennsylvanian swamps. There are more than 30 well-defined layers or beds of coal below the surface of this region, and seven of these have been extensively mined. Less well understood are the petroleum deposits, now greatly depleted, that have been found in the Pennsylvanian rocks. Petroleum is also organic in origin, but it has often migrated from its source to rock layers of other ages. While more than half the oil that has been produced in Illinois was pumped from Pennsylvanian rocks, it does not follow that so great a proportion originated there.

Surrounding the Pennsylvanian rocks just described are those deposited at the very beginning of that geological epoch. These consist of thick and strong (massive) sandstones that resist both weathering and erosion and thus tend to stand as rugged ridges when they are surrounded by weaker rocks. Sandstone consists mostly of the mineral silicon (quartz), a very hard substance that does not readily form a deep soil and that contains few of the minerals needed for the growth of most crops. In Illinois, this rock outcrops along a quite narrow band in the extreme south.

Succeeding these sandstones outward from a center near Centralia, are the oldest of the Mississippian rocks (+350 to +300 million years old). This geological era was one during which deep, warm, and quiet seas usually covered the Midwest. The rocks that formed when the seas were present are made up, for the most part, of lime (calcium carbonate, $CaCo_3$), and consist mostly of the "skeletal" remains of small, often near-microscopic,

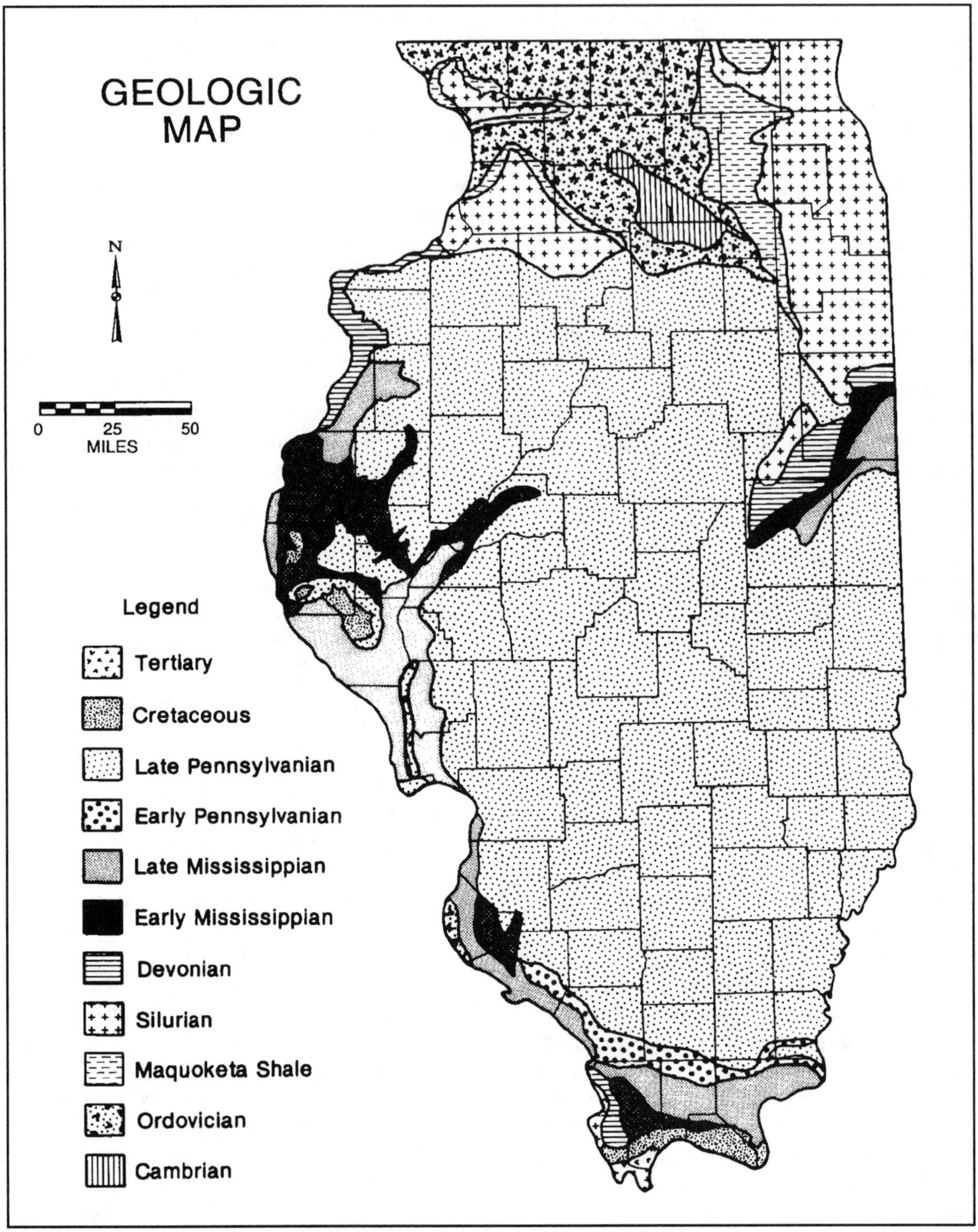

FIGURE 2-2. Bedrock surface of Illinois. (Source: Illinois Geological Survey.)

marine plants and animals. Some are easily cut by saws, make excellent building stone, and are a convenient source of lime used in making quicklime, dentifrices, face powder, and agricultural fertilizers. These "pure" limestones are easily weathered and eroded, and they are alkaline rather than acidic. They frequently produce lowland plains and often are responsible for "karst" topography. Pure limestones are so easily dissolved by water that the seepage of rainfall downward through fissures in the rock and along bedding planes (where one "layer" or "bed" of rock succeeds another) often results in caverns. When these underground rooms and passages become extensive, their roofs often collapse; the result is a "dimpled" surface whose funnel-shaped depressions may contain small, nearly round lakes. The Pennyroyal "Plateau" (Figure 2-1) has such a karst landscape. Note that this particular formation is scarcely present in Illinois, but is widespread in Kentucky. The best-developed karst landscape in Illinois is in southern Calhoun County, north of St. Louis.

Among the uppermost Mississippian rock layers is still another massive sandstone closely resembling the oldest of the Pennsylvanian rocks described above. Like those formations, it is represented by a narrow band in the extreme south, parallel to its neighbor (not shown on Figure 2-2).

Much of the rest of the state is floored by late Mississippian limestone (Figure 2-2). Although some of these are relatively "pure," the greater number include other minerals as well. The most significant of these minerals is magnesium; when a great deal of it is present, the rock is called dolomite. Dolomite is quite resistant to weathering and erosion, not easily dissolved, and likely to produce either a high plain or a range of low hills. Few of the rocks in this area are true dolomites; they are best described as "dolomitic limestones," with characteristics intermediate between dolomites and the pure limestones described earlier.

The only areas of Illinois not yet described are in the far north. Most of the surface here is underlain by a dolomite (Niagara) of Silurian age. Because it has so much magnesium, it has resisted erosion. In the northwest it makes a pronounced ridge along which runs US Route 20, and outside the state it is responsible for the Door Peninsula (northeast of Green Bay, Wisconsin), Mantoulin Island, and the escarpment that forms Niagara Falls. Just below the Niagara formation (and outward toward the northern boundary of Illinois) is a thick layer of very weak shale, called the Maquoketa (for a town in Iowa) or New Richmond (for a town in Indiana). It has the same characteristics that predominate in Cretaceous areas. In southern Wisconsin, the shale is succeeded northward by Galena-Platteville dolomites of Ordovician age, a layer of very weak sandstone (the St. Peters), still another dolomite (the lower Magnesian limestone), and finally by a great expanse and thickness of early Paleozoic (Cambrian) sandstones.

Below all these Paleozoic rocks is the Precambrian "basement"—the complex (usually very hard), metamorphic (changed), crystalline (recrystallized by heat and pressure) rocks whose surface (if we were to strip away all the layered Paleozoic sediments) would form a gently undulating plain over the interior of the United States. These rocks (and the surface) are indeed exposed in northeastern Minnesota, northern Wisconsin, and over much of southern and eastern Canada.

DEVELOPMENT
OF BEDROCK GEOGRAPHY

Bedrock geography may be a term that falls strangely on the ear, because the study of the earth's mineral crust is usually the province of geology. But geographers are concerned with patterns of distribution on the

FIGURE 2-3. Niagara dolomite caps the crest of a mound in the driftless region of northwestern Illinois. (Photograph by A.D. Fentem.)

earth's surface, and as they explain and interpret these patterns, they are inevitably involved with sister sciences whose subject is the thing or event that makes the pattern.

Figure 2-2 clearly shows a regular, concentric pattern that indicates the operation of general processes. Spatial patterns develop through time, and in the paragraphs to follow, we will trace that development, beginning with the situation at the dawn of the Paleozoic (ca. 600 million years ago). The processes that took place during that immense span can be generalized as

1. *Sedimentation*—the deposition of sediments in or at the edge of water bodies
2. *Warping*—the gentle bending of the earth's crust into shallow basins scores or hundreds of miles across and separated by equally broad swells or domes
3. *Erosion*—the stripping away of much of the accumulated sediment by running water to produce a near-uniform plain

While this division is useful for understanding, it should be remembered that these processes overlapped to a considerable degree.

Figure 2-4 shows the geological column for Illinois—the sequence and timing of the sedimentations which built up the layers of rock we have been describing. We do not know how long these depositions continued nor how thick the deposits eventually became. With but one exception, the upper layers—if they were ever present—have been eroded away and washed into the sea. The exception consists of

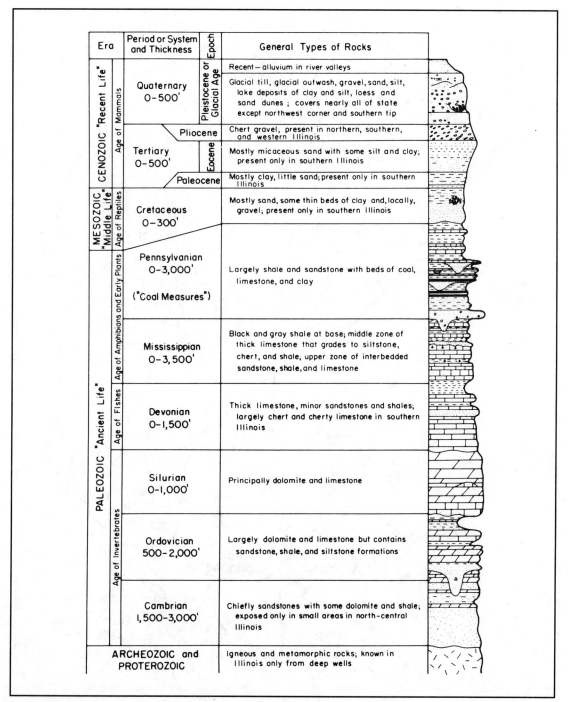

Era	Period or System and Thickness	Epoch	General Types of Rocks
CENOZOIC "Recent Life" — Age of Mammals	Quaternary 0–500'	Pleistocene or Glacial Age	Recent — alluvium in river valleys
			Glacial till, glacial outwash, gravel, sand, silt, lake deposits of clay and silt, loess and sand dunes ; covers nearly all of state except northwest corner and southern tip
	Pliocene		Chert gravel, present in northern, southern, and western Illinois
	Tertiary 0–500'	Eocene	Mostly micaceous sand with some silt and clay; present only in southern Illinois
	Paleocene		Mostly clay, little sand; present only in southern Illinois
MESOZOIC "Middle Life" — Age of Reptiles	Cretaceous 0–300'		Mostly sand, some thin beds of clay and, locally, gravel; present only in southern Illinois
PALEOZOIC "Ancient Life" — Age of Amphibians and Early Plants	Pennsylvanian 0–3,000' ("Coal Measures")		Largely shale and sandstone with beds of coal, limestone, and clay
Age of Amphibians and Early Plants	Mississippian 0–3,500'		Black and gray shale at base; middle zone of thick limestone that grades to siltstone, chert, and shale, upper zone of interbedded sandstone, shale, and limestone
Age of Fishes	Devonian 0–1,500'		Thick limestone, minor sandstones and shales; largely chert and cherty limestone in southern Illinois
Age of Invertebrates	Silurian 0–1,000'		Principally dolomite and limestone
Age of Invertebrates	Ordovician 500–2,000'		Largely dolomite and limestone but contains sandstone, shale, and siltstone formations
	Cambrian 1,500–3,000'		Chiefly sandstones with some dolomite and shale; exposed only in small areas in north-central Illinois
ARCHEOZOIC and PROTEROZOIC			Igneous and metamorphic rocks; known in Illinois only from deep wells

FIGURE 2-4. Geologic column for Illinois. (*From Guide to Rocks and Minerals of Illinois,* Illinois State Geological Survey, Urbana, 1959.)

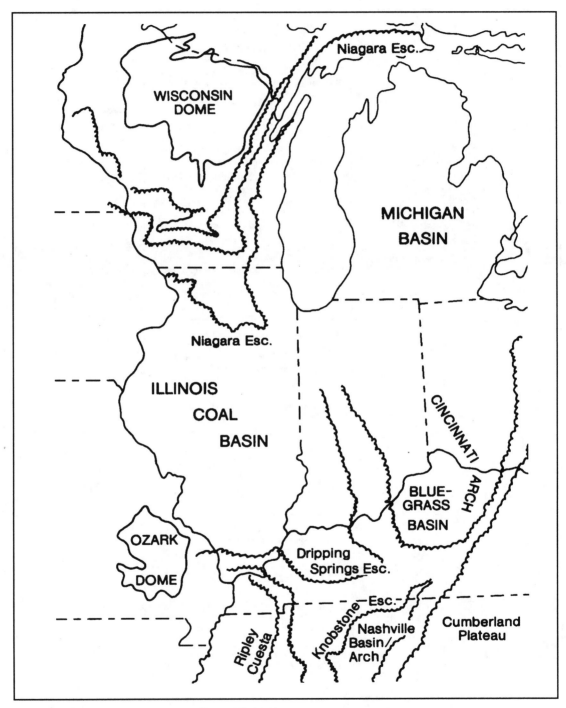

FIGURE 2-5. Basins and Domes in the Middle West.

the Cretaceous sediments that lie unconformably (that is, with missing layers in between) upon much older sediments (Figure 2-2).

At the same time that the sediments were accumulating upon the Precambrian basement (and perhaps continuing afterward), the earth's crust was bending very gently and slowly to form the broad, shallow basins shown in Figure 2-5. The thickness of the sediments in the basins (geosynclines) naturally became much greater than those deposited on domes (geanticlines). The two western domes (the Ozark and the Wisconsin) are separated by a "sag" in the Precambrian rocks, and a similar depression separates the upwarpings in Ontario from those in Kentucky-Tennessee. Between these two major axes are two broad, shallow basins, one of which is deepest in Michigan and the other whose lowest point is in southeastern Illinois. Similarly, the two basins are separated by a broad swell called the Kankakee Arch.

The last step in understanding the general concentric arrangement of bedrock types in Illinois is to consider the effects of long-continued erosion. To aid in visualizing the relations among sedimentation, warping, and erosion, Figure 2-6 has been prepared. The end result of erosion on a stable landscape is to bring about a nearly horizontal surface. Since the rock layers are *not* horizontal, but slope gently downward into the basins, the erosional surface slices across the bedrock surface. Note that the youngest surface rocks are at the centers of the simple and idealized basin shown in Figure 2-6, while the oldest appear near the centers of the domes. (The Precambrian basement rocks peep through the sediments in the Ozarks, in north-central Wisconsin, and in Ontario).

In the real world, of course, perfectly geometric features almost never occur, and the midwestern geanticlines and geosynclines are not truly circular. It is worthwhile to compare the surface bedrock map (Figure 2-2) with the cross-sections shown in Figure 2-6 and in Figure 2-7 until you can see, in your mind's eye, the three-dimensional image that relates the vertical and horizontal views. The bedrock map reveals that not only were the downwarped basins not circular, but also that the warping and bending were accompanied by some wrinkling of the crust, and sometimes even breakage or faults. The wrinkles (called anticlines when they are sharply upward) duplicate, in their results, what occurs on a large scale with the geanticlines.

The more important of these structures may be seen on Figure 2-8, where the surface as it would exist on top of one of the rock layers is shown. The most significant anticline from an economic point of view is the LaSalle, because it formed a "trap" roofed by nonpermeable shale, which confined petroleum deposits to a narrow band. Almost all the oil found in Illinois before 1935 came from this structure. It had little effect on topography, however, because the younger rocks that were exposed by its erosion were almost identical to those that surround it. Dashed lines mark the crest of the Du Quoin and other anticlines that produced a great deal of oil after the mid-1930s.

A very large monocline in the vicinity of the Illinois River in west-central Illinois also has had important consequences. In this case, nearly horizontal bedrock plunges suddenly into the Illinois Coal Basin (eastward) at a rather steep angle. The results are

1. The flat-lying bedrock (the western Illinois Platform) may have helped to preserve an upland plain
2. The coal measures in the Pennsylvanian rocks stayed near the surface and therefore can be easily reached by strip mining over a large area.

With the Illinois River providing an inexpensive means of sending coal to the large

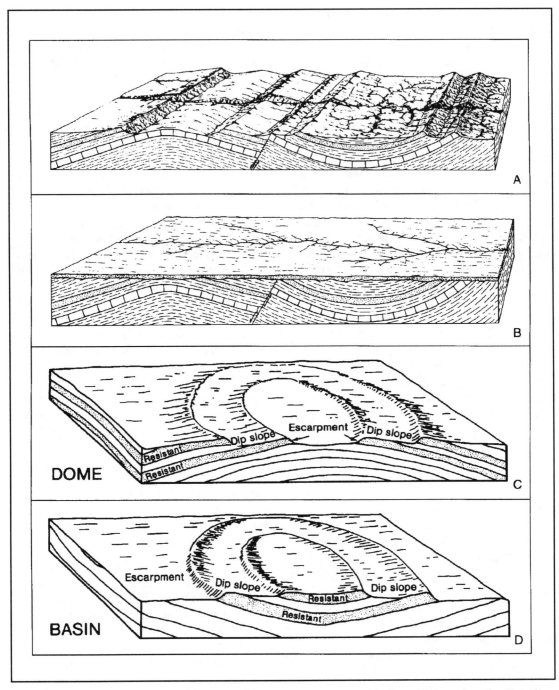

FIGURE 2-6. Erosion Landscapes of basin and dome structures. A and B from *Photogeology*, by V. Miller, 1961. C and D from *Physical Elements of Geography* , 4th Edition by Finch et al. Copyright 1957. Reprinted by permission of McGraw Hill, Inc.

Chicago market, the counties of Fulton, Peoria, and Knox have become important for strip coal mining.

More dramatically, the faults named Sandwich, Savanna, and Cap Au Gris provide sudden contrasts in either or both bedrock and topography. Not only did the earth's crust fracture along these lines, but segments of it slipped vertically relative to each other, sometimes for hundreds of feet. Thus, what is accomplished by gently tilted bedrock over hundreds of miles can be duplicated in a few thousand feet. This can be seen on the bedrock map, where Ordovician sediments are found side by side with those of Cambrian age along the Sandwich fault. Figure 2-6B shows how such a condition comes about.

Folding and faulting have produced the beautiful scenery in Calhoun County and at Pere Marquette State Park, north of St. Louis. Here Mississippian and Ordovician bedrock are brought side-by-side at the surface to make contrasts in landform scenery, including the karst topography of the southern tip of the county. Palisades Park, north of Savanna,

owes its vertical rock walls to both folding and faulting which elevated a weak shale to the level of the Ice Age Mississippi with dolomitic (Ordovician) limestone above. As the weak shales were eroded, the massive limestone blocks they supported came tumbling down to form the cliffs which are the park's attraction. A number of closely spaced faults in the southeast have altered that scenery as well. There is still a great deal of instability in the crust of the Middle West, and earth tremors are relatively frequent. Within historic time, a massive movement and quake produced Reelfoot Lake in western Tennessee.

THE SHAPE OF THE BEDROCK SURFACE

The previous discussion has brought the development of landforms in Illinois to the condition that seems likely to have existed sometime in the Tertiary—a condition similar to that idealized in Figure 2-6B. Just as natural large-scale patterns are seldom geometric,

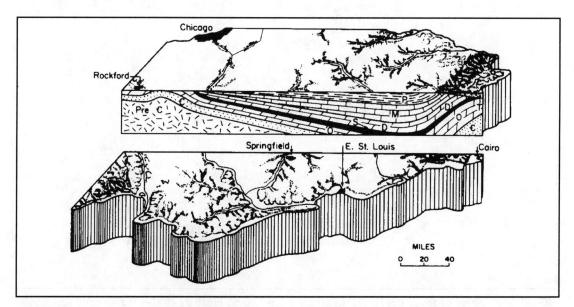

FIGURE 2-7. Geologic Profile of Illinois. (From *Guide to the Geologic Map of Illinois,* Illinois State Geological Survey, Urbana, 1961.)

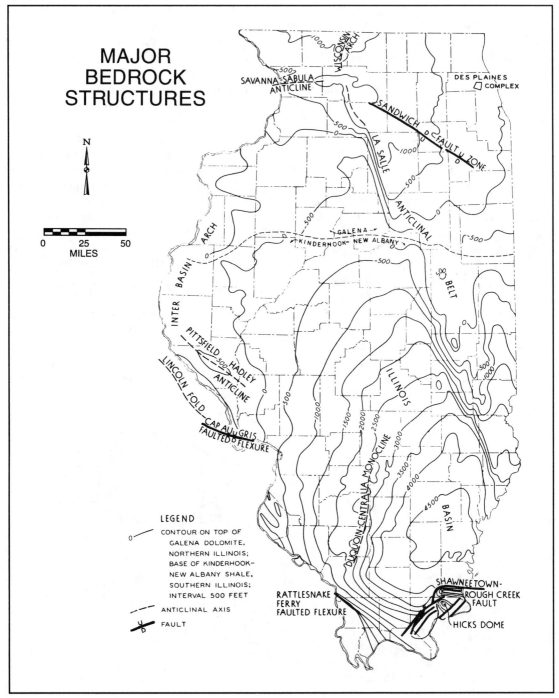

FIGURE 2-8. Major bedrock structures. (Adapted from Leland Horberg, *Bedrock Topography of Illinois*, Illinois State Geological Survey, Urbana, 1950.)

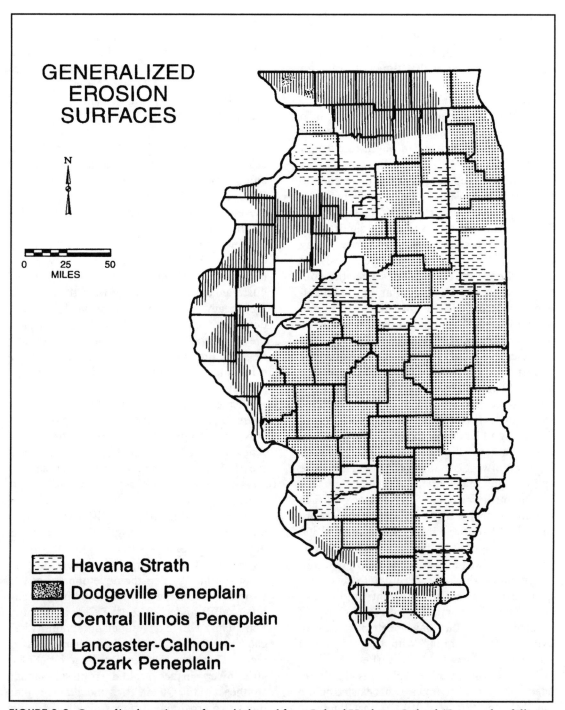

GENERALIZED EROSION SURFACES

N

0　25　50
MILES

▨ Havana Strath
▨ Dodgeville Peneplain
▤ Central Illinois Peneplain
▥ Lancaster-Calhoun-
　　Ozark Peneplain

FIGURE 2-9. Generalized erosion surfaces. (Adapted from Leland Horberg, *Bedrock Topography of Illinois,* Illinois State Geological Survey, Urbana, 1950.)

broad surfaces have never been perfectly flat; nevertheless, that condition (called a peneplain when it results from erosion) was probably approximated at that time. In order for such a condition to be maintained, the earth's surface must remain stable. If the level of the sea into which streams flow is lowered, or if the land is uplifted above the sea, the streams will flow more swiftly and begin to cut new valleys.

It is at this juncture that the difference in the erodibility of the bedrock types becomes important. The softer rocks (especially the Pennsylvanian shales and weak sandstones) are rather quickly attacked and carried away as sediments to the sea. The more resistant rocks (the massive sandstones near the contact between the Mississippian and Pennsylvanian systems and the dolomites of Silurian and Ordovician age) tend to remain as upland belts (as idealized in Figure 2-6). Fairly recent uplift and renewed erosion have produced the "cuestaform" hilly belts found in the extreme south and northwest of the state. The Shawnee Hills is the name given to the two belts of sandstone, the Pennsylvanian (Dripping Springs) and Knobstone escarpments, as they cross southern Illinois. The Silurian outcrop in the northwest is made of Niagaran dolomite and is called the Niagara Escarpment.

One further aspect of the bedrock surface requires attention, for it appears that there are four roughly defined levels of elevation in the state. On Figure 2-9 their relative locations are shown. Each level may be a remnant of a peneplain as shown in Figure 2-6B. To suggest how such an arrangement might have come about, we will describe the series of repeated events that have been posed as an explanation, beginning with the oldest and outermost surface, the Dodgeville.

The development of peneplains is closely related to the pattern of drainage that existed before the Ice Ages (Figure 2-10). At that time, the Mississippi River followed roughly the route of

the present Illinois River and was joined near present Beardstown by a very large river which we can think of as the ancestral Ohio. When we remember that peneplanation requires a long period of stability in the earth's crust, it is not surprising that the process was interrupted more than once before it could be completed.

After an uplift of the land relative to the sea, the major rivers gain renewed strength and begin cutting new valleys in the flat surface. As the valleys become deeper, however, the energy of their streams diminishes, and finally there is no further downcutting; the master streams then begin to widen the floors of their valleys. At the same time, new tributaries begin to form and erosion advances upstream or "headward" (toward the heads of the stream) until they too can no longer cut downward; consequently, they begin to widen their own valleys. The peneplain comes into being as the widening valleys approach each other and coalesce to form a new and lower surface.

Just such a surface, the Dodgeville, appears to have developed throughout the Midwest. Subsequently, however, there was renewed uplift, a rejuvenation of streams, and the development of a new surface (the Lancaster)—again working its way outward from the juncture (confluence) of the major streams. This cycle embraced all of Illinois, but outside the state, the Dodgeville surface—a gently rolling plain—still exists and the tributaries are still deepening their narrow valleys.

In similar manner, the development of the Lancaster surface was interrupted by still another uplift and surface (the Central Illinois Peneplain), which developed quite rapidly because the Pennsylvanian rocks there were so weak. There are two smaller "patches" of a still lower surface (called strath lowlands) southeast of the Quad Cities and near Havana; these two erosional surfaces may be of quite recent date. Finally, the valley floors of the

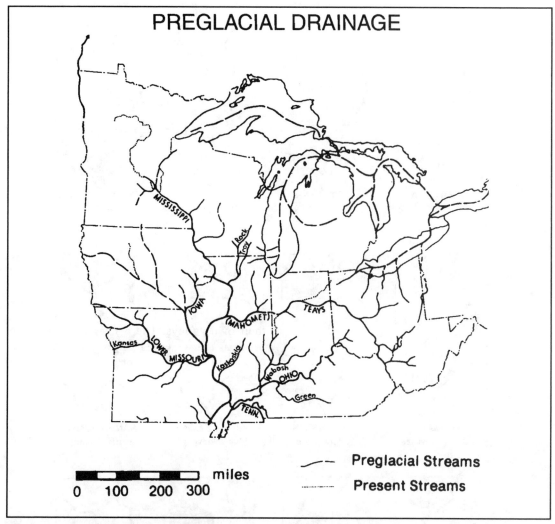

FIGURE 2-10. Preglacial drainage patterns in the Middle West. (Adapted from Leland Horberg, B*edrock Topography of Illinois,* Illinois State Geological Survey, Urbana, 1950.)

streams in the southern third of the state, such as the Kaskaskia and the Big Muddy, are so wide and flat that they, too, may have developed during this fourth, and last, erosional cycle. The idea of peneplanation in the Midwest is rejected by some geomorphologists, who think that the "four story" topography is more related to the resistance of rocks than to erosional cycles as described.

These dissenters would say that the widespread upland surface in western Illinois is there because the rocks are hard (dolomitic) and horizontal.

A million or two years ago the physical landscape of Illinois would have looked much as that of southwestern Wisconsin does now. Its terrain would have been much rougher than it now is, and flat or nearly flat land

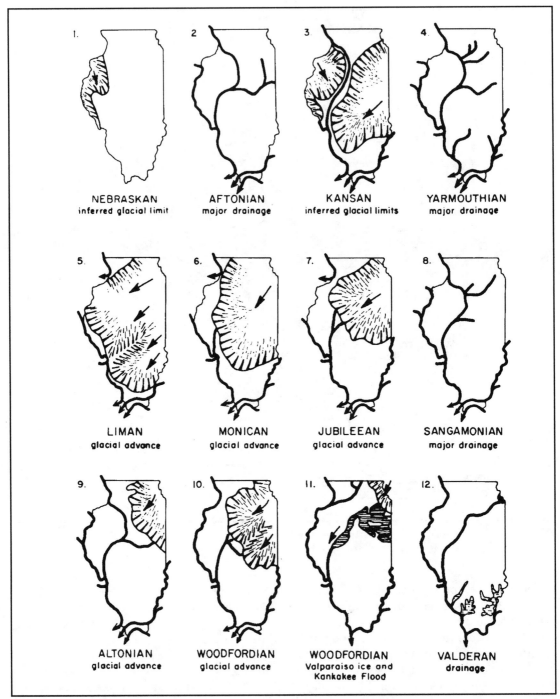

FIGURE 2-11. Glacial stages and drainage in Illinois. (From H.B. Willman and J.C. Frye, *Pleistocene Stratigraphy of Illinois,* Illinois State Geological Survey, Urbana, 1970.)

would have been relatively rare. Although its finer features have been obliterated, and its drainage pattern only approximates the present one, today's surface is much influenced by both rock types and the preglacial topography, as we shall presently see.

THE PLEISTOCENE

Although the glacial epoch occupied only a very small fraction of geological time, its results for present landforms are very great—partly because of its recency. There has been but little time since the last ice advances—a few thousands of years—and glacial features are still much in evidence. It is now widely believed, indeed, that the Ice Age has not ended and that all of human civilization belongs to a brief interval between major ice advances.

More than 90 percent of Illinois was overridden by glaciers during the past 100,000 years, and the last of the glacial ice was still present in the northeast less than 15,000 years before the present.[2] Earlier glaciations, reaching back more than a million years, have had their most obvious manifestations wiped out by time or by later ice invasions. We will first indicate what the general effects have been, and then point out differences that result either from the manner of glaciation or the length of time that has elapsed since glaciation occurred. Figure 2-11 shows the most important stages and substages in the glaciation of the state, as well as the currently used names. From time to time it will be necessary to refer back to this sequence of maps in order to clarify the discussion that follows. Figure 2-12 is a summary which shows the distributions of the last ice sheets to cover each area of the state.

The initial effect of glaciation was to add to, rework, and redistribute the regolith (unconsolidated earth materials overlying the bedrock). Whether by the weight and passage

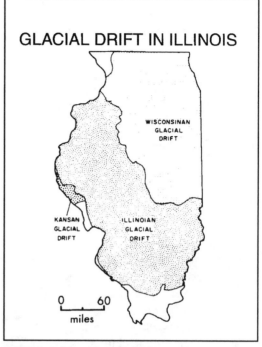

FIGURE 2-12. Glacial drift in Illinois. (From *Guide to the Geologic Map* of Illinois, Illinois State Geological Survey, Urbana, 1961.)

of the ice itself, or by the vast amounts of running water from melting ice, these earth materials filled in valleys and formed new surfaces even higher than the crests of the bedrock hills. The new surface was much smoother than the terrain it replaced and it was virtually free of stream valleys.

Differences from place to place in the present glaciated surfaces depend upon three factors:

1. The thickness of the unconsolidated materials (called drift when they are of glacial origin)
2. Differences in the behavior of the ice itself at different periods
3. The length of time since the ice sheets retreated

The last of these is particularly important, because the drift was so easily eroded that even a

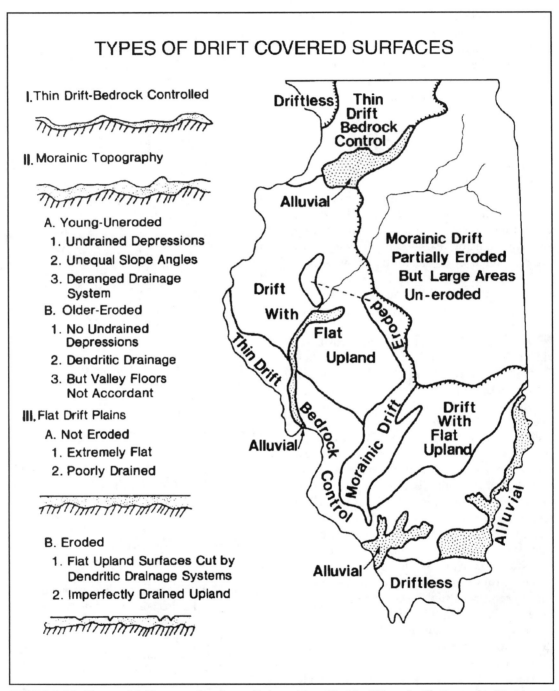

FIGURE 2-13. Types of drift covered surfaces. (Adapted from P. MacClintock, *Physiographic Divisions of the Area Covered by the Illinoisan Driftsheet in Southern Illinois,* Illinois State Geological Survey, Urbana, 1939.)

few thousands of years witnessed a great deal of erosion and alteration of the landscape.

TYPES OF DRIFT COVERED SURFACES

Figure 2-13 depicts the major types of drift-covered surfaces to be found in the state and the accompanying diagrams indicate the relations between and among thickness, age, and original drift characteristics. After distinguishing between driftless and drift-covered areas, the most important distinction is between thin and thick drift within the portions of the state once covered by Illinoisan but not by Wisconsinan ice sheets. The *upland* of the thin-drift areas in the north-central and south-central regions of the state appear little differ-

ent than they would have had glaciation not occurred; nevertheless, hills have been smoothed and rounded and slopes are longer and more gentle than they otherwise would have been. In the south-central region, much of the original glacial material is now collected in the valleys of streams and this helps to account for their very wide and very flat floors.

Within the areas covered by thick Illinoisan drift, the original surface was, unlike those of most regions of glacial deposition, exceptionally flat. No one is quite sure of the reasons for this flatness, but two explanations have been suggested. Ice movements during the Illinoisan glaciation was from northeast to southwest (Figure 2-11). In the thick-drift regions, the Illinoisan ice was building and flowing across rocks that were mostly shales;

FIGURE 2-14. Headward erosion by postglacial streams on the flat upland of the Galesburg Plain. (Photograph by A.D. Fentem.)

the clayey drift that resulted may have been plastic enough to be flattened by the weight of overlying ice.

It is more likely that the flatness was related to the manner in which the ice retreated after the Monican advance (Figure 2-11). Ordinarily, glacial retreats are halting and interrupted by readvances. During periods when the ice margin is nearly stationary, the ice continues to flow, carrying with it earth materials. Melting at the margin removes the ice there and results in the accumulation of the earth materials in long hummocky ridges that mark the stationary edge of the ice. These low ridges are called *moraines*—end moraines when they mark the farthest advance in a glacial period and recessional moraines when they mark pauses during a glacial retreat. It is possible that the Monican retreat resulted from an abrupt climatic change which was not reversed, and that the ice wastage and retreat was rapid and nearly uninterrupted. If that were true, recessional moraines would be nearly absent. Furthermore, a rapid retreat would have produced an enormous amount of extremely turbid (dirty) water; sediments might have been deposited from these waters over wide areas and helped to smooth the land surface. Even when recessional moraines are rather abundant on the Illinoisan Drift Plains, as they are, for example, east and northeast of St. Louis, they are not very prominent because erosion has subdued them during the long period (more than 100,000 years) since their formation.

During the long period since the Illinoisan ice sheets last retreated, a great deal of the originally flat surface has been destroyed and converted into an extensive system of stream valleys. Nevertheless, there are broad areas of that flat surface still present on the Illinoisan Drift Plains, and even when valleys are quite close to each other, the narrow divides between them are almost perfectly flat.

The shapes of the land on the Wisconsinan drift surface are in rather sharp contrast to

these of the Illinoisan plain. Curving parallel moraines are the most conspicuous features (Figure 2-15). These broad belts of low hills are often thousands of yards across and extend for many scores of miles. In the south, where the drift is oldest, there has been time enough to establish a fairly well integrated drainage system, and there were few natural lakes when Europeans first arrived. However, rainwater is removed from the lowlands between the moraines very sluggishly, and many areas were wet or covered with standing water during much of the year. Vast temporary lakes developed after heavy rains.

To some observers, the southern part of the Wisconsinan glaciated area seems quite flat; nevertheless, even the areas between the morainic ridges usually have a very gently rolling surface as compared with the extreme flatness of the upland surface in the Illinoisan plain. Another difference between the two regions is the vertical location of the flattest land. It is the *lowest* land between moraines that has most of the flat land in the Wisconsinan areas of thick drift; it is the *highest* areas between stream valleys which embrace most of the flat land in the Illinoisan areas of thick drift.

The northern and northeastern reaches of the Wisconsinan drift have landforms that are distinctive because ice retreats and advances were even more frequent and because the latest of these occurred little more than 12,000 years ago. High morainic ridges are quite close to each other. The stream drainage system has not had time to develop completely, and there were, therefore, numerous lakes, swamps, and peat bogs when Europeans first saw the area.

As the ice retreated from northeastern Illinois, huge volumes of meltwater accumulated in the Great Lakes; the concentric moraines of northeastern Illinois served as temporary dams, impounding the waters of an expanded Great Lakes system which spilled over into

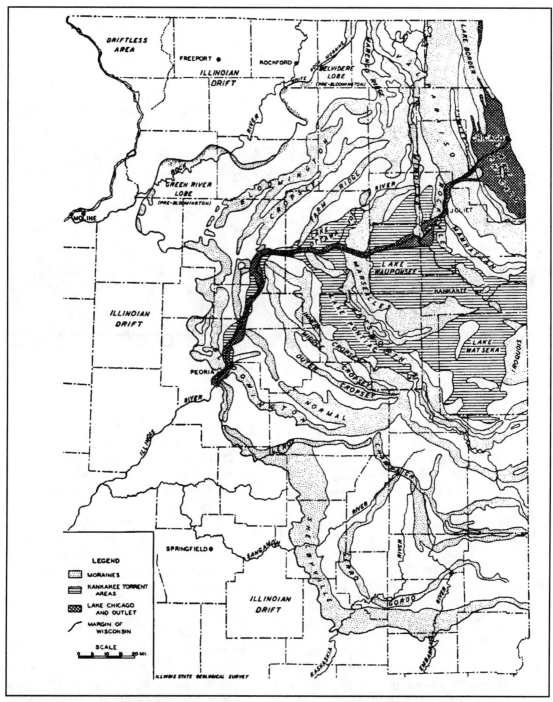

FIGURE 2-15. Wisconsinan glacial features in Illinois. (From M.M. Leighton, George E. Ekblaw, and Leland Horberg, *Physiographic Divisions of Illinois*, Illinois State Geological Survey, Urbana, 1948.)

FIGURE 2-16. Undulating surface of the Cerro Gordo Moraine near Champaign-Urbana. (Photograph by A.D. Fentem.)

the areas between the moraines and made huge shallow lakes (Figure 2-15). These lake bottoms accumulated sediments and became low and very flat plains. Most of the land where Chicago now stands was covered by the enlarged Lake Michigan.

One of the most spectacular events toward the end of the Wisconsinan glaciation was the Kankakee torrent. Meltwater escaping from an ice tongue occupying what is now Lake Michigan, and from still further east (glacial Lake Erie), spilled into Illinois from the east along a path south of Lake Michigan, forming temporary lakes: Watseka, Wauponsee, Pontiac, and Ottawa. After breaching the moraine near Marseilles, Illinois, the flood was directed westward toward the present sharp bend in the Illinois River at Hennepin. The upper Illinois River was entrenched (cut

downward) during this flood. Downstream from Hennepin, the gigantic flood of water deposited great quantities of sediment in the valley itself. And because the waters were so high, the tributary streams were backed up in their valleys and formed temporary slackwater lakes where still more deposition took place. Finally, as the course of the flood across the lake floors became more concentrated and the currents more swift, huge sand and gravel bars were formed.

OUTWASH AND VALLEY TRAINS

As we have just seen in the case of the Kankakee Flood, the effects of Pleistocene glaciation on landforms are not limited to the areas covered by the ice sheets. In the following paragraphs, large areas of the state whose landforms have been partly shaped by outwash and valley train will be identified. "Outwash" is the general term used to identify material carried away from ice margins by moving meltwater. "Valley trains" occur when meltwaters become concentrated along valleys and deposit the earth materials they hold in suspension along those valleys. The surface of areas covered with outwash is, of course, generally quite flat immediately after deposition, as are all water-lain deposits.

Two areas that received large quantities of outwashed materials were the Havana Strath Lowland above Beardstown, and the Green River Lowland south and east of Rock Island and Moline. Portions of both areas were covered by water from time to time and became lake bottoms, but much of their character can be attributed to the vast amounts of earth materials washed into them from melting glaciers. On occasion, the deposited material consisted of fine sands. Before vegetation became firmly established on the sands, strong westerly winds whipped them into migrating sand dunes. Although these dunes are now stabilized, their shapes reveal to us their origin thousands of years ago.

Much of the meltwater escaped down the Wabash River Valley and thence through the Ohio or the Cache River gap (an earlier path for the Ohio River parallel to and north of its present course). So much valley train was deposited in the Wabash and Ohio valleys that it dammed the mouths of the tributary streams in Illinois and converted them into lakes, further widening and flattening those valley floors (Figure 2-13).

DRAINAGE CHANGES

Figure 2-17 represents the bedrock surface below the glacial drift as envisioned by Leland Horberg after detailed reading of hundreds of well logs.[3] Although a number of refinements have been made in charting buried bedrock valleys since Horberg's monumental investigation, the valley systems which are shown by dashed lines give an essentially accurate picture of the drainageways that were present before glaciation but have since been abandoned and filled with glacial drift. The buried valleys are themselves of great interest because they often serve as sources of groundwater. Sand and gravel deposits (usually old valleys trains) found in these buried valleys can contain a large amount of water, and this water can move freely through the buried alluvium into shallow wells that penetrate it.

There are also important landform results of drainage diversions. New valleys—those formed since the Late Pleistocene—were entrenched during short periods of time and are therefore deep in relation to their width, while preexisting valleys usually have bluffs that are quite far from each other. Preexisting valleys that are still present, but whose streams were permanently diverted to other courses by the later ice invasions, may provide convenient pathways for railways or for canals. New valleys, forced into their present courses by ice invasions, make it easier to build bridges and dams; earlier, they were fordable, and hence helped to direct the paths of early settlement and commerce. The "ford" in the town name Rockford reveals the early significance attached to a rockbottom ford on the Rock River; it helped to develop one of the state's leading cities.

In some areas of Illinois, local stream drainage changes are the most important geomorphological (landform developmental) events that have occurred; in others, these diversions have had far-reaching consequences for the present human geography. Some of the most important effects will be described below.

By far the most significant change was the permanent diversion of the Ancient

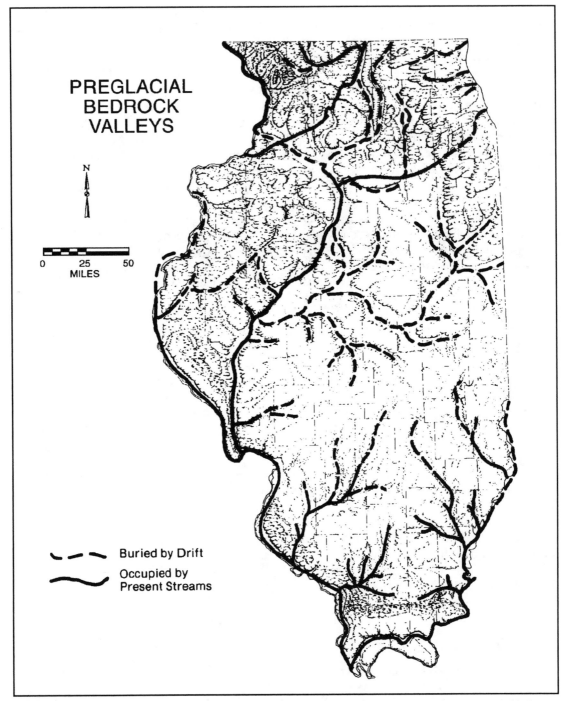

PREGLACIAL
BEDROCK
VALLEYS

N

0   25   50
MILES

- - - Buried by Drift

⌒ Occupied by
Present Streams

FIGURE 2-17. Preglacial bedrock valleys. (Adapted from Leland Horberg, *Bedrock Topography of Illinois*, Illinois State Geological Survey, Urbana, 1950.)

Mississippi. Until about 20,000 years ago that river flowed from a point south of Savanna to the present big bend in the Illinois River at Hennepin through what is today called Meredosia Channel. That great valley, 300 feet deeper than at present, became the route of the Hennepin (Illinois-Mississippi) Canal. Although the canal has not been used for commerce for a long time, its right-of-way, locks, and bridges are visible history. Furthermore, a Chicago-Upper Illinois River-Hennepin Canal-Upper Mississippi Waterway to Dubuque and beyond is often proposed. Such a route, feasible from an engineering standpoint, would of course divert Mississippi River commerce eastward to the southern end of Lake Michigan. The economic effects on Mississippi River cities such as St. Louis and Memphis would be profound.

Eastward from the Hennepin bend on the Illinois the valley is new (cut during the Kankakee Flood); southward from there, the river is an underfit stream, flowing through a wide floodplain. The old (formerly Mississippi) and new valleys together unite the Great Lakes system (via the Cal-Sag Waterway and the Des Plaines and Chicago River tributaries) with the Gulf of Mexico. The canalized river route—the Illinois and Michigan Canal—is the only connection between the Great Lakes-St. Lawrence and Mississippi systems. This fact helps to explain why Illinois—despite its location in the heart of Mid-America—is often foremost among the states in foreign trade.

Vast tonnages of grain and soybeans leave Illinois via Chicago on Lake Michigan and New Orleans on the Gulf by this route. Coal

FIGURE 2-18. The Illinois River upstream from its sharp bend at Hennepin is narrow and occupies a postglacial valley. (Photograph by A.D. Fentem.)

FIGURE 2-19. The first locks on the upper Illinois River were constructed at Lockport, thus uniting the Great Lakes and the Mississippi waterways for barge traffic. (Photograph by A.D. Fentem.)

from southern Illinois, petroleum and sulfur from east Texas and Mexico, copper and tin from Bolivia and Chile, and alumina (from the Caribbean and northern South America via Gulf Coast processors) are among the products that pass along this route and help to determine the industrial character of the Calumet region of Chicago and the Metro East area opposite St. Louis. The worldwide markets of the earthmoving equipment industry at Peoria are brought closer by this waterway. The Port of Chicago, developed around Lake Calumet and connected with the main waterway by the canalized Calumet River, is the place where foreign flag vessels meet domestic barges in Illinois.

Because the river flows in a new valley above Hennepin, the scenery of the upper Illinois valley is quite different and includes the picturesque bluffs at Starved Rock. The valley narrows abruptly above the bend, and barge sizes must be changed at that point. Another stretch of young and strikingly different valley occurs along the Mississippi downstream from the point where the river was finally diverted into its present course by the Woodfordian ice advance (Figure 2-11). As the ice advanced from the east, it successively closed off the channels that could accommodate the waters flowing southeastward through the Meredosia Channel. Finally, at present Cordova, the rising waters spilled over southwestward and cut a narrow gorge (Cordova Gorge) to Rapid City. The river here is quite narrow, and before locks and dams were built this stretch was marked by swift water. A similar reach occurs upstream from Keokuk, Iowa, and marks the site of a major hydro-power dam—the only significant one in the environs of the state.

As may be seen on Figure 2-17, a part of the upper Rock River occupies a new (and therefore scenic) valley between Grand Detour and Bryon, and there is a similar narrows in the Illinois River below Peoria.

SURFACE DRAINAGE CONDITIONS

Except for the farmers who must directly contend with environmental conditions in making a living, most citizens are only intermittently aware that the natural conditions of drainage in Illinois are so poor that in the past they posed a serious problem in using the land. The poor drainage has been overcome or controlled only through great efforts and huge capital expenditures. The flooding of flat-lands along stream courses is common to almost all regions with humid climates, but in Illinois most of the uplands suffered poor drainage as well. For the most part, sluggish and incomplete drainage resulted from the recent glaciations.

No matter what the shape of a land surface at a given time in earth history, natural processes will produce, within a relatively brief period (scores or hundreds of thousands of years), a drainage system competent to remove quickly all the precipitation from the surface and convey it to the seas. Glaciation, however, is a catastrophic process that not only obliterates the pre-existing system, but also deposits earth materials in a topsy-turvy fashion and frequently results in slopes that converge in depressions lacking outlets. Most of the present natural lakes in Illinois are in

FIGURE 2-20. Typical drift surface and glacial lake plain on the Wisconsinan Till Plain of the Wheaton Morainal Country. (Photograph by A.D. Fentem.)

FIGURE 2-21. A common sight on the Bloomington Plain near Mt. Pulaski—standing water after spring rains. (Photograph by A.D. Fentem.)

the northeast, from whence the glacial ice most recently retreated. Although glaciated upland surfaces in Illinois were sometimes very near to being flat, there are undulations almost everywhere. In addition, the floors of former glacial lakes were flattened by the accumulation of sediments, as were the valley trains adjacent to the southern rivers and in the Havana and Green River strath lowlands.

The conditions of drainage encountered by early settlers were influenced by both the terrains just described and by the length of time since glaciation. Floodplain, valley train, and lake basin drainage is accomplished by engineering on a large scale. It requires surveys by civil engineers, the digging of an artificial system of drainage ditches, and the building of levees for deflecting floodwaters to human-constructed basins and channels where

they can be contained. In Illinois, however, the poor drainage of the recently glaciated uplands was a problem encountered and solved by individual farmers and groups of neighbors. And it was a problem with which they were at first ill-prepared to cope.

Two regions of the state, in particular, had to attack the drainage problem early: the west-central area of thick Illinoisan drift, and the Wisconsinan drift region in the east-central part of the state. In both instances, a high proportion of the land was useless until drained. While the northeast embraced large fractions of poorly drained land that included numerous lakes and bogs, these were relatively small and fragmented; the slopes of the closely spaced moraines, as well as sandy expanses (with rapid under drainage), provided early sites for farming. Southern

Illinois, despite its thin drift, also had limitations imposed by poor drainage, but these were brought about, in large measure, by soil characteristics that prevented water from percolating downward.

In the west, poor drainage was more easily ameliorated than in the east. Here the unusual flatness of the Illinoisan drift surface was frequently interrupted by the ramifying tributaries of a stream system that had been developing for perhaps 100,000 years. The solution turned out to be the installation of networks of tiles made from fired clay. Most farmers in the region initially were unfamiliar with their use, and tiles were not at first readily available. Also, digging a mile of ditches and installing thousands of tiles to drain a forty-acre plot was no picnic. Nevertheless, the fact that there was usually near at hand a gully heading the natural drainage system made the projects feasible. In effect, the natural system was completed by an artificial underground system. If we could "X-ray" western Illinois, we would see that nearly all the flat upland is now tiled.

In the "Prairie Province" of east-central Illinois, the low-lying stretches between the great moraines were, during the wettest part of the year, little better than marshes. There are early travellers' accounts of standing water stretching almost unbroken for miles in spring. The younger drainage system here is much less complete, and the shallow valleys much further apart than in the west. The heads of all the rivers draining the region focus—and their tributaries radiate—from a point north of Champaign-Urbana. Tiling and hand-dug ditches would not suffice, neither could drainage be undertaken by individuals without vast wealth or large acreages of land.

FIGURE 2-22. Corn damaged by standing water in Coles County. (Photograph by A.D. Fentem.)

FIGURE 2-23. A field in western Illinois with a recently installed tile network. (Photograph by Scott Miner.)

Illinois, unlike Indiana and Ohio, did not pass enabling legislation providing for public surveyors or for compacts organizing drainage districts. The effects on settlement and agricultural development were profound. Poorer settlers tended to be excluded because they could not afford the large expenditures necessary to make farms profitable; one large farmer devised a ditching plow drawn by 65 oxen! Landownership plots became unusually large while tenant farming and pioneering went hand and hand. Settlement was doubtless impeded and delayed. The land was not completely brought into production until the twentieth century—partly because of these circumstances, and also because the marshlands bred mosquitoes which in turn subjected the

pioneers to frequent bouts with "the ague," as malaria was then called.

LANDFORM REGIONS

Having surveyed the distribution and development of the various aspects of landform, we will now turn to a regional description summarizing the relations among the elements which give particular character to places.[4] On Figure 2-24, boundaries have been drawn about areas that have similar landform characteristics. A classification of the regions themselves should help to make the differences and similarities among regions more clear (Figure 2-25).

Since the driftless areas have escaped the most dramatic effects of Pleistocene events,

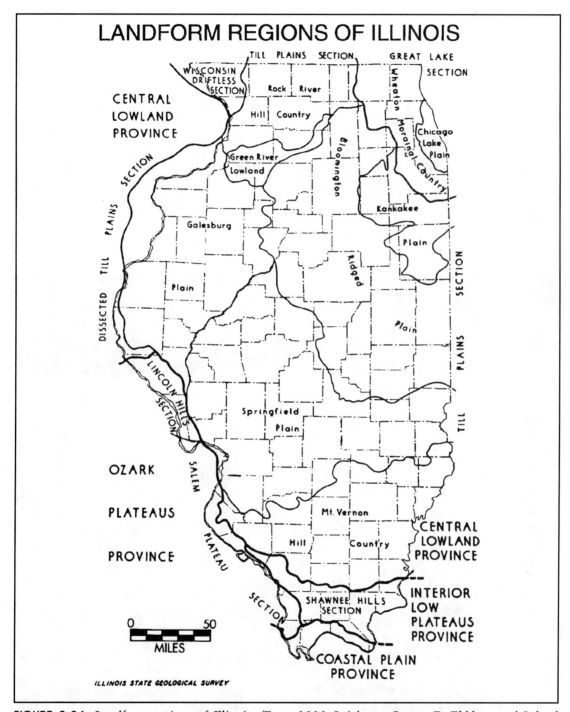

FIGURE 2-24. Landform regions of Illinois. (From M.M. Leighton, George E. Ekblaw, and Leland Horberg, *Physiographic Division of Illinois,* Illinois State Geological Survey, Urbana, 1948.)

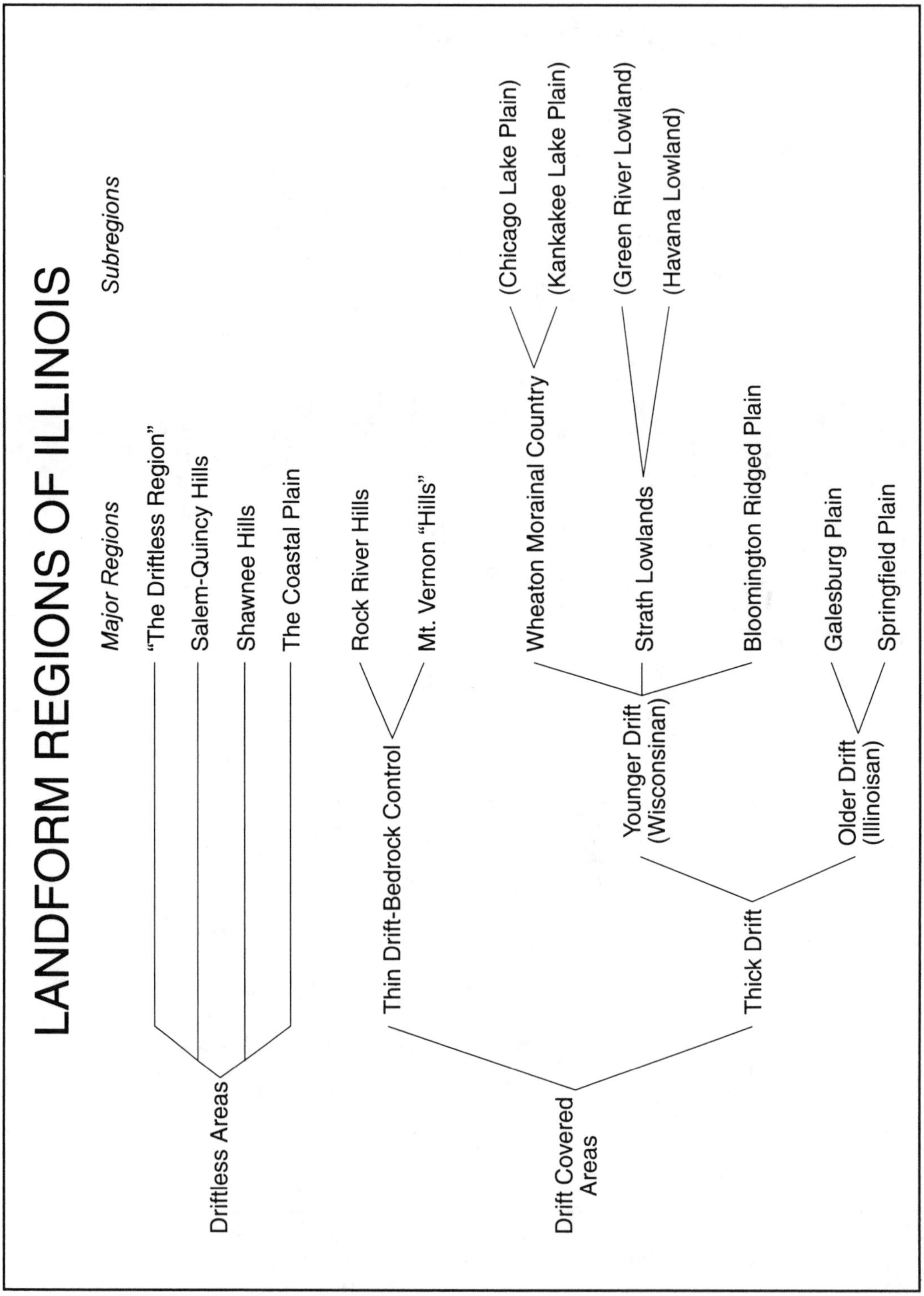

LANDFORM REGIONS OF ILLINOIS

Major Regions   *Subregions*

Driftless Areas
- "The Driftless Region"
- Salem-Quincy Hills
- Shawnee Hills
- The Coastal Plain

Drift Covered Areas
- Thin Drift-Bedrock Control
  - Rock River Hills
  - Mt. Vernon "Hills"
- Thick Drift
  - Younger Drift (Wisconsinan)
    - Wheaton Morainal Country
      - (Chicago Lake Plain)
      - (Kankakee Lake Plain)
    - Strath Lowlands
      - (Green River Lowland)
      - (Havana Lowland)
    - Bloomington Ridged Plain
  - Older Drift (Illinoisan)
    - Galesburg Plain
    - Springfield Plain

FIGURE 2-25. Classification of the landform regions of Illinois.

they have the fewest complications in their developmental histories and are probably the most scenic parts of the state because their angular features have not been obscured by glacial erosion and deposition. Both the Driftless Region (a name generally used to describe the adjacent unglaciated portions of Illinois, Iowa, Minnesota, and Wisconsin) and the Shawnee Hills are beautiful, if not very productive.

The greatest local relief (difference in elevation between the highest and lowest points within small areas) in the state occurs in the *Driftless Area* where the Mississippi cuts through the Niagara Escarpment, that low and nearly continuous range of hills that dominates the region and marks the outcrop of Niagaran dolomite. The north-facing slopes of the hills generally drop quite sharply for a hundred feet or so and are then gently concave until they pass through the underlying shale and reach the Galena-Platteville dolomitic limestone, a beautiful buff-colored stone that has been used in the construction of attractive old houses and country schools in the area. On the southwestern "backslopes," elevations decline much more slowly, but these gentle slopes are interrupted by narrow valleys that deepen with increasing distance from the crest of the escarpment. The escarpment should not be pictured as a sharply defined feature; in fact, it is quite "ragged" and extends bold promontories toward the north. Considerable fragments have also become detached from the retreating face of the escarpment and stand as buttes or "mounds" on the gently undulating (Lancaster) plain that stretches northward far into Wisconsin. Some of these picturesque hills are many tens of miles away from the escarpment itself and bear names such as Scales Mound and Charles Mound

FIGURE 2-26. Outliers of the Niagara Escarpment, called "mounds," rise above the Lancaster surface. (Photograph by A.D. Fentem.)

FIGURE 2-27. A view along the Niagara Escarpment showing a peninsula of that cuesta in profile. (Photograph by A.D. Fentem.)

in Illinois; Sinsinawa Mound in Iowa; and Belmont, Platte, and Blue Mounds in Wisconsin. They are exceptionally scenic and interesting, and afford magnificent views from their flattish crests.

Finally, the latest cycle of erosion has cut deep and narrow valleys into the Ordovician rocks of the Lancaster Peneplain, which are themselves quite attractive. One of these, the Apple River valley, steeper and deeper than most because of its youth (waters in a tributary of the Rock River were impounded behind an advancing ice dam in the vicinity of Stockton, Illinois, and spilled over into the Mississippi drainage during the Pleistocene), has become a tourist attraction. One of its branches has become a state park, while the other is being developed as a private venture. The wide floodplains of these streams in their lower courses have developed, to a large ex-

tent, since settlers came upon the scene, cleared and cultivated the hills, and initiated lead mining. Galena River was once wide and deep enough to allow steamboats to dock bow-to-shore at the town of Galena; now their passage alone could not be accommodated.

The *Shawnee Hills* may be even more impressive in their natural scenery. Like the Driftless Region, this area is traversed by a bold and prominent escarpment with elevations exceeding 800 feet and with precipitous slopes on its outfacing (southern) flank. Magnificent views are to be seen from secondary roads that follow the crest. Near the western end a prominent hill, Bald Knob, not unlike the mounds of the northwestern corner of the state, rises more than 200 feet higher; its even crest may be, like those of the northern mounds, a remnant of the oldest (Dodgeville) erosional surface.

Southward from their highest elevations, the Shawnee Hills descend like a series of giant stairsteps (each step marking the outcrop of a gently inclined and resistant rock stratum) to the floodplain now occupied by Cache River and the Ohio River, which surrounds—like a sea about an island—the low and gently rounded Cretaceous Hills. South of Carbondale, Giant City State Park has been created about house-sized sandstone blocks piled topsy-turvy along the foot of the escarpment. When the Illinoisan ice advanced from the north—moving upwards along the backslope of the Pennsylvanian Escarpment, its meltwaters were impounded by the higher elevations there, creating a great "head" of water that was then forced, through fissures

and along bedding planes in the rock, to escape along the escarpment face, which is also known as the Dripping Springs escarpment. Soft shales just below the massive sandstones were by that means removed from beneath, and the sandstone blocks slid and toppled into their present positions. Local faulting is responsible for still other impressive scenic features in the Shawnee Hills region.

Along the western extremity of the state there are narrow uplands also lacking glacial features and bordered on one or more margins by the bluffs of the major rivers, the Illinois and the Mississippi. This *Salem Upland-Quincy Hills* region is essentially a relatively narrow ribbon of gently undulating, limestone supported, and highly elevated

FIGURE 2-28. Erosion-control terraces constructed on the limestone topography of the Quincy (Lincoln) Hills. (Photograph by A.D. Fentem.)

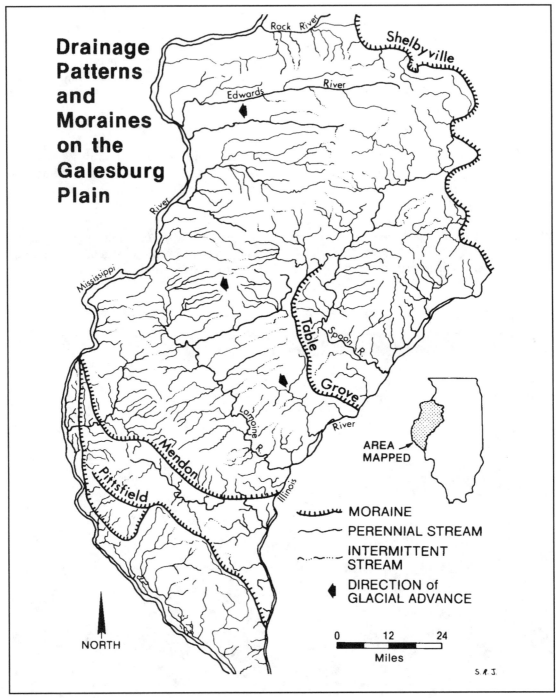

FIGURE 2-29. Drainage patterns and moraines on the Galesburg Plain. (Adapted from Fred Caspall, "Parallel Drainage in West-Central Illinois," Western Illinois University, Macomb, 1965.)

FIGURE 2-30. From the air, the absence of stream valleys on the Bloomington Ridged Plain is shown by rectangular fields and nearly 100 percent cultivation. (Photograph by A.D. Fentem.)

plain that extends ridges to the west (in the north) and both east and west (in the south) in a herringbone pattern. These ridges are separated from each other by short but steep re-entrants and terminate on the floodplains. The ribbon of upland (the Lancaster Peneplain) broadens to the north in Pike and Adams counties and narrows to hundreds of yards wide in the south between the two rivers before ending abruptly at the Cap Au Gris fault. Southward from there the hummocky karst topography described earlier occupies part of Calhoun County.

Two landform regions of the state have features only partially subdued by glaciation. Despite the similarity of their genesis, they are quite different in appearance. The *Rock River Hills* are very much like what the Driftless Region would be without the Niagara Escarpment, which ''peeks'' through the drift only occasionally in the extreme west. The region consists, in its major aspect, of a broad and gently undulating plain, which is thought by some to represent the Lancaster erosional surface. A thin layer of drift modulates, but does not obscure, its erosional origin. Occasionally, there are small patches of exceptionally flat upland that probably mark the sites of former glacial lakes. The region's scenic beauty is provided by the stream valleys that interrupt the surface. Tributaries of the Rock River, chiefly the Pecatonica, have cut valleys up to 250 feet deep. Because the rock is hard, the valleys are narrow and have steep and prominent sides.

The *Mount Vernon Hill Country*, on the other hand, has almost no relief features of note and only rarely does local relief reach as high as 100 feet. There are three reasons for this contrast with the Rock River Hills. First, the surface is but little higher than the master streams (the Mississippi, the Ohio, and the Wabash) so that valleys were deepened but little and only very slowly. This situation allows valleys to become exceptionally wide, even as they are incised. Second, the weak rocks and the veneer of glacial drift were soft and easily eroded—still another reason for the extreme width of the valleys. Third, the valleys have probably been filled with a great amount of outwash during glacial melting. Perhaps half the total surface, then, consists of the nearly featureless flats along the water courses. Between these wide floodplains, the very low and gently sloping hills are usually no higher than the tallest trees. The effect on the traveller is one of great monotony. Here and there, especially northeastward from St. Louis, are subdued remnants of Illinoisan moraines, but these too rise only a hundred feet or so above the gently sloping uplands. In the Wabash valley, the valley train was so voluminous that it obliterated the upland over large areas, leaving only occasional island-like hills to protrude through the glacial-fluvial materials.

Having experienced similar histories, the *Galesburg Plain* and *Springfield Plain* are quite similar in their geomorphic landscapes. Both have thick Illinoian drift and are characterized by a high proportion of unusually flat land. In both cases, the flat land is located at the upper end of the elevation range; small

FIGURE 2-31. "Swell and swale" topography typical of young drift surfaces. (Photograph by A.D. Fentem.)

FIGURE 2-32. The Cal-Sag Channel, the canalized route of one branch of the Kankakee Torrent, skirts the southern end of a moraine remnant called Blue Island. On the site of Blue Island is the modern city with that name. (Photograph by A.D. Fentem.)

and scattered moraine remnants stand on top of the high flats in both regions, and both have dendritic (treelike) stream systems that are actively reducing the high flats to sloping land.

Despite these many similarities, there are grounds for distinguishing between the two. The differences are in local relief and in the forms of the stream valleys. The valleys of the Galesburg Plain are deeper by 60 or 80 feet. (Spoon River and Crooked Creek valleys are about 160 feet deep in their middle courses.) These valleys also have steeper sides and narrower floodplains than does, for example, the Sangamon valley of the Springfield Plain. When we recall that western Illinois has a higher surface and one floored by stronger

(mostly Mississippian) rocks, these differences are expectable. In order for the stream valleys of the Galesburg Plain to approach their base levels, they have had to deepen more quickly, while erosion acting on the valley sides has been operating for the same length of time. And valley sides that expose dolomites will retreat more slowly than will those that develop largely on shales.

The Galesburg Plain has two other minor characteristics that lend some distinction. The first of these is a large area of exceptionally straight and parallel valleys; the second consists of a sprinkling of small, shallow, and roundish basins. The Buffalo Hart (now called the Table Grove) and Mendon moraines shown on Figure 2-29 appear to mark a stage in the retreat of the Illinoisan ice sheet during 'which the thickness of the ice was unusually small and the melting continuous and rapid. Geographer Fred Caspall has demonstrated that the surface in this region bears remarkably straight and parallel low ridges and alluvial (water lain) deposits trending northeast-southwest at right angles to the ice margin.[5] Scattered shallow depressions in this same region are thought to mark the sites where icebergs ran aground. Deposition of sediments from glacial meltwater might have built the surrounding plain a little higher. It is possible also that these depressions, which would have become shallow lakes and marshes later on, could have attracted grazing animals, probably buffalo, and that their trampling and destruction of vegetation exposed the basins to wind erosion.

As has been previously described, the area covered by thick Wisconsinan drift—the *Bloomington Ridged Plain* and the *Wheaton Morainal Country*—owe their peculiar character to the recency of their glaciation and to the manner in which the glacial materials were deposited. Both areas have a distinctive pattern of curving low ridges (the moraines), a gently undulating surface, and somewhat sketchy and incomplete drainage systems. On the Bloomington Plain local relief is everywhere low and there are no deep valleys to lend interest to a somewhat monotonous landscape. Early explorers, impressed by the near absence of trees, often likened its appearance to the sea. (Those who live there may be compensated for the dearth of scenic excitement by their knowledge that most rural square miles have a value of well over a million dollars). The bolder and more closely spaced moraines of the Wheaton Morainal Country provide a more exciting and varied natural landscape. Not only are the ridges steeper on their flanks, but their crests are sometimes broken by conical knolls and sharp depressions (kames and kettles). The courses

of the major streams have been directed by the curving ridges so that the lowlands between them have become stream valleys as well. Since the drift was topsy-turvy to begin with, and because the time elapsed since deposition has been so short, the area is dotted by small lakes, marshes, and bogs. The materials that make up the drift include a great deal of sand and gravel, and this too helps to vary the natural environment; within short distances, areas that are low, flat, and wet alternate with those that are high, steep, and dry.

Two lowlands within the area covered by the Wisconsinan ice, the *Kankakee Lake Plain* and *Chicago Lake Plain*, are the beds of former lakes. They are very flat, lower than surrounding regions, and are almost devoid of stream valleys. Among the most conspicuous relief features are railroad embankments and other features of human construction. Nevertheless, some variety is imparted by differences in materials, as well as by fluctuations in Pleistocene lake levels and by the Kankakee Flood, an event described earlier.

The deposition of silts and sands in Glacial Lake Kankakee subdued the already smooth surface of the Kankakee plain. Part of the area was modified by the flowage of a vast amount of water across the plain toward the end of the brief period during which all the interconnected lakes were being lowered after the breaching of the Marseilles Moraine. The swiftly moving water transported sands and gravels, which were deposited as gigantic bars as the torrent finally ebbed. As seen from above, the pattern of bars resemble that which might be seen on the bottom of a washtub after extremely dirty water has been drained from it by "pulling a plug" at one edge.

The Chicago Lake Plain is varied by two kinds of features—beach ridges (gently curving and parallel sandy ridges) that mark former beaches, and remnants of moraines that now stand as "islands" above the former lake bottom. One of these, Blue Island, is the site of the town with the same name. As in the Kankakee Plain, the materials of the lake floor range from fine clays to sands and gravels. Both the bogs in the lower areas and the sands were ideal for producing vegetables and before these foods were imported from great distances, truck farming was an important industry here. The beach ridges, because they were dry, helped determine early routes of travel and continue to be used as major throughfares today.

There are two additional landform regions that, like the lake plains, are lower in elevation than their environs and also lack stream valleys—the *Green River Lowland* and *Havana Strath Lowland*. The similarities of their features are matched by similarities in their origins. Both have flattish bedrock floors and both have accumulated huge amounts of glacial-fluvial materials (outwash) that contain fairly large quantities of sand. Viewed from the prominent Bloomington Moraine, the Green River Lowland appears to be almost featureless. Green River itself has been so modified that it resembles a system of drainage ditches more than it does a natural stream. From the floor of the basin the imposing wall of the bounding moraine stretches along the southeastern horizon and from this vantage point one becomes aware of small "hills," a few tens of feet high, and may note that their southeastern slopes have frequently been chosen as sites for farmsteads. In the far west, these features sometimes coalesce into fairly extensive tracts of low hills that resemble giant waves. They are in fact former sand dunes whose migrations across the lowland ceased when vegetation became established on them. Their soils are so poor and excessively drained that they are almost useless for crops and are usually relegated to scrub timber or poor pasturage. The hills are favored for building because they make possible the

avoidance of wet basements and muddy feedlots that prevail elsewhere. Irrigation is practiced to some extent on the sandier areas. The sand hills disappear to the east where extensive peat bogs once existed, and in the narrow panhandle some of the best soils in the state are devoted to seed corn. The Havana Strath is much smaller, but very similar in its alteration of sandy relict dunes and intervening flats. Here the accumulation of coarse drift between the Wisconsin terminal moraine and the Illinois River is so great that it can contain a huge amount of easily accessible ground water, some of which is now used to irrigate about 160 farms in the area.

NOTES

1. Most of the information about landforms has been gleaned from publications of the Illinois State Geological Survey, which have been comprehensively compiled and indexed in H.B. Willman, Jack A. Simon, Betty M. Lynch, and Virginia A. Langenheim, *Bibliography and Index of Illinois Geology through 1965,* Bulletin 92 (Urbana: Illinois State Geological Survey, 1968).

2. The most comprehensive survey of ice age events and consequences in Illinois is in H.B. Willman and J.C. Frye, *Pleistocene Stratigraphy of Illinois,* Bulletin 94 (Urbana: Illinois State Geological Survey, 1970).

3. The pioneering work of Leland Horberg is the most important source of knowledge about the bedrock surface. See his *Bedrock Topography of Illinois,* Bulletin 73 (Urbana: Illinois State Geological Survey, 1950).

4. The landform regions described are based on M.M. Leighton, George E. Ekblaw, and Leland Horberg, *Physiographic Divisions of Illinois,* Report of Investigations No. 129 (Urbana: Illinois State Geological Survey, 1948).

5. Fred Caspall, "Parallel Drainage in West-Central Illinois" (M.S. thesis, Western Illinois University, 1965).

3

THE PHYSICAL ENVIRONMENT: CLIMATE, VEGETATION, AND SOILS

Arlin D. Fentem
Western Illinois University

To consider the climate of a single state comprising only a tiny fragment of a continent is to risk loss of perspective and comprehensive understanding of variable and changing atmospheric patterns. For the global atmosphere is a restless unitary system; nothing occurs within it that does not ramify throughout the system and hence affect all areas of the earth. The word *climate*, itself, refers to an average of all the meteorological events that occur in the atmosphere over an area throughout a period of years. When it is remembered that a single day at a midwestern town may have a temperature range greater than the average difference between the coldest and warmest months at Springfield or Chicago, it may be seen that the generalization we call climate is capable of obscuring as much as it reveals. And the climate for one year may be quite different from that of the next. For these reasons, the discussion of Illinois climate will begin with the presentation of climatic norms, and then go on to a consideration of dynamic (changing) conditions in the atmosphere that have brought them about.

THE CLIMATE OF ILLINOIS

Temperature and precipitation, the primary ingredients of climate, both vary markedly over Illinois, not only because the state is almost 400 miles (or six degrees of latitude) from its northern border to its southern tip, but also because of its situation in relation to air masses and fronts (boundaries between air masses), to major storm tracks which move along the fronts, and to the Caribbean Sea and the Gulf of Mexico—the sources of most atmospheric moisture in Illinois.

Temperature

Two maps in Figure 3-1 show the seasonal distributions of temperature over the state, while a third map in the series relates those to the most significant measure for crop growth—the growing season. For so small a state, the differences between the northern and southern extremities are truly remarkable. As suggested by Figure 3-1, heating bills are nearly twice as high at Rockford as at Cairo. Equally as significant for human needs and

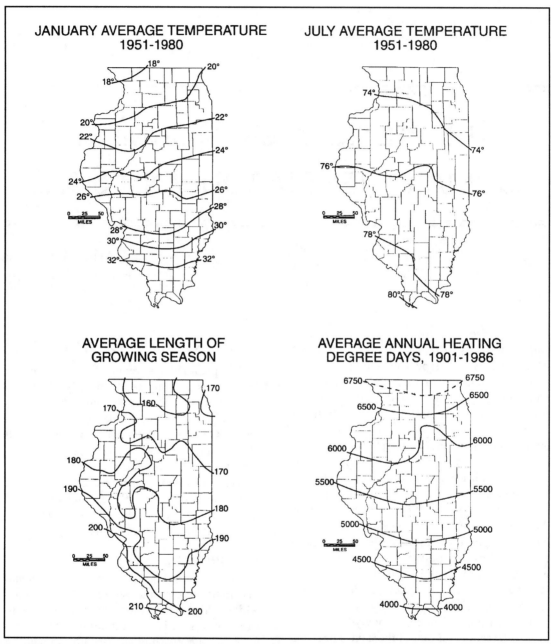

FIGURE 3-1. Temperatures, heating degree days, and growing season. Average temperatures for 1951-1980 compiled from U.S. National Oceanic and Atmospheric Administration, *Climatological Data: Illinois Annual Summary,* 1990. Heating degree days from William Easterling, et al., "The Appropriate Use of Climatic Information in Illinois Natural Gas Utility Weather Normalization Techniques," Illinois Department of Energy and Natural Resources, *Report of Investigations* 112, Illinois State Water Survey, 1990. Growing season data adapted from *Water Resources and Climate,* Section 1 of *Atlas of Illinois Resources,* Illinois Department of Education and Registration, 1958.

TABLE 3-1. Selected Climatic Data for Rockford, Springfield, and Cairo, Illinois.

City	Summer Precip.	Winter Precip.	Jan. Temp.	July Temp.	Annual Range	Days with Snow	Days with Snow Cover	Days of Growing Season
Rockford	11.4"	5.8"	18°F.	73°F.	52°F.	34	80	160
Springfield	9.9"	6.3"	25°F.	76°F.	51°F.	20	40	188
Cairo	9.9"	10.5"	33°F.	78°F.	45°F.	10	20	210

use, the range in the growing season made the northern quarter of the state very risky for growing corn (before hybridized seeds and grain dryers) while permitting cotton to be grown at the foot of the Pennsylvanian escarpment in the south. The types of agriculture practiced—from dairying in the north to "plantation" cotton in the south—were begun, at least in part, in response to these climatic differences.

Two conditions, "continentality" and relatively high latitudes, combine to bring about great ranges of temperature for locations in the interiors of large land masses. "Continentality" refers to the fact that land heats more rapidly than the seas when it is receiving more solar energy than it is radiating to outer space (during the summer half-year) and cools more rapidly when incoming energy is less than that lost at the outer limit of the atmosphere. The higher the latitude (distance north or south of the equator), the greater the seasonal difference in the length of day and in the angle at which the sun's rays meet the earth's surface. Since both direct rays and long days increase the energy received from the sun, colder winters and relatively hotter summers accompany increasing latitude. The high annual ranges of temperature in Illinois between the averages of the coldest and warmest months (from 43°F in the south to 55°F in the northwest) are thus shared with the interior of the country generally. The higher ranges in the northwest result from both its

higher latitude and (especially) from its greater distance from the moderating influence of the sea.

Comparison of the maps in Figure 3-1 and reference to Table 3-1, however, shows that there are still other important controls of temperature differences. The *difference* in annual temperature range from north to south is mostly the result of differences in the winter temperature. Although both the January and July maps use an interval of two degrees F between isotherms (lines connecting points with equal temperatures), there are twice as many of these lines on the January map as on the one for July. The rates of change (temperature gradient) are -1°F per 66 miles northward at the height of summer and -1°F per 26 miles in midwinter. A part of this discrepancy may be attributed to the effects of latitude; in July the day length at Rockford is about a half-hour longer than at Cairo, but in January the sun is above the horizon at Cairo for the longer time. The more important reasons for the steeper winter gradient, however, are connected with another control of climate—air masses and the fronts that exist between them. The air mass source regions affecting Illinois are shown on Figure 3-2. Most of the weather we experience is associated with the polar continental (cP) and tropical maritime (mT) air masses and the interactions between them that take place when they are in contact over or near Illinois.

Fronts are boundaries between air masses. During the First World War it was discovered that when air masses with different humidities

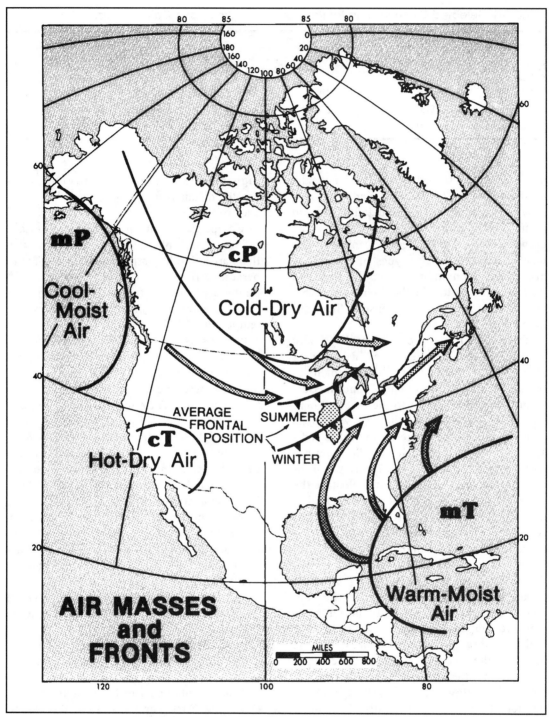

FIGURE 3-2. Air masses and fronts. (Adapted from Robert E. Gabler (ed.), *A Handbook for Geography Teachers,* National Council for Geographic Education, 1966.)

and temperatures meet, they resist mixing and tend to maintain their internal characteristics just as two spreading blobs of gelatin on a glass plate would do when they met. Both polar and tropical air masses are "fed" from above by air settling from high in the atmosphere and the volumes of descending air are constantly changing; therefore, one of the two air masses in contact is usually advancing at the expense of the other. To meteorologists, the analogy with shifting battle fronts in Europe seemed clear, and the boundaries were dubbed "fronts" for that reason.

On Figure 3-2 the positions of the polar front are approximated for both summer and winter, and this feature of the map illustrates why the winter temperature gradient is so steep. Quite often, the front in winter divides Illinois so that the cold and dry polar air lies over the north and warm, moist tropical air over the south. The streamlines that approximate the directions of air movement are another indication of the same situation—the air streams in the two air masses are indeed "advancing" against each other.

To visitors from other climates, one of the most surprising (and distressing) features of the Midwestern winter is the rapid day to day alternations in temperature and moisture characteristics that result from the to-and-fro migration of the polar front. The immediate cause of these abrupt changes is the passage of cyclonic storms along the front. When the continent is viewed from a few hundred miles in space (as is done by sensors in artificial satellites) the polar front boundary appears as a line. One way to conceive a cyclone is as a wave travelling along this boundary—just as a "wave" travels along a loosely suspended rope when it is sharply shaken at one end. A typical wave is several hundred miles from trough to crest (south to north) and even wider in its east-west dimension. In winter, the crest (northerly apex) of a typical wave (cyclone) might progress from Denver to New England in about three days. And during that three days, the people of Peoria might experience one cold and cloudy day as the storm approached, a rapid rise in temperature as the "warm front" of the wave passed the city, a day of unusually high temperatures, and, finally, what we call a "cold wave" as the "back" side of the wave passed by and placed central Illinois behind the "cold front" and again in the polar air mass. (See Figure 3-3.)

It frequently happens that the bulge of cold dry air behind an advancing cold front has so much strength that it advances more rapidly than the leading (warm) front. The wave may become so steepened that a huge "bud" of the polar air mass bursts through the front as a great, roughly circular dome of extremely cold and dry air occupying much or all of the eastern United States. These buds or domes are called anticyclones and bring with them bright, sunny, and extremely cold weather. Recalling the analogy with military movements, such an event is similar to a massive "breakthrough" by the "forces of the north" which begins as a major flanking movement on the "right" while giving up ground on the "left."

One final explanation for the steeper winter temperature gradient may be understood by referring to Figure 3-4 and Table 3-1. These illustrations reveal that the number of days with significant snowfall increases more rapidly with distance northward in the "top" half of the state than in the south (the isolines are more closely spaced from about the latitude of Peoria northward), and that the number of days with snow on the ground is distributed in much the same way. The effect is to make the northern part of the state colder than it otherwise would be. The sunshine that penetrates the earth's atmosphere and reaches the surface includes a high proportion of short-wave lengths (in the violet and ultraviolet range of the visible light spectrum). When the surface that receives the light is dark, most

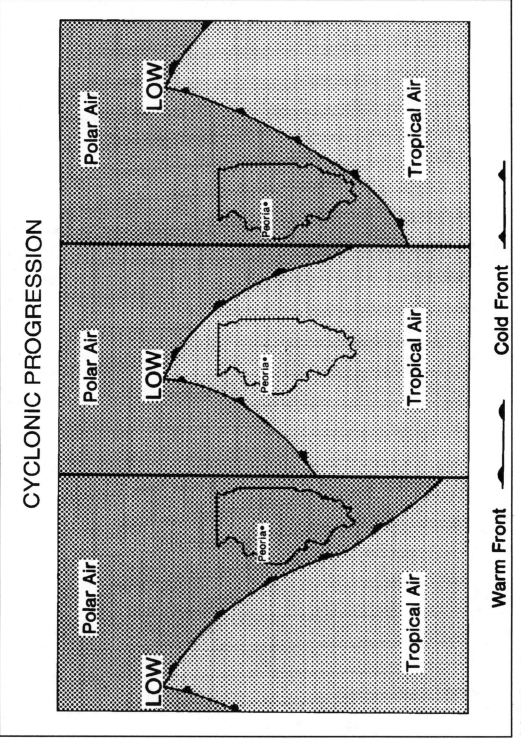

FIGURE 3-3. The progression of a typical cyclone.

of the solar energy is "absorbed" and is converted to heat. The warm earth surface radiates energy just as the sun does (although at much lower levels), but in this case the energy is transmitted outward through the atmosphere by much longer waves. These waves cannot pass through the atmosphere without being absorbed by the molecules that make up the gases in the air (especially water "gas" or vapor and carbon dioxide) and by pollutants such as dust and smoke particles. This uninterrupted passage of incoming radiation, its absorption at the surface, its conversion to long-wave radiation, and its eventual "entrapment" is called the "greenhouse effect." (In greenhouses, the glass panes greatly reinforce the effect of the atmosphere itself.) When snow covers the ground, however, the incoming solar radiation is reflected rather than absorbed. This is the reason that snow appears "white;" it is "returning" all the light that it receives. When a surface appears black, it is absorbing all the energy. The reradiated long waves are invisible, although some of them will register on specially prepared (infrared) photographic films. The short waves of energy reflected from the snow-covered surface pass as easily through the atmosphere (without warming it) as they did on their inward journey, and are lost to space. Thus, on a great many days each winter (twice as many at Rockford as at Springfield), the sun is heating the atmosphere much more efficiently in the south than in the north—even on days when the amount of solar energy received is the same.

Precipitation

Figure 3-4 and Table 3-1 show seasonal and annual distributions of precipitation over Illinois. Although there is considerable variation from place to place, the decrease in precipitation northward is compensated somewhat by decreased evapotranspiration. Because different crops have greatly different water requirements, it is never possible to say that any region has an optimum supply of moisture. Corn and soybeans, however, dominate the field crop agriculture of the state to an unusual degree, partly because of their suitability to Illinois' soils. If every year were an average year, the distribution of precipitation in time and space would be close to ideal for these crops. Indeed, the suitability of the physical environment in Illinois for farming is matched by few places on earth. Most of the environmental difficulties experienced by Illinois agriculturists occur during occasional drought years, or because rains fall at inopportune periods during the growing season.

The decrease northward in annual precipitation is entirely the result of the steep gradient during winter (Figure 3-4); the summer rains are actually a little greater along the northern border than in the southern half of the state. Both the winter and summer patterns are best understood in relation to the general circulation of the earth's atmosphere and to the polar fronts that help produce the contrast in seasonal temperature gradients already described.

On an average day, the atmosphere over Illinois contains about two trillion gallons of atmospheric moisture and approximately five percent of that falls as precipitation. An even lesser amount is contributed by the surface of Illinois to the atmosphere through evaporation and the transpiration of water vapor by vegetation. Most atmospheric moisture, including even that in the interiors of continents, is evaporated from the seas. Because Illinois is situated so far eastward in North America, air transported from the mT source region over the Atlantic and the Gulf of Mexico supplies a high proportion of its atmospheric moisture.

The probability of precipitation for a particular place depends in part upon whether or not Atlantic-Gulf air is present. The United States is situated in the Polar Westerlies, a

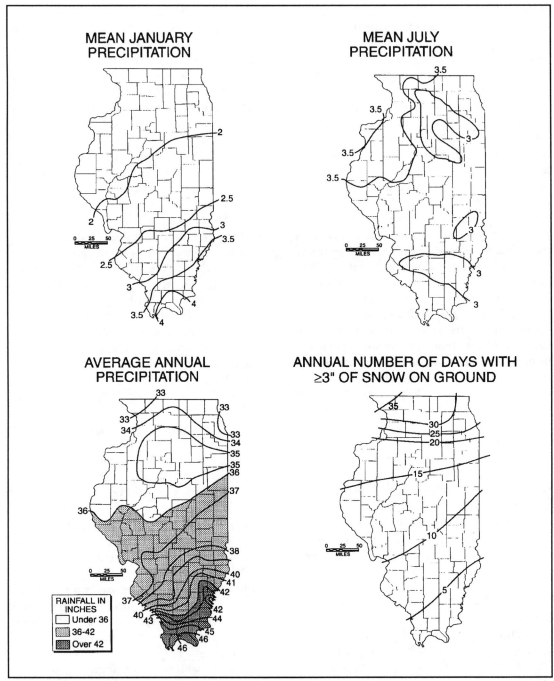

FIGURE 3-4. Precipitation and snow cover. Precipitation data compiled from U.S. National Oceanic and Atmospheric Administration, *Climatic Data: Illinois Annual Summary*, 1990. Snow cover adapted from Stanley A. Changnon, Jr., *Climatology of Severe Winter Storms in Illinois,* Bulletin 53, Illinois State Water Survey, 1969.

latitudinal zone in which prevailing winds, both at the surface and aloft, are from westerly points. The energy that drives the Westerlies (which are but one component of the planetary wind system) is the steep gradient in temperature northward. As the foregoing discussion of temperature gradients points out, this gradient is much less steep in the summer hemisphere, and the Westerlies are therefore much weaker then. Consequently, continentality— the heating of air over continents—tends to make summers a great deal hotter than average for given latitudes. Gases expand when heated and become less dense; air during the summer, therefore, tends to drift into the interior of the continent from the (relatively) cooler seas. At the same time, the Westerlies, along with the other planetary winds, shift northward. The result is (on average) a slow drift of warm, moist air up the Mississippi Valley and overspreading Illinois. The surface Polar Front boundary between polar and tropical air masses in summer is usually located north of Illinois. In winter, on the other hand, the Polar Front frequently bisects the state, so that the prevalence of the moisture-rich air mass is much greater at Carbondale than at Rockford.

Almost all precipitation is brought about by cooling large volumes of the atmosphere; cooling *of this kind,* in turn, is almost always a consequence of great vertical displacements. Although air may be cooled in other ways, temperature drops that produce thick clouds result from the expansion of parcels of air when they are lifted from near the surface (where they are compressed by the entire weight of the overlying atmosphere) to higher levels (where the weight and densities are less). If upward-moving air is cooled to a critical temperature—the "dew point"—the condensation of water vapor into water droplets begins, clouds form, and precipitation may result. One of the "triggers" of upward-moving columns of air is the heating that takes place at the base of the atmosphere by radiation from the earth's surface on hot summer days. When moist tropical air overspreads the state, as it does in summer, air mass thunderstorms may occur in one place about as well as in another, which is one reason why summer precipitation varies so little over the state.

The waves (cyclones) along the Polar Front, which are so important in understanding the changeability in temperature, are also closely connected to precipitation patterns.[1] While a detailed explanation of cyclonic precipitation is beyond the scope of this text, the following discussion is intended to draw a general picture of the processes involved and to relate them to the planetary circulation. The great stream of air flowing from west to east in these latitudes may be thought of as a fluid moving in a gigantic "river" at speeds which increase with altitude. For reasons too involved to discuss here, this stream of air circles the North Pole in a series of huge waves so that its path across the continent curves gently. One of these waves is shown over the United States in Figure 3-5. The speed of this river of air is greater at some places than at others, and is greatest over places where the temperature gradient (which supplies the energy) is also greatest. That location is, of course, the Polar Front—especially in winter when the air arriving from Canada may be well below 0°F. This narrow zone of very high velocities near the lower limit of the stratosphere has been named the Polar Front Jet Stream.

Within the Polar Jet Stream, the surface cyclones interact with the upper air waves to produce surges of greater wind speed that travel through the long waves. Air moving southeastward along the stream lines is compressed as it changes direction (following the wave form) to the east, subsides (sinks) in the atmosphere, and tends to bring with it an outbreak of polar continental (cP) air. As the air passing through the surge turns northeastward, its velocity slows and the air begins to spread out (diverge), creating lower density and favoring the movement of air upward from the surface (Figure 3-6).

In the earlier description of cyclones, the movement of air at the surface was described as currents advancing against the fronts. This *convergence* of surface air toward the crest of the cyclonic wave is greatly accelerated if a Polar Front Jet Stream surge is passing over the surface front, because the zone ahead of the speed surge produces divergence at upper levels to make "room" for converging and rising air within the cyclone. The horizontal picture of surface winds and fronts presented earlier is accurate, but a cyclone may also be conceived as a huge, rotating mass of air with strong upward movement. In many ways it

resembles the eddies that form in strongly-moving (and especially) curving streams of water. Both the waves at upper levels and the counterclockwise circulation of air near the surface in cyclonic storms result in part from the rotation of the earth, from frictional or "shear" effects as streams of air encounter each other, and from turbulence caused by irregularities in the earth's surface. In the United States, the chief source of surface modification on air streams and frontal positions is the Rocky Mountains.

The development of waves (cyclones) on the Polar Front is thus seen to be closely related to the path of the Polar Jet Stream, and the Jet Stream is sometimes said to "guide" the cyclones in their courses across the continent. There is considerable stability in the

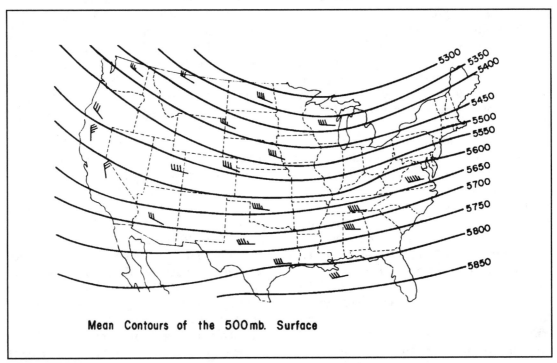

Mean Contours of the 500 mb. Surface

FIGURE 3-5. A wave over the United States. From TROPOSPHERIC WAVES, JET STREAMS, AND UNITED STATES WEATHER PATTERNS by Harmon. Copyright 1971 by Association of American Geographers. Reprinted by permission.

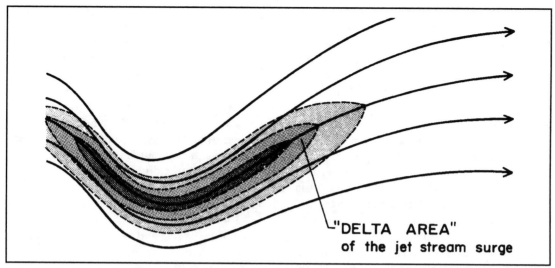

"DELTA AREA"
of the jet stream surge

FIGURE 3-6. A segment of the polar jet stream. From TROPOSPHERIC WAVES, JET STREAMS, AND UNITED STATES WEATHER PATTERNS by Harmon. Copyright 1971 by Association of American Geographers. Reprinted by permission.

large wave pattern at high altitudes over the United States. This persistence of the Jet Stream position, and therefore of "*preferred*" storms tracks, helps to explain why similar weather patterns often continue for days or weeks (say drought over the Northern Plains and violent storms and heavy rain in the Ohio Valley). When the upper air waves shift their positions, the persistent weather patterns are then broken and replaced by others. Long-range weather forecasting is grounded in attempts to assess probable upper air wave patterns and Jet Stream locations.

Most winter precipitation occurs with the passage of cyclonic storms. We have seen that, in a general way, this is true because huge volumes of air are being lifted toward regions within the upper atmosphere where density is becoming less through expansion (divergence) ahead of Jet Stream surges. *Within* the moving cyclone, some regions receive more precipitation than others. When other conditions are equal, maximum precipitation occurs near the center where convergence and lifting are greatest. The other major

controls are the fronts themselves. Since the warm and cold air masses have quite different densities, boundaries between them are maintained both at the surface and aloft. The advancing cold air mass to the west often travels more rapidly than the storm itself (the back of the wave is steepening). When this happens, the warm, most mT air is forced upward vigorously and precipitation tends to be heavier (though shorter-lived) just ahead of, along, and immediately behind the cold front.

The heavy concentration of winter precipitation in the southern part of the state may be understood in relation to cyclone tracks. Ordinarily, a semipermanent long wave in the upper atmosphere loops far southward over the United States, with the Jet Stream imbedded in it. A favored zone for the development of cyclonic storms is just east of the Rockies, because the mountains are so high and so continuous that they tend to maintain the boundaries between the polar maritime (Pacific) air to the west and polar continental and/or tropical maritime air to the east.[2]

Cyclones developing in southern Colorado and southward to Texas tend to follow the southward-looping Jet Stream to the middle of the continent and then to turn sharply to the northeast (recurve) so that the centers of the cyclonic eddies then move up the Ohio Valley, often passing through southern Illinois in a northeasterly direction (Figure 3-8). It is just after the recurving takes place, as the cyclones turn toward the northeast, that the Jet Stream surges are likely to intensify and to overtake the cyclone (Figure 3-6). This interaction intensifies the storm and increases precipitation. Seventy percent of severe winter season storms in Illinois are "Texas Track" cyclones, both because of their great intensities and because their centers passed through or near Illinois. In summer, on the other hand, the centers of lows usually pass well north of the Wisconsin border so that cyclonic precipitation is no longer concentrated in the southern part of the state.

One further mechanism for inducing precipitation which favors southern Illinois is orographic lifting. Only in the extreme south is there a terrain barrier to the movement of surface air currents, the Shawnee Hills. While the escarpment crest is only a few hundred feet above the surrounding countryside, it appears to trigger a slight increase in rainfall.[3]

Climatic "Insults" and Climate Change

Unusual events and periods in the environmental system are more significant to people and to the environmental complex itself than their short-lived character seems to indicate. People design a cultural complex—house

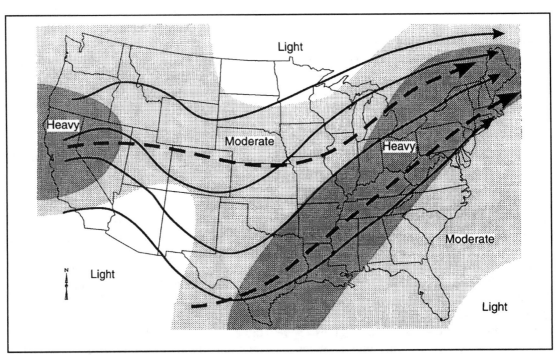

FIGURE 3-7. Continental air flow and precipitation. Dashed lines represent cyclone tracts and continuous lines mean constant pressure at upper levels. From TROPOSPHERIC WAVES, JET STREAMS, AND UNITED STATES WEATHER PATTERNS by Harmon. Copyright 1971 by Association of American Geographers. Reprinted by permission.

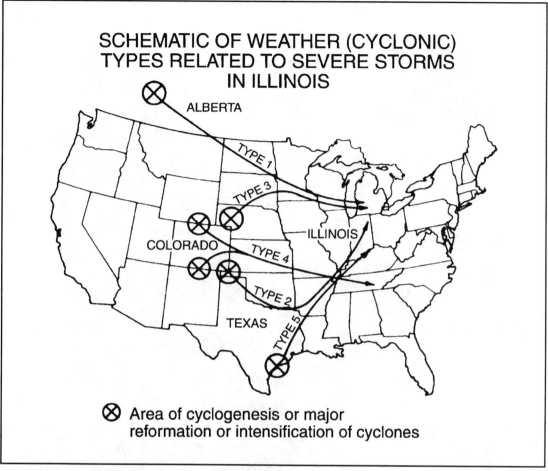

SCHEMATIC OF WEATHER (CYCLONIC) TYPES RELATED TO SEVERE STORMS IN ILLINOIS

ALBERTA

TYPE 1

TYPE 3

COLORADO

ILLINOIS

TYPE 4

TYPE 2

TEXAS

TYPE 5

⊗ Area of cyclogenesis or major reformation or intensification of cyclones

FIGURE 3-8. Types of cyclone tracts. (Adapted from Stanley A. Changnon, Jr., *Climatology of Severe Winter Storms in Illinois*, Bulletin 53, Illinois State Water Survey, 1969.)

types, space-heating systems, transportation facilities, cropping regimes, etc.—that copes rather well with usual conditions, but they and their constructions suffer severe damage when unusual weather produces floods, strong winds, ice glazing, deep snows, rainless periods during the growing season, or wet fields at planting time. One storm may produce more erosion than results from all the others during a year; thus, even the shape of the land surface may have been largely sculpted during catastrophic weather events.

Nine years in ten might have enough rainfall to support a forest vegetation, but episodic droughts might prevent young trees from ever being established.

Tornadoes have been the most destructive to human life of all storm types and have killed more than a thousand people in Illinois since 1916. They almost always occur along with violent thunderstorms, which themselves may be widespread and cause much damage. The probability of tornadoes at any point in Illinois is less than that in seven other states,

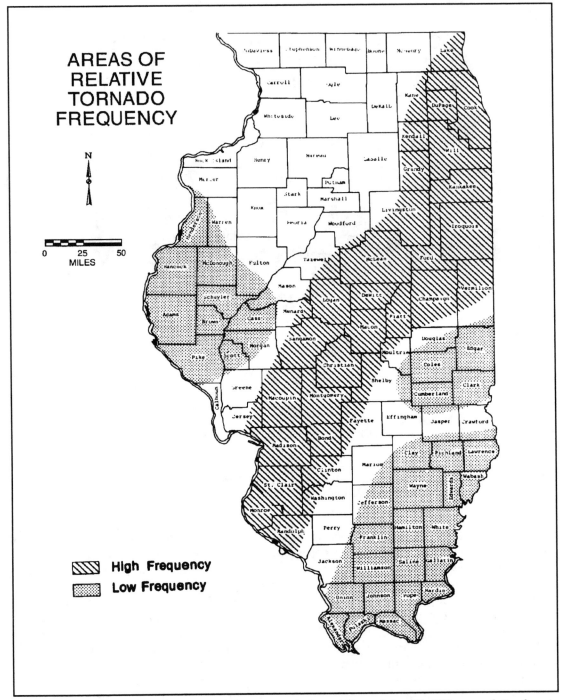

FIGURE 3-9. Areas of relative tornado frequency. (Adapted from John W. Wilson and Stanley A. Changnon, Jr., *Illinois Tornadoes,* Circular 103, Illinois State Water Survey, 1971.)

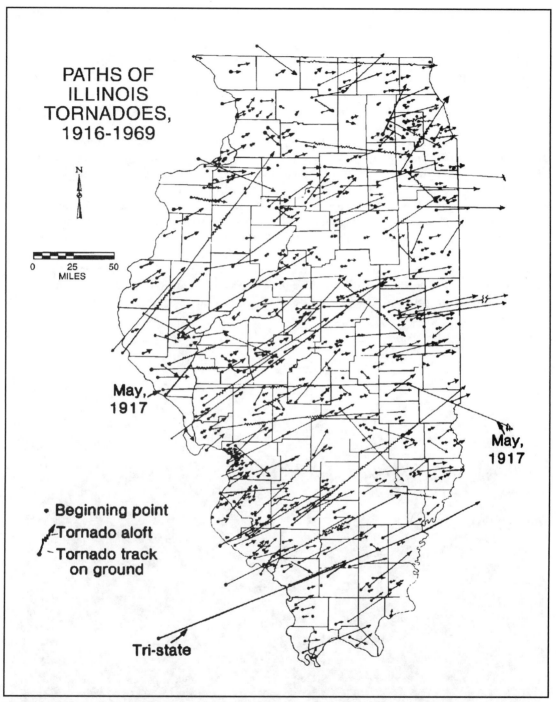

FIGURE 3-10. Paths of Illinois tornadoes, 1916–1969. (Adapted from John W. Wilson and Stanley A. Changnon, Jr., *Illinois Tornadoes*, Circular 103, Illinois State Water Survey, 1971.)

but Illinois ranks first in the number of tornado-caused deaths and second, after Oklahoma, in property damage for the period 1916-69.[4] Tornado damage is usually limited to a very small area (less than four square miles, on average). Nearly 1,500 years would pass before every area of the state experienced a tornado passage if the 1916-69 frequency were maintained. It is common for news reporters to refer to "tornado alley" and to give the impression that there is a zone of high danger in the state (Figure 3-9), but the near random distribution shown by Figure 3-10 indicates that a longer record will be necessary before the calculation of probabilities at points will mean very much. More than two-thirds of tornado-caused deaths occurred on either May 26, 1917, or March 18, 1925. The average length of tornado tracks on the ground in Illinois is less than 14 miles and the average diameter 185 yards; the killer tornadoes of 1917 (101 deaths) and 1925 (603 fatalities) were continuously on the ground for 283 miles (188 in Illinois) and 219 miles respectively, and reached maximum diameters of 0.5 and 1.0 miles. The 1925 storm, the Tri-State Tornado, is by far the most destructive tornado on record anywhere.

A tornado is a violently-rotating funnel of air that descends from a thunderstorm cloud to the ground. The greater the violence of thunderstorms, the greater the probability that they will spawn tornadoes. Most midwestern tornadoes develop in squall lines—series of thunderstorms ahead of, along, or just behind a cold front—in well-developed cyclones. The same conditions in the upper atmosphere, a speed surge through the Jet Stream, that

FIGURE 3-11. Damage from a 1991 tornado in Macomb, Illinois. (Photograph by Thomas B. Williams.)

favors intense cyclones also increases the likelihood of tornadoes. Another element favoring tornadoes is warm and moist air, because thunderstorms are more frequent and more intense when there is a great deal of water vapor present. When water gas (water vapor) condenses, great quantities of heat energy are released. The heating of air in rain clouds causes it to expand and to continue its upward movement; once condensation occurs, cloud development tends to be self-sustaining. Spring is the favored season (63 percent occur between March and May) for tornadoes in Illinois because it offers the most favorable *combination* of cyclone development *and* abundance of warm and moist air masses. In summer, tropical maritime (mT) air is abundant but frontal activity diminishes. A secondary tornado maximum occurs in the fall as conditions favorable for cyclogenesis intensify. Finally, in winter, cyclones increase in frequency and intensity but tornadoes are relatively rare because the presence of mT air is less frequent and its moisture content much lower.

Tornadoes are so dramatic and so frightening in their violence that a great deal of effort has been devoted to their forecasting. Radar can sometimes detect their characteristic "hook" shapes, but more often echoes from heavy rain areas obscure them. About all that forecasters can do is to warn the public when there are conditions favorable to tornado development approaching a region. Once an intense squall line or single storm has been identified on radar screens, radio and TV announcements warning of the approach of the storms can be broadcast. This "tornado warning" can be supplemented by tornado alerts, which are reports of ground sightings of tornadoes or funnel clouds. If possible, the alert is accompanied by an announcement of the direction of movement. Most tornadoes move in directions close to northeast, so that a sighting alone can be more useful if one keeps a map of the local area available.

There is some promise for another warning system based upon sferics, the excess of electrical activity within thunderstorms having funnel clouds. Tornado-breeding thunderstorms seem to have almost continuous lightning (10-100 flashes per second) at a center high in the cloud that can be detected by radio receivers tuned to 150 kilohertz. So far, the necessary combinations of radar detectors (for the thunderstorms), receivers, and direction-finders have not been developed into a remote warning network.

Severe winter storms cause far greater economic loss than do tornadoes and nearly as many directly-caused deaths. Indirect losses are probably several times greater than measurable property damage and death tolls imply. The best record of indirect losses for a single storm was assembled for one that crossed northern Illinois on January 26-27, 1967. Reported physical property damages totaled nearly $22 million and indirect economic loss was calculated to be $174 million. Automobile accidents due to ice and snow, transportation tie-ups that closed or reduced the efficiencies of businesses and public institutions, heart attacks that resulted from overexertion, and the costs of rescue operations and snow removal all contributed to the death toll of 56 and economic losses of nearly $200 million.

Between 1900 and 1960, 293 winter storms either produced at least six inches of snow at some point in Illinois or caused glazing (icing) conditions to be reported by 10 percent or more of Weather Bureau substations. The winter of 1977-78, which saw 18 severe storms, was estimated to have cost at least one billion dollars and resulted in a significant number of deaths and injuries. The social and institutional impacts of this unusually severe winter are difficult to assess. They range from more potholes in streets and delays in disposing of court cases to 27 percent more births than normal in Chicago during the ensuing

FIGURE 3-12. A snowplow at work in Rock Island County following a winter storm. (Photograph by Allen Englebright.)

October and November, 1979. One of the perceived results of that record storm year was a shift in the political leadership of Chicago after it was widely condemned by residents for its inability to clear streets and highways of record snowfalls.[5]

Colorado-Texas Track cyclones moving northeastward toward the lower Great Lakes constitute more than two-thirds of these storms. The typical severe storm moves in a direction 30° south (to the right) of northeast and maximum snowfalls occur in an elongated (3:1) core about 75 miles north and closely parallel to the path followed by the cyclone center. A model (generalization) developed from 304 storms indicates snowfalls of more than six inches in an area 156 miles by 52 miles (about 8,000 square miles or 15 percent of the state's area). Locally heavy snowfall may be anticipated when a fairly strong easterly wind is blowing as snowfall begins. If the wind direction is maintained for several hours or very slowly backs toward the northeast, the storm center is likely to pass not far south of the observer and to place him or her in the core of the heaviest snowfall.

The typical student reading this chapter probably has a somewhat different impression of Illinois' climate than does its elderly author. The past twenty years has been characterized, not only by falling temperatures, but also by heretofore unprecedented variability. The three worst winters on record: 1977-78, 1978-79, and 1981-82, all occurred within a five-year span and the winter of 1977-76 was nearly as severe. Each of the three record-setting years saw 18 major winter storms affecting the state, compared with the 1901-80 average of six per year.

Long and severe droughts were frequent occurrences in the 1930s and early 1940s, but succeeding decades had largely escaped devastating dry periods until the very severe drought of 1988-89. As may be seen on Figure 3-14, it is difficult to discern any long-term trend in precipitation. Perhaps the most startling feature of that chart is the great annual variation it reveals; year-to-year differences of 15 and even 20 inches per annum are not uncommon, and some years have had twice as much precipitation as others.

New precipitation records—and unprecedented floods—made 1993 a memorable year for many parts of Illinois and neighboring states. Some weather stations received more than twice their average annual rainfall in that year. Peoria already had recorded 155 percent of its normal annual rainfall by the end of July in 1993. Floodwaters following the rain dev-

astated cropland, property, and entire towns in the Mississippi valley. Crops on one million acres were lost in Illinois, and many towns and cities along the Mississippi and its tributaries were at least partially inundated as levees failed. Valmeyer, a village of 900 people near the Mississippi in Monroe County, was so completely destroyed that residents chose to rebuild the entire community on higher ground nearby. Transportation also was disrupted for weeks and months. All bridges across the Mississippi between St. Louis and Burlington were closed for a time, and barge traffic on the river above St. Louis was halted for two months. For people and towns in the floodplains, full recovery from the floods of 1993 may require years.

Amid our concerns about global warming, it may surprise some to discover that Illinois' climate has become cooler since reaching a peak about 60 years ago (Figure 3-15). By the 1980s, mean annual temperatures were about

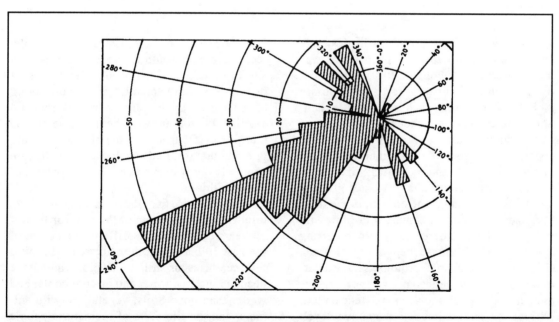

FIGURE 3-13. Direction of storm movement across Illinois sorted by number per 10-degree interval. (From Stanley A. Changnon, Jr., *Climatology of Severe Winter Storms in Illinois,* Bulletin 53, Illinois State Water Survey, 1969.)

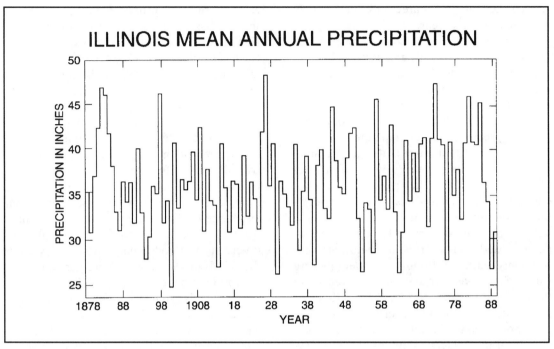

FIGURE 3-14. Mean annual precipitation for Illinois, 1878-1989. (Peter Lamb, ed., *The 1988-89 Drought in Illinois,* Illinois Water Survey Research Report 121, Illinois Department of Energy and Natural Resources, 1992.)

4° F cooler than they had been in the early 1930s, and the growing season of 1992 was the coolest ever recorded. It would be extremely unwise, however, to extrapolate global patterns from such a miniscule sampling of the earth's atmosphere.

During the past 20 years it has become apparent that "spells" of weather are typical of our climate. "January thaws" and October's "bright blue weather" are not just figments of our imaginations, and weather events do not occur in a random fashion with respect to time. There really are recurring cycles of drought and sequences of wet years. Climatologists and meteorologists have determined that each persistent weather type is associated with an equally persistent pattern of upper air circulation, and they have developed some insights into what conditions may affect the planetary wind system during both long and short runs of time. For example, recent droughts south of the Sahara in Africa and in India seemed to be signaled by a sudden increase in the amount of the north polar area covered by ice and snow (as seen on satellite photography). The increase in the meridional (from equator to pole) temperature gradient that resulted may have supplied more energy to the system and altered the pattern of flows within it.

Returning to this theme serves as a reminder that the climate of the past (or future) 10 years could be quite different than that of the past 100, and that the climate of Illinois a few centuries or millennia ago might have been distinctly warmer and drier or cooler and wetter than now. Soils, vegetation, wild animals, and the life style of prehistoric people in the Midwest may have been greatly influenced by former climates. In particular, neither the natural vegetation nor the soils of the state are congruent with the present climate; it may be that they are relicts of earlier and different meteorological conditions.

NATIVE VEGETATION

Of all the unusual features encountered by the explorers and settlers of the Midwest, none so excited their curiosity or posed so many practical problems as did the Prairie Peninsula (Figure 3-16). This vast sea of tall grasses, extending as a wedge from the eastern boundary of the semi-arid Great Plains, stretched all the way past the tip of Lake Michigan into present Indiana and had outliers as far as western New York State. To the scientific mind, which had learned to associate grasslands with dry climates, these tall and coarse grasses were not only different from those found elsewhere, but were also a puzzle because they grew in humid environments that seemed clearly capable of supporting dense forests. To the agriculturalist, the prairies were both boon and bane, although at first encounter the difficulties they imposed far outweighted the blessings that would later become evident.

Since only vestiges of the original prairie remain, it might seem that the grassland could be dismissed with only a cursory examination in this chapter. Its original distribution, however, is closely tied to other elements of the physical environment, and its presence influenced both agricultural settlement and the types of farming that developed. Its most important consequence was to make soils much more productive than they otherwise would

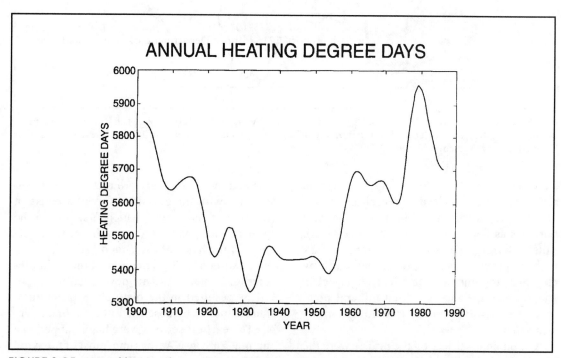

FIGURE 3-15. Annual heating degree days, 1900–1990. (From William Easterling, et al., "The Appropriate Use of Climatic Information in Illinois Natural Gas-Utility Weather Normalization Techniques," Illinois Department of Energy and Natural Resources, *Report of Investigations* 112, Illinois State Water Survey, 1990.)

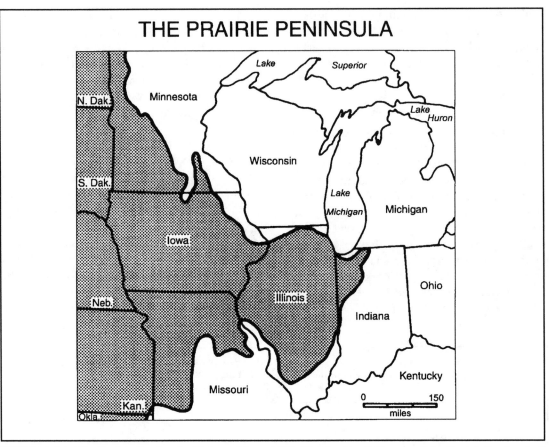

FIGURE 3-16. The Prairie Peninsula. (Adapted from H.E. Wright, Jr., "History of the Prairie Peninsula," in Robert E. Bergstrom (ed.), *The Quaternary of Illinois,* Special Publication 14, University of Illinois Department of Agriculture, 1968.)

have been. The very extensive systems of fine roots that characterized the prairie grasses were readily incorporated into the upper soil horizons as decayed organic matter (humus). This material imparts the characteristic dark color to prairie soils, improves their tilth by clumping together the smallest of particles, and reduces the rate at which mineral plant foods are dissolved and carried downward by rainwater.

The pioneer settlers on the prairie were the products, both in Europe and in America, of woodland environments, and they had developed an appropriate technology and system for making woodlands productive. Their encounter with the grassland forced a pause in the tide of settlement, which first skirted the prairie margins and eventually reached its core in the heart of the Grand Prairie about a generation late. This delay, and the technologies and tenure systems developed in the interim, resulted in the distinctive agricultural landscapes to be described in a later chapter.

The earliest settlers were handicapped in a number of ways. At the time, most of them did not have the steel moldboard plow invented by John Deere at Grant Detour and put into mass production about 1847 at Moline. The tough sod of the prairie plains could not be turned over effectively by the cast iron plowshares then available. Poor drainage and prairie soils were usually associated, and the combined capital expenses of sod breaking, ditching, and tiling were beyond the financial resources of most. It was also harder to find wood for fuel and housing, particularly on the wide expanses of the Grand Prairie in eastern Illinois. Even "navigation" could be a problem. James Caird, a Scots traveller, essayed a journey from Springfield to Decatur in 1858 and has left this description:

. . . we stood across the great plain which stretched out before us. The horses struck without hesitation into the long coarse grass, through which they pushed on with very little inconvenience, although it was in many places higher than their heads. It was not thick, and parted easily before them;. then sweeping under the bottom of our wagon it rose in a continuous wave behind us as we passed along. The surface of the ground was firm and smooth. We had fixed our eye on a grove of timber on the horizon as our guide, and drove on for about an hour in a straight line, as we believed, towards it. But stopping now and then to look at the soil and the vegetation, we found that the grove had disappeared. Without knowing it we must have got into a hollow, so we pressed on. But after two hours' steady driving we could see nothing but the long grass and the endless prairie, which seemed to rise slightly all round us. I advised the driver to fix his eye upon a cloud right ahead of us, the day being calm, and to drive straight for it. Proceeding thus, in about half an hour we again caught sight of the grove, still very distant, and the smart young American driver "owned up" that he had lost his way.[6]

Finally, the wet prairie was accurately associated in the popular mind with hazards to health, which were thought to be caused by "miasmas"—noxious vapors believed to be included in the night mists which developed over the poorly-drained hollows of the plain. The chills and fevers of "ague," as they called the disease, are now known to have been caused by mosquito-borne malaria.

GRASSLANDS AND WOODLANDS

The species making up the tall grass prairie can be seen today only in abandoned pastures, in unkempt cemeteries, along rural roads and railroad rights of way, or in nature preserves. Fortunately, the native vegetation patterns can be reconstructed with a high degree of accuracy. In the case of the Grand Prairie, some lands were not plowed until about 1900, and interviews with original settlers could be conducted as late as 1920. Surveyor's notes made during the original rectangular surveys prescribed by the Northwest Ordinance (1787) may also be consulted.

The map of vegetation prepared by R. C. Anderson in 1970 (Figure 3-17) shows the fragmented nature of the grassland distribution. Except in the extreme northeast and the far south, the grassland dominated in most areas, but it was broken wherever the land was sloping. When it is remembered that flat land prevails throughout most of the state, and that most sloping land is found on the valley sides of streams which are actively invading the higher tracts of flat land, the dendritic pattern of the forested areas becomes understandable. In eastern Illinois, the crescentic moraines also bore forests on their steeper slopes. Flat land and grassland, then, usually occurred together; the major exceptions were the floodplains of streams, where mixed hardwood forests occupied large tracts. Upland forests, which occur in better drained situations, were usually composed of mixed oaks and hickories in the northern part of the state. In the south, however, the variety of deciduous trees increased dramatically and included even species usually associated with the southern United States, such as tupelo (black gum), red gum, and cypress.

Figure 3-17 also reveals what at first seems surprising: The most coherent and unbroken tracts of prairie occur, not in the west closest to the steppes of the dry plains states, but east of the Illinois River near the tip of the peninsula. This seeming paradox is explained by the general absence of well-defined valleys in the Bloomington Ridged Plain; since the Wis-

consinan ice retreated only a short time ago, there has not been time enough to develop a well-defined drainage pattern.

Formal attempts to explain the "anomalous" prairie date back to the early 1900s, and a number of published works could be cited by E. N. Transeau in his sweeping review of 1935. By that time, it was becoming clear that the tall grass prairie had invaded from the west and that it was present in so many and varied environmental situations that a *general* explanation must be sought.[7]

For some, the present climate of the region seemed a sufficient causal agent, and it was Transeau who demonstrated, for both average and "extreme" years, some distinctive climatic characteristics that might help maintain the tall-grass prairie: lower humidity in midsummer; lower precipitation during drought years; greater variability in climate; decreased precipitation across the southeastern border in winter; and decreased snowfall northeastward in winter.

John Borchert extended this analysis by relating drought year climatic characteristics to the relative prevalence of air mass types. He concluded that:

1. The axis of the Prairie Peninsula coincided with the greatest average transport of air from the base of the Rocky Mountains
2. The greatest eastward transport of air from the mid-continent occurs during drought years
3. Storm tracks are displaced away from the grassland during those same years[8]

"Rocky Mountain" air, because of its long trajectory over land, and because it has lost moisture in being lifted over the western Cordillera, is a poor source of atmospheric moisture as compared with the mT air that has been displaced southeastward. Deprived of both moisture and the cyclonic triggers of

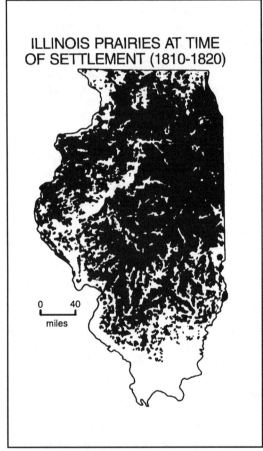

FIGURE 3-17. IIllinois prairies at time of settlement (1810-1820), based on original land survey records. (Adapted from R.C. Anderson, "Prairies in the Prairie State," *Transactions of the Illinois Academy of Science,* 63 (1970).)

FIGURE 3-18. Rare survivors of trees planted in rows along the highways of east-central Illinois during the drought years of the 1930s. (Photograph by A.D. Fentem.)

precipitation, the grasslands might develop a distinctive, and drier, climate. Such a climate could result from an increase in the strength of the Westerlies. Indeed, a change in the average curvature or location of the long waves in the atmospheric circulation could have profound effects on the distribution of heat and moisture over the continent. Did such a climate exist in the past, and did it "create" the prairie wedge? Transeau thought a "late post-glacial dry period with more widespread drought conditions and more prolonged drought" was indicated,[9] but many scholars remained unconvinced.

Carl Sauer was the best-known spokesman for an alternative hypothesis, and was vociferous (as were his many disciples) in attributing the prairies to fire—and particularly to fires set by humans in pursuit of big game. The invasion of people from Asia across the land bridge that connected North America and Asia during the last ice age presaged, in his view, not only the extinction of a wide range of large beasts, but also the displacement of post-glacial woodlands from what is now the North American grassland.[10]

Evidence for past vegetations and climates comes from a variety of sources, the most useful of which may be counts of plant fossils, pollens, and seeds preserved in bogs. Appropriate sites are few, pollens may be blown great distances, and dating is sometimes difficult. Nevertheless, the accumulating evidence is now so strong that a consensus exists for a drier and warmer post-glacial epoch that began around 9000 years before the present and ended about 6000 B.P.[11] Evidence from eastern Illinois records a deciduous forest replacing an earlier spruce-tundra vegetation by 9000 B.P. and the beginnings of prairie replacement on uplands at about 8300 B.P.[12]

Careful reconstructions of the extent and thickness of glacial ice, of near-continent ocean temperatures, and of late-glacial drainage changes, have provided data for computer simulations of climates during the Holocene. These suggest higher evaporation-precipitation ratios in the grassland and a 4°F higher temperature than now during that warmer-drier period.[13]

Although the evidence for a climatic explanation for the prairie now seems overwhelming, few doubt that prairie fires were frequent or that they played an important role in maintaining a grassland vegetation. Figure 3-19

shows a species (Eastern Red Cedar) that is highly vulnerable to fire invading a bluestem pasture. Protected from fire, this hardy tree may often be seen in Illinois colonizing environments that range from swamps to barren dry bluffs—often in the same general area. Present efforts to recreate and maintain prairies for their aesthetic and ecological values regularly employ deliberately set fires as a management tool.

Motorists traveling in northerly directions along bluffs flanking major rivers in the state may see still another type of natural grassland—the "hill prairies." Many bluff promontories with southwesterly exposures bear patches of native grasses, surrounded by trees, near their crests. Although rarely exceeding two or three acres, their original cumulative extent is thought to have been

FIGURE 3-19. Fire susceptible red cedar has invaded native bluestem grasses in Hancock County (Galesburg Plain). (Photograph by Fred C. Caspall.)

at least 75 square miles. Their vegetation of bunch and prairie grasses is thought to be a response to the xeric conditions induced by exposure to the afternoon sun and prevailing summer winds (southwest) in situations where permeable loess overlies bedrock.[14]

SOILS

Few residents of Illinois any longer have direct contact with the soil. More than 80 percent of the state's vast agricultural output is produced by farmers who, together with their families, could easily be accommodated within a metropolitan area the size of Springfield. Nevertheless, the quality of these soils—and the maintenance of their fertility—remains as vital as ever. Illinois soils are among the most productive in the world and produce vast surpluses of feed-grains and soybeans for the rest of the nation and for export. The "agribusiness" that the soil supports dwarfs the 8 billion dollar value of direct farm sales in 1993.

The soil occupies a distinctive niche within the total environmental complex. Not only does it express in its character the interaction of all other physical elements, but in it are rooted, ultimately, all forms of terrestrial life. Since humans develop their most intimate contact with the earth's surface through the medium of soils, they have lavished far greater attention to the changing character of soils from place to place than they have to the other elements of the natural world. More than 10,000 soil series have been recognized and mapped in the United States alone, and many shades of difference within the series themselves are recognized in detailed county mapping.

Ordinarily, the soil develops from a parent material—such as bedrock, glacial drift, alluvial sands, or windblown silts—over very long periods of time. The interrelations of slope, exposure to wind and sun, temperature, moisture, natural vegetation, and parent materials are expressed in weathering (physical and chemical) and biological processes within the soil body itself. In a humid lowland environment such as that of Illinois, the soil may be conceived as a continuous "blanket" of varying thickness and includes any materials that support plant life. Describing and subdividing this layer are rather difficult tasks because the soil extends in all three dimensions and because its character varies in all directions.

The purpose of this introduction to Illinois soils is to understand the broad regional differences in soil character and quality as responses to natural processes. In order to do this effectively, it is necessary

1. To identify the characteristics that affect utility
2. To have a basic understanding of the processes that produce differences in soil character
3. To describe the methods by which the broad regional variations are generalized (summarized) from the complex mosaic of "soilscapes" that are typically found within even small territories

Soil Characteristics

If a soil unit is undisturbed for a sufficiently long time it usually becomes arranged vertically into three major (and some minor) horizons, each one roughly parallel with the soil surface. One of the important reasons for horizon development is the vertical transfer of materials from place to place within the unit induced by gravity, capillary action, and biological activity. The upper, or A horizon, has the most active population of organisms, is often, therefore, marked by an accumulation of dark organic matter, and has usually lost some of its soluble materials and smallest particles through movement downward into the B horizon. This horizon, in turn, is marked by the accumulation of materials from above, usually by stronger (often reddish) color, and frequently by distinctive prismatic or blocky structure. Finally, the C horizon consists of the weathered rock material from which the upper two horizons have been formed. When we describe soils in relation to their utility, we are most often describing the A horizon, for it is in this zone that field crops are seeded and very largely grow. It is often true, however, that the other two horizons influence, both directly and indirectly, the vigor of plant growth.

An elemental soil characteristic is texture, the mix of particle sizes that results when a soil sample is totally disaggregated and dispersed in a liquid. The range may include quite large objects, but most particles are less than 5 millimeters in diameter; the largest are designated as sand, the middle range as silt, and the finest as clay. Mixtures of all sizes are called loams. The selective diffusion of the finer (clay-sized) particles downward is called eluviation and increases the density of the B horizon (which is then said to be illuviated). The transfer of minerals in solution from the A horizon is called leaching, and these minerals often reappear as salts in the B horizon. Both processes tend to result in horizontal layers (called pans) that either greatly retard or totally impede the percolation of water, with unfortunate results for crop production because soils with pans are alternately waterlogged after rains or excessively dry during rainless episodes. (Available stored soil moisture is limited to the shallow zone above the pan.)

A very obvious soil character is structure, or the arrangement of particles into aggregates or clumps that are usually separated from each other by air or water-filled porespaces. Structure is observed most often as tilth or "handling consistence," the "feel" of soil between the fingers and its resistance to crushing. Anything which "clumps" particles together "improves" structure because it affords more passages for water and air, increases infiltration and percolation of water, and "crushes" into more friendly seedbeds. Aggregation is aided by an abundance of calcium and humus, or decomposed organic matter. These often combine into a gluey substance which aggregates the fine particles.

Organic matter at the surface and humus at greater depths are present in most Illinois soils; however, it tends to be much more abundant in areas that were naturally covered by grass. Here the extremely dense fine-root system of grasses is protected from direct exposure to the weathering and decomposes beneath the surface. When humus is abundant, leaching, eluviation and pan formation are slowed or stopped, while tilth, fertility, and resistance to drought and waterlogging are improved.

The processes that form the A horizon are largely subtractive; fine particles are carried downward and minerals are taken into solution and percolated into the B horizon. The effectiveness of this latter process (leaching) depends not only upon the amount of rainfall and the texture and structure, but also upon soil acidity. If soil water percolating downward is acidic, it takes minerals into solution much more effectively. The results for the human use of soil are two:

1. Minerals that serve as plant foods are carried downward beneath the reach of cultivated plants
2. The soil acidity itself hinders most plant growth.

Finally, the more acidic a soil is, the more acidic it tends to become. Calcium, a very common mineral element (and one easily obtained from limestone), is quite alkaline and helps to maintain a balance between extremes of alkalinity and acidity in soils. But calcium is also very easily dissolved, and thus removed, which makes soil water even more acidic and calcium depletion even more effective.

When we review the processes that form soils in humid areas like Illinois, we may note with some surprise that (with the exception of humus formation) they all seem to produce soils with lower potential for crop growth the longer they operate. This greatly simplified version of soil formation provides a key to understanding broad regional differences in Illinois soil quality. It applies, of course, only to areas where soil formation has been in progress for thousands of years, and not to those areas that receive periodic deep deposits of materials through flooding or that have very rapid erosion or very poor drainage.

Illinois and the central Midwest have soils of quite exceptional fertility for four reasons, the most important of which are their parent materials and their youth. Glacial drift and wind-deposited dust (loess) predominate as parent materials, so that the long process of developing soils through the weathering of bedrock is replaced by the alteration of materials that have already been finely divided and have been subjected to so many alterations and so much mixing that they are likely to provide the "raw materials" for a wide range of soil elements. The short time Illinois soils have been in place under conditions similar to the present means that the most negative results of soil development are still far in the future. The areas covered by the prairie have more humus, less leaching, and better structure than otherwise would have been the case; in effect, the aging of soils under prairie vegetation was retarded. Finally, the soils with the very highest fertility—grassland soils in semiarid climates—suffer from moisture deficiencies that do not plague Illinois agriculture.

Loess, Time, and Fertility[15]

The parent material for a high proportion of upland soils is loess, a thin blanket (sometimes more than one) of windblown dust that was laid down over the entire state before, during, and after the Wisconsinan glaciation. The identified loess sheets range in age from the Sangamon (post-Illinoisan) interglacial to the Peorian (ca. 20,000 B.P.), and loess is probably still accumulating. As a parent material, the more recent loesses are quite youthful and rich in desirable minerals.

To understand the processes that have produced a major proportion of present soils, we will begin by describing the soils that would exist in an area soon after it had been blanketed by several feet of Peorian loess. Even though no soil development had taken place, the parent material itself would be highly productive of crops. One of several reasons for high fertility is the mineral composition of fresh loess, which includes a variety of plant nutrients and a very high content of calcium carbonate (often more than 40 percent by weight). The high calcium carbonate ($CaCo_3$) content in turn helps produce a desirable pH (a measure of alkalinity). Secondly, thick and freshly deposited loess contains a mixture of particle sizes, most of which are in the desirable silt-sized range. The irregular shapes of these particles results in an arrangement (structure) that provides an abundance of pore space and hence rapid infiltration of water and excellent tilth. Finally, since no horizons have developed, the pans which often impede underdrainage in mature soils are missing, and when topsoil is lost by erosion, the soils beneath are equally as productive as those lost.

The map of loess thickness (Figure 3-20) and the graphs in Figures 3-21 and 3-22 that show changes in the present character of loess from one place to another reflect both differences in the original materials and modification following deposition. The pattern in Figure 3-20 seems to indicate that the sources of the windblown dust were river valleys, valley trains, outwash deposits, and the beds of former lakes. The uplands southeastward from major river valleys, from the Havana

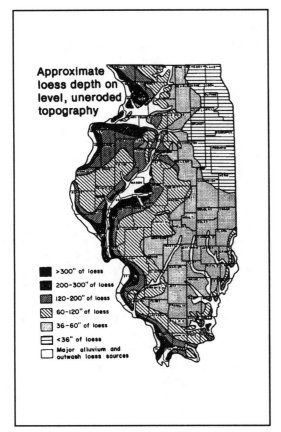

FIGURE 3-20. Approximate loess depth. (Adapted from J.B. Fehrenbacker, et al., in Robert E. Bergstrom (ed.), *The Quaternary of Illinois,* Special Publication 14, University of Illinois Department of Agriculture, 1968.)

Strath, and from the Green River Lowland all have very thick accumulations. Even more interesting is the area of shallow deposits centering in Western Illinois—downwind from a narrows on the Mississippi caused by an Ice Age stream diversion. (This "narrows" was chosen as the site for a major power dam at Keokuk in 1918.) It is very likely that most of the loess came from such nearby areas, that it was largely deposited in winter, and that the immediate sources were those broad expanses where abundant glacial meltwater was present and deposition was occurring in summer.

During the ensuing winter, the meltwaters must have ebbed, exposing large areas without vegetation and therefore very subject to wind erosion. The graphs showing decreases in loess thickness and particle size with increasing distance from bluffs all imply the accuracy of this reconstruction of events and hint at a close relation with soil quality.

It was earlier suggested that there is a tendency for soils in Illinois to deteriorate with age; it is important to realize, however, that soils age much more rapidly in some situations than in others. In the areas with several feet of loess accumulation, the process has been slow indeed; hence, the calcium content remains virtually the same as it was in the beginning. It is probable that water from the rains of summers succeeding those dusty winters ten and twenty thousands of years ago was quickly saturated with "lime" and thus was incompetent to leach calcium from the annual deposit. In fresh roadside cuts near river bluffs today, small nodules of marl (calcium carbonate) can be seen.

Figures 3-21 and 3-22 demonstrate two characteristics of loess deposits in Illinois that are closely related to the distribution of soil productivity—particle size and thickness. Both of those, in turn, are primarily dependent upon distance from the sources of the loess and the direction of the winds that deposited it. Generally, although not always, prevailing winds appear to have been northwesterly.

The largest (and heaviest) fractions in the dust clouds were deposited first—the materials on the bluffs flanking the source areas are often sandy and today have a billowy appearance. Some indeed, might be called dunes. With increasing distance, the particle sizes become progressively smaller, resulting in finer-textured parent material. During subsequent soil development, the finest particles, as we have seen, are eluviated (carried downward) to lower soil horizons,

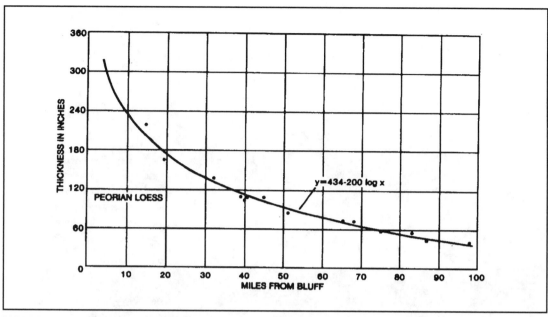

FIGURE 3-21. Relationship of loess thickness and distance from bluff. (Adapted from Guy D. Smith, *Illinois Loess—Variations in its Properties and Distribution: A Pedologic Interpretation,* Bulletin 490, University of Illinois Agricultural Experiment Station, 1942.)

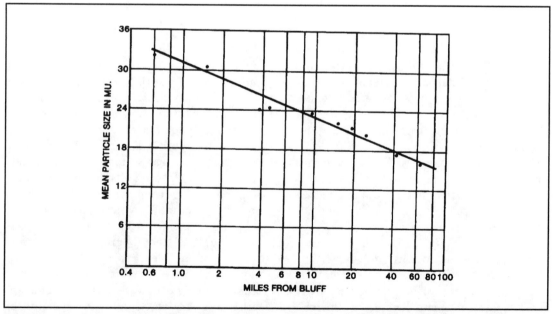

FIGURE 3-22. Relationship of loess particle size and distance from bluff. (Adapted from Guy D. Smith, *Illinois Loess—Variations in its Properties and Distribution: A Pedologic Interpretation,* Bulletin 490, University of Illinois Agricultural Experiment Station, 1942.)

FIGURE 3-23. A brook has cut deeply into thick loess between the Illinois and Mississippi rivers in the Illinois Peninsula. (Photograph by A.D. Fentem.)

eventually impeding drainage as they make those horizons denser. Frequent and prolonged saturation, in turn, depletes calcium, increases acidity, and leads to precipitation of minerals in the increasingly dense and impermeable B horizon.

The thinning of the loess with distance has similar consequences, but for a different reason. During the long interval between the Illinoisan and Wisconsinan glaciations, uplands everywhere developed a mature soil termed the *Sangamon paleosol*. This "old soil," now buried beneath the loesses (or young Wisconsinan drift), must have become acidic and infertile, its upper horizons leached, its lower ones impermeable and tending to "perch" water above its dense lower horizons, or pans. It was upon this paleosol that the wind blown

loess descended. In thin loess areas, the wet and acidic old soil beneath served as a "sink," rapidly depleting the available calcium and incorporating mineral salts in an increasingly dense hard pan. The areas shown on Figure 3-25 are those whose A horizons have lost virtually all their calcium. Locally, these soils are called "crayfish soils" in recognition of the high perched water tables which provide a favorable environment for these crustaceans. Geographers call them "grey prairie" soils because of the bleached appearance of the uppermost horizon which, in newer and fresher soils further north, is characteristically very dark-colored. A vegetation of prairie grasses appears to retard leaching, but not to halt it.

Glacial Drift, Time and Fertility

There is an important exception to the relation between thickness and fertility. On Figure 3-20, the area covered by the young Wisconsinan Drift Sheet is outlined, and it may be seen, by the sudden thinning of loess at its boundary, that earlier loesses have been obliterated, as has been the Sangamon soil. In its absence, the recent loess is highly productive. Once it is accepted that the passage of time is an important factor in soil formation, it is not surprising that *in general* young drift—specifically that of Wisconsinan age, is quite productive, even when loess deposits are thin or absent.

Generally, transitions from highly productive to infertile soils are gentle, although they may be disturbed somewhat by local conditions such as topography. One very dramatic contrast, however, is found where the southern boundary of younger drift (at the Shelbyville Moraine) abuts against the claypans developed from thin loess overlying thin Illinoisan drift. The tier of counties north of the boundary produces twice the value of agricultural products as does the row just south of the moraine.

The Distribution of Soil Fertility

Almost any rural landscape in Illinois presents a complicated mosaic of soil types that contrast sharply with each other in fertility. Such a landscape is shown on Figure 3-26, which represents an area in the transition zone between the highly productive soils of western Illinois and the grey prairie of the southern counties. The information shown here is just

FIGURE 3-24. Cultivation has resulted in the removal of several feet of easily eroded loess surrounding fenced pasture near Bluffs in the Illinois Valley. (Photograph by A.D. Fentem.)

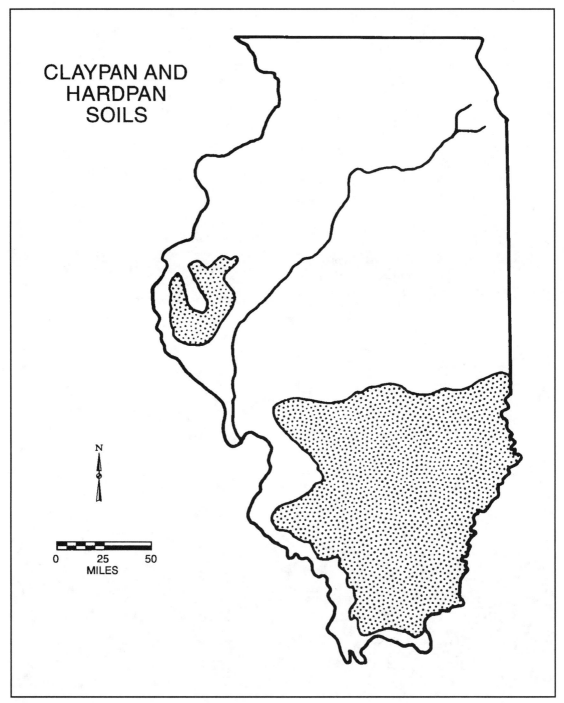

FIGURE 3-25. Claypan and hardpan soils in Illinois.

the kind that someone bidding on a farm in Montgomery County needs, but only eight of the 35 soil series found in the county and of the 375 that occur in the state are present in the sample area. Clearly, some way of examining soil qualities in a more general way (and mapping them at smaller scales) is needed.

The means for such analyses are provided by soil sample information collected by the U.S. Soil Conservation Service for its *Water and Soil Conservation Needs Inventory*. On average, these data provide a two percent random sample by county, and they can be extrapolated so as to make estimates of the total distribution of soil series in each county. A second step in arriving at estimates of soil productivity by county is to consult yield records for each soil series, a painstaking task that has been completed for a ten-year period.

The yield records come from farms that have had soil samples taken and whose operators report management practices. It is thus possible to determine, for a particular soil, the yield of each of the major crops under high, medium, and low levels of management. In compiling the data from which Figure 3-27 was made, Paul Mausel used only corn yields expected under a "high level of management."[16] Detailed study, however, showed that in general the variation in yields among the major crops by soil series was very similar.

The effects of topography are clearly seen in the low yields in the Shawnee Hills and in the Driftless Area in the extreme northwest, as well as along the Illinois River. The southerly portion of the Wisconsinan Till Plain, where the drift is old enough for reasonably good surface drainage, young enough to have

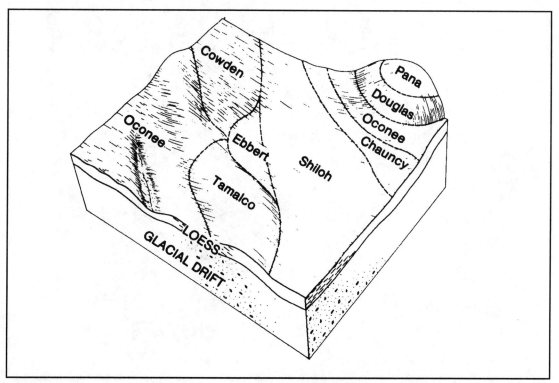

FIGURE 3-26. Generalized landscape of soil types. (Adapted from C.E. Downey and R.T. Odell, *Soil Survey of Montgomery County, Illinois,* U.S. Government Printing Office, 1969.)

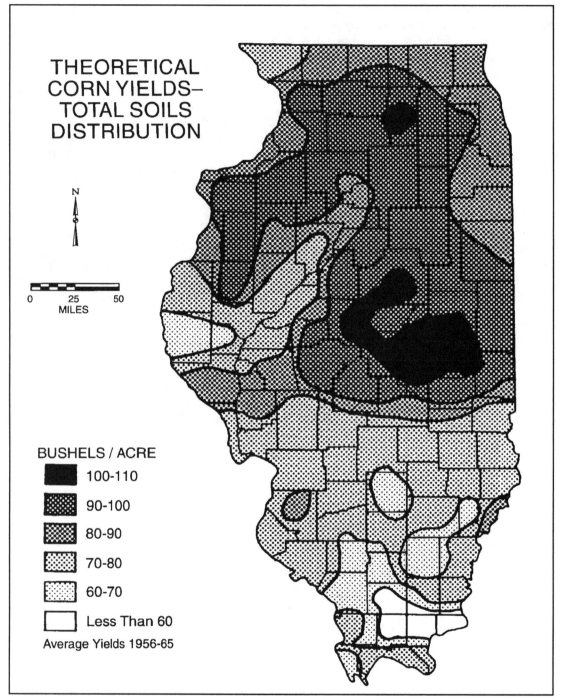

FIGURE 3-27. Theoretical corn yields—total soils distribution. Figure adapted from THE PROFESSIONAL GEOGRAPHER, 23. Reprinted by permission of Blackwell Publishers.

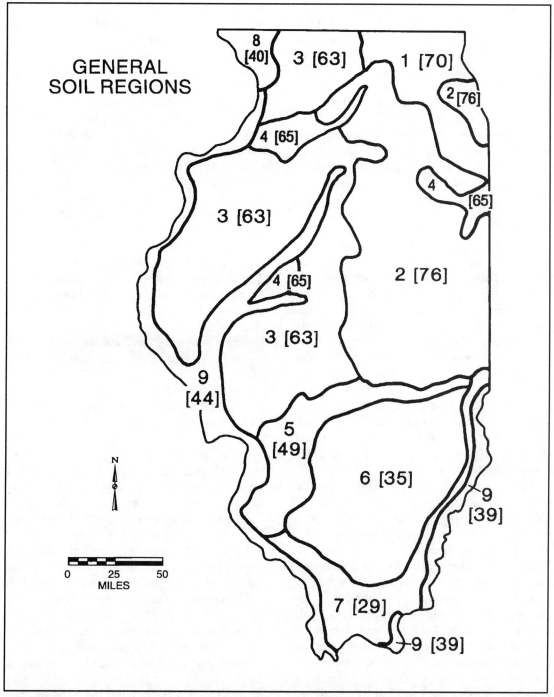

FIGURE 3-28. General soil regions of Illinois. (Adapted from Paul W. Mausel, "General Subgroup Soil Regions of Illinois," *Bulletin* of the Illinois Geographical Society 11 (1969).

escaped excessive leaching, and remote enough from major streams to have avoided deep dissection, has the best soils. The rapid transition from this peak to much lower productivity southward testifies to the importance of fresh and unleached materials as young drift and rapidly accumulated loess is succeeded by thin loess. The differences from place to place in total agricultural production are even greater than those in "theoretical corn yields" because counties with unproductive cropland usually have much greater proportions of pasture, woods, and wastelands.

Soil Regions

If we were to pinpoint a particular spot on Figure 3-27 and try to predict the productivity of a field from the average productivity of all fields in the county where it is located, we might make a very bad guess. If we take the further step of grouping counties and portions of counties into regions of similar soils associations, we will be even further from the reality of individual fields. Nevertheless, it is as important for understanding to see the larger picture as it is to distinguish between counties and among individual plots of land. Soils regions, such as those appearing on Figure 3-28, help to simplify the great complexity revealed by detailed soils study.

Modern soil scientists, in developing the soil classification system now used in the United States, recognized ten orders into which any soil sample in the world could be placed. Five of these orders are present in Illinois. The five orders may be subdivided into 47 suborders worldwide, further broken down into great groups (numbering 206), subgroups, families, and—ultimately—the soil series (of which there are approximately 10,000). Detailed examination by Paul Mausel revealed that 65 subgroups were represented in the state, but that only 34 of these made up as much as 3 percent of any of selected representative counties. The final step

in regionalization was to examine soil maps at various scales in order to draw boundaries that enclosed associations of subgroup soils in particular areas of the state (Figure 3-28).[17]

Although areal classification of this kind furthers our understanding of the natural world, the practical interests of most people are focused upon the capabilities of soils for producing crops. For that reason, the soil productivity data amassed by the Soil Conservation Service were again consulted by Dr. Mausel in assigning productivity levels to the subgroups (as he had already done in estimating county productivities). Since the proportion of each region that is covered by each subgroup can be measured, it is possible to estimate productivity ranges for each region. The bracketed figures shown within each region on the map are very general estimates of the differences in expected yields of the major crops under low levels of management. The numbers are index numbers only and should not be translated into yields as measured by bushels or some other unit. The low level management index has been chosen in the hope that it will better reflect *natural* productivity. The disparities in productivity among regions are much less when high level management indexes are used. (This could mean that technological and other human-created inputs are becoming more important in determining returns from varied soil resources.)

The soils of an area reflect the interactions of the other physical environmental elements; therefore, the general soils regions delineated on Figure 3-28 may also be thought of as environmental regions. Region 1, in the northeast, embraces much variety because of the recency and complexity of glacial events there. Closely spaced clayey moraines with steep slopes are bordered by aprons of sandy outwash and alternate with old lake plains and bogs. Despite the unpromising nature of some of the parent materials and the near absence of loess, the fertility of these soils is generally quite high, both because of their youth (ca. 12,000 years) and because natural prairies were widespread. The crests of the morainic ridges are sometimes forested and often pastured; in recent decades they have been attractive to exurbanites seeking rural homesites with prospects overlooking the pleasantly rolling countryside.

Most of Cook County is included in another region because its relief is lower, its parent materials are more uniformly derived from outwash and old lake sediments, and its native vegetation was more consistently made up of prairie grasses. The remainder of region 2 is essentially the older part of the Wisconsinan Till Plain where moraines are broad, widely spaced, and gently sloping. Since stream dissection is minimal, slopes are almost everywhere gentle. Add to these favor-

able auspices a virtually unbroken prairie and the advantages of youth, and the very high natural fertility is readily accounted for. The rather thin loess blanket accumulated quickly; it has been suggested that organic matter and loess might have accumulated simultaneously in this region. The great thickness of the dark (organically stained) horizon in the Brunizem soils might have resulted from development upward as well as downward.

Region 3 is the domain of relatively thick Illinoisan Till overlain by thick loess. The most favored areas are the broad interfluves between streams which had the advantages of nearly flat surfaces, fresh and nearly unleached loess, and prairie vegetation. The best of these soils compare favorably with those in the region just described; however, the greater age of parent materials, the thinning of loess

FIGURE 3-29. A relict dune in the Havanna Strath Lowland; in the past the dune migrated southeastward and buried the thick A-horizon of a soil rich in humus (foreground). (Photograph by A.D. Fentem.)

near the southeastern margins of the region, and —most of all—the greater proportion of forested and steep slopes resulting from 70,000 years of stream erosion create a greater range, and lower overall level, of potential productivity.

Smaller areas, like those set aside as region 4, are incorporated within the regions just discussed; the Green River Lowland, Havana Strath, and Kankakee Lake Plain, however, are large enough for separate consideration. All three areas have a high proportion of sandy and medium-textured parent material deposited by glacial meltwater as outwash or lake sediments; all were predominately flat and were largely covered by prairie. In general, fertility is quite high, but some of the coarser materials are subject to drought, especially where sands were blown into dunes; still other areas require ditch drains. Ground water for irrigation is sometimes easily accessible, and supplemental irrigation is used successfully to produce grains, potatoes (Green River Lowland and Havana Strath), and green beans (Havana Strath).

Region 5 is transitional between the two major soil orders in Illinois as the result of increasing dissection of the surface and the thinning of the loess blanket. The soils here are more leached and their B horizons are usually characterized by clay accumulations.

Region 6 has the least productive soils to be found on essentially flat surfaces; even the best of the soils here are lacking in organic matter, are droughty, excessively wet during rainy episodes, and acid. Soybeans—better adapted to survive drought than corn and capable of maturation even when planting is delayed into summer by waterlogging—have made these soils more useful since their introduction during World War II. This region is the only one where soybean acreage exceeds that of corn.

The low productivity of region 7, on the other hand, is attributable to steep slopes and forest vegetation as well as to the passage of time and a warm, wet climate. Region 8, The Driftless Region, is likewise characterized by steep slopes. Finally, the areas along the major streams are most difficult to describe because they embrace great extremes—from extensive areas of floodplain now reclaimed and exceptionally fertile to excessively sandy areas elsewhere, and from thin soils on the upper slopes of valley sides to deep accumulations of productive loess on flattish but narrow interfluves. Problems of drainage, flooding, weed control (weed seeds arrive with floodwater), and erosion all help to limit usefulness, although some areas of floodplain are exceptionally productive.

CONCLUSION

This two-chapter survey of Illinois' physical environment has been comprehensive, but it has viewed the state at some far remove. In effect, we have seldom been closer to our subject than we might have been from one of our orbiting satellites. If we were to observe a county as closely as we have the state, we would find that the same level of complexity exists there. Still closer examination brings us to landscapes we may see in their entirety with our own eyes. Between the county scale and our own backyard, we can make our observations with the aid of aerial photographs or topographic maps at scales ranging from 1 inch:1 mile to 1 inch: 2,000 feet. Field trip leaflets describing and interpreting more than a hundred local physical situations are available from the Illinois Department of Registration and Education. Few activities are more interesting or rewarding than applying the knowledge and principles of earth science to the study of our home territories.

NOTES

1. The discussion of the role of planetary circulation in Illinois weather as adapted and simplified, with apologies, from J.A. Harman, *Synoptic Climatology of the Westerlies: Process and Patterns.* Washington, D.C., Association of American Geographers, 1991.

2. See Stanley A. Changnon, Jr., Climatology of Severe Winter Storms in Illinois. *Bulletin 53,* Illinois State Water Survey, 1969.

3. Douglas, Jones; Floyd Hugg; and Stanley Changnon, Jr., Causes for Precipitation Increases in the Hills of Southern Illinois, Illinois State Water Survey, *Report of Investigations 75,* Urbana, 1974.

4. John W. Wilson and Stanley A. Changnon, Jr., *Illinois Tornadoes,* Circular 103 (Urbana: Illinois State Water Survey, 1971), p. 4.

5. Stanley Changnon; David Changnon; and Phyllis Stone, Illinois Third Consecutive Severe Winter: 1978-79. Illinois Institute for Natural Resources, *Report of Investigations* 94, 1980.

6. Sir James Caird, "1858: Birdseye View of Illinois, Prairie Farming, and the State Capital," in *Prairie State,* ed. Paul M. Angle (Chicago: University of Chicago Press, 1968) p. 325.

7. E.N., Transeau, "The Prairie Peninsula," Ecology 16 (1935), pp. 423-37.

8. John Borchert, "Climate of the North American Grassland," *Annals of the Association of American Geographers* 40 (1950), pp. 1-44.

9. E.N. Transeau, op. cit.

10. Carl O. Sauer, "A Geographic Sketch of Early Man in America," *Geographical Review* 34 (1944), pp. 529-73.

11. See George Jacobson, Thompson Webb III, and Eric Grimm, "Patterns of Rates of Vegetation Change during Deglaciation in eastern North America," Chapter 13 in *Geology of North America*, Vol. K-3 North America and Adjacent Oceans during the east Deglaciation (Geological Society of America, 1987) and Thompson Webb III, P. Bartlein and John Kutzbach, "Climatic Change in eastern North America during the past 18000 years," Chapter 20, Ibid.

12. J.E. King, "Late Quaternary Vegetational History of Illinois," *Ecological Monographs* 51 (1981): 43-62.

13. Webb, Bartlein, and Kutzbach, op. cit.

14. Evers, Robert A., "Hill Prairies of Illinois," *Bulletin* of the Illinois Natural History Survey, 26(5), 1955.

15. A pioneer in exploring these relationships was Guy D. Smith. See his *Illinois Loess—Variations in Its Properties and Distribution: A Pedologic Interpretation,* Agricultural Experiment Station Bulletin 490 (Urbana: University of Illinois, 1942). A summary of recent findings is provided by J. B. Fehrenbacher, I. H. Jansen, and K. R. Olson, *Loess Thickness and Its Effects on Illinois Soils*, Agricultural Experiment Station Bulletin 782 (Urbana: University of Illinois, 1986).

16. Paul W. Mausel, "Soil Quality in Illinois - An Example of a Soils Geography Resource Analysis," *Professional Geographer* 23 (1971), pp. 127-136.

17. Paul Mausel, "General Subgroup Soil Regions of Illinois," Bulletin of the Illinois Geographical Society 11 (1969), pp. 70-81.

4

HISTORICAL GEOGRAPHY

Alden Cutshall
University of Illinois at Chicago

In 1818, Illinois became the nation's twenty-first state and began a record of impressive development. Over the years it attained a position of leadership in agriculture. By 1860 the Prairie State was the leading producer of corn, wheat, and oats. Today Illinois leads the nation in corn again (having lost the leadership to Iowa for a period of years), soybeans, Swiss cheese, onion sets, pumpkins, and horseradish. It is one of the world's

FIGURE 4-1. The old state capitol building in Vandalia. The seat of government in Illinois was transferred from Kaskaskia to more centrally located Vandalia in 1820 and then to Springfield in 1839. (Photograph by R.E. Nelson.)

great centers of industrial production, its factories fueled in part by the state's coal and petroleum resources. In the heartland of America, it is a crossroads state, and Chicago, its major city, has become the world's foremost rail center, a leading center for highway transportation, and the greatest inland seaport. O'Hare Field in Chicago is the world's busiest airport. In 1992, Illinois had a per capita income of $21,608, which was seventh highest among the 50 states. The state has come a long way since the first permanent settlement at Shawneetown in 1809, and since the capital was moved upon attaining statehood from French-flavored Kaskaskia to more centrally situated Vandalia (Figure 4-1).

THE INDIAN ERA

Even before there was a State of Illinois or an Illinois territory, what is now Illinois was Indian land, with many different tribes or groups at different times and places within the area. Early Illinois Indians lived with the environment, not against it. Their ancient village sites are widely distributed. There are ashes of burned-out camp fires along the bases of bluffs and outlines of long-gone

105

settlements once enclosed with palisades. Their crudely drawn pictures and rough carvings have been found on several rock outcrops, but only a few are left, normally beneath a protecting rocky ledge. Most of those that remain are only faded fragments, generally splotches of a single color. There are rounded stone pits or more irregular depressions where the Indians pounded corn or other food products. Bone needles suggest the manner in which they made their clothes. Remains of camp refuse and piles of flint chips indicate where they made weapons or implements. Illinois, from Chicago to Cairo, is rich in prehistoric Indian lore.

Most striking of the early Indian remains are the mounds, more than 10,000 of them constructed by prehistoric people in Illinois. And most striking of the mounds are those at Cahokia where there are over 120 on the site of an early Indian city. The largest one, Monk's Mound, is said to be the largest prehistoric earthen construction in the world.

A dense Indian population once lived in the American Bottoms, that broad flood plain area to the east of the Mississippi River from Alton to Chester. This was one segment of a prehistoric culture that flourished along several midwestern rivers (Mississippi, Illinois, Wabash, Ohio, and Tennessee). A string of "towns," possible satellites of Cahokia, extended upstream along the Illinois Valley, and others downstream as far as Memphis. Ninety percent of the mounds probably were used as burial places. Dickson Mounds, south of Peoria, was a major burial ground and is now a state museum. It was of the same age and culture as Cahokia, and probably was a

FIGURE 4-2. One of the more prominent Indian mounds at Cahokia. The largest, Monks' Mound, is pictured in Figure 7-3. (Illinois Geographical Society)

satellite of its larger southern neighbor. Archaeological explorations at the Koster site (Kampsville) have unearthed remains of twelve successive Indian settlements, the oldest dated 8,500 years ago.[1]

According to archaeological finds, the Cahokia site was first inhabited about 700 A.D. The city of Cahokia covered about six square miles and had a peak population of about 20 thousand. It was primarily an urban area; the residential section was quite extensive. Houses were arranged along what appear to have been streets or around open plazas. The inhabitants hunted, fished, gathered wild plant foods, and practiced gardening. The main agricultural fields were probably outside the city. These people were dependent upon a well-developed agricultural system with corn, beans, and squash the principal cultivated crops.

Only a culture that relied on the growing of plants for food could have supplied the apparently large population and comparatively permanent settlements of these people. The type of culture helps to explain the flood plain or alluvial terrace location of their settlements; these sites offered a more fertile, more friable soil than that which existed on the nearby uplands.

For reasons that are most unclear, the prehistoric Cahokian culture came to an end about the time Columbus reached the New World. A gradual decline in population began around 1300 A.D. and by 1500 A.D. the site had been abandoned. Around 1300 A.D. there was a climatic change throughout much of the southern part of the United States. This may have adversely affected the food supply and made it increasingly difficult to feed the large population. Other contributing factors may have been the depletion of natural resources in the surrounding woodland areas, revolution, disease, and war. Possibly the number of people simply increased to the point of overpopulation with respect to the available

resource base. Whatever the cause or causes of abandonment, Cahokia represented the highest achievement of prehistoric Indian civilization north of Mexico, and it represents but one of several vanished civilizations that peopled parts of Illinois prior to the time of recorded history.

At the time of European penetration into the Illinois region, many groups of Indians resided within the area. During the eighteenth century the Potawatomi, the Kickapoo (a portion of the Piankeshaws), and the Illini tribes lived within the present limits of Illinois. Early accounts of the Indian population indicate that tribes of the Miami family inhabited most of the lower Wabash Valley. At that time these people occupied much of the present basins of the Wabash, Maumee, and Miami rivers to the east of Illinois, and at a slightly later date were as far west as Peoria. The Shawnee had come northward from their more southern habitat and occupied the area at the junction of the Wabash and the Ohio as well as downstream from this confluence to the Mississippi. The friendliness of the Shawnee in this area, holding the key position with respect to river transportation, probably contributed much to encourage and foster the early settlement of whites in southern Illinois and adjacent Indiana.

Other names of Indians that appear in early history are the Kaskaskia (primary along the Mississippi), the Ottawa, and Chipewa. The Potawatomi lived around the shores of Lake Michigan, but penetrated well beyond the lake area. In the northern portion of present Illinois were the Winnebago, Fox, Sauk (Sacs), and Iroquois. Both the Fox and Sauk roamed the forested hills of the Driftless Area in large numbers prior to 1800 and normally lived by collecting, farming, and hunting; in later years some of them shifted to lead mining. (The Galena area was listed as the nation's principal lead mining region in the early 1800s, although the first mining there by

Europeans occurred in the 1820s. Apparently lead was extracted in the vicinity by Indians as much as a century earlier.) Whatever the group of Indians, however, their life-style was attuned to the environment.

In many ways the Illinois Valley can be considered the core area of Indian occupation in the 1700-1800 period of time. The owners of that valley at the time of statehood were the Illini, or Iliniwek, who may have occupied much of the present state's area in the seventeenth century. They were essentially prairie Indians and lived largely by the chase. The vast areas of prairie grass and the wooded areas along the rivers and streams apparently harbored abundant animal life to supply their needs. Their dependence on game forced them to move widely on hunting trips. Consequently, they did not build permanent homes. The "towns" of the Illini were clusters of wigwams. Their social institutions were primitive. At this period in history they were quite willing to have Europeans live in their midst. They had been the largest tribe in the Illinois Valley region, and at one time were probably the strongest tribe in the midcontinent. Their chief food was maize (corn), but they also cultivated beans, squash, and other vegetables. Apparently the debilitation of their society was a result of several factors. In random order, some of them were wars with the Iroquois; the sudden introduction by Europeans of gun powder and iron into their stone age culture, and its attendant change in their life-style; their use of English rum and French Brandy when they had no inherited resistance to alcohol; and their susceptibility to the European diseases of measles, small pox, and tuberculosis. In most cases, the arrival of the European was followed by a decrease in Indian population and a weakening and debilitation of Indian society.

The Illini have been described as handsome, brave, tall, strong, fast, proud, and affable. They were also called idle, jealous, dissolute, and thievish. Perhaps all the adjectives were applicable, not only to the Illini but to other tribes of the area as well.[2]

THE SEQUENCE OF EARLY SETTLEMENT

The early history of modern Illinois is one of European conquest, at least in a general way. Initially, the activities of the English and the Dutch were confined to the Atlantic coast; the Spanish concentrated on Florida, the Southwest, and the West Indies; and the French were most interested in the St. Lawrence Valley. The English had difficulty in penetrating deeply into the interior, partly because of the ferocity of the Indians, especially the Iroquois. The Spanish focus was southward; they were much more interested in the gold of Mexico and Peru than in the less spectacular resources of interior North America. French explorers and fur traders were the first Europeans to visit what is now the American Midwest, but they did not come as settlers to develop the country.

Specifically, the first white men to penetrate the heart of the American continent were a priest and a mapmaker. Father Jacques Marquette, a French Jesuit missionary came to New France in 1666. He studied the Indian languages and worked among the Ottawa tribes up the Upper Great Lakes. Louis Jolliet (Jolîet), a French-Canadian fur trader and mapmaker, met Marquette at Sault Sainte Marie in 1669. Four years later the two men led the historic expedition from Green Bay up the Fox (Wisconsin Fox) and down the Wisconsin rivers to the "great river of the west," the Mississippi. They continued downstream well to the south of present-day Memphis until they were convinced that the river led to the Gulf of Mexico. The small party then returned north, ascended the Illinois and Des Plaines and portaged across the low, swampy

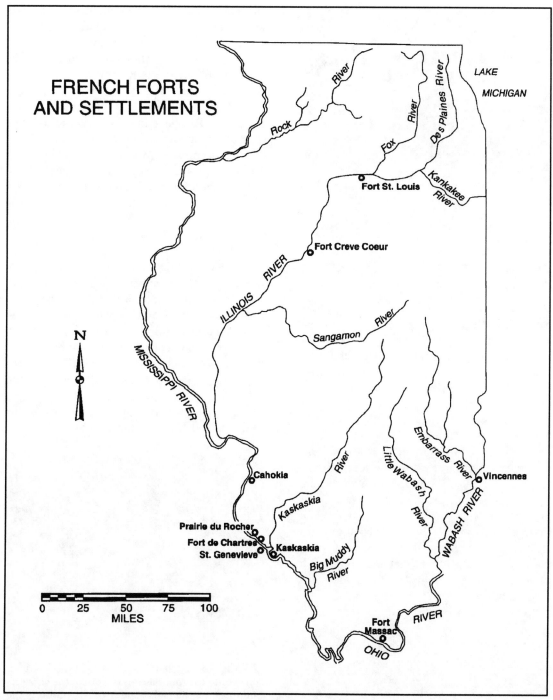

FRENCH FORTS AND SETTLEMENTS

FIGURE 4-3. French forts and settlements. These small villages and forts, all located on major rivers, were the earliest European settlements in Illinois and vicinity.

FIGURE 4-4. Fort de Chartres was one of the early eighteenth century French strongholds on the Mississippi River in southwestern Illinois. The restored gate of the fort is pictured here. (Photograph by R.E. Nelson.)

continental divide into the Chicago drainage basin, following approximately the alignment of the old Illinois and Michigan Canal and present Chicago Sanitary and Ship Canal. Thus, they became the first explorers to cross the present state of Illinois and to see its entire western border.

This expedition, together with the one by Robert Cavelier, Sieur de La Salle, almost a decade later, laid the foundation for a series of forts and fur trading posts between Detroit and New Orleans. La Salle was a woodsman with many friends among the Indians, a convincing conversationalist, and an international promoter—a truly unusual combination of talents. He envisioned the economic potential of the great Middle West and proposed to establish a large colony of fur traders and farmers in the heart of the Mississippi Valley. As a result of his vision, and his ability to persuade people and raise funds in France, a number of French forts and settlements was established (Figure 4-3). Fort Crevecoeur, constructed on the bluffs above the river near Peoria in 1680, was the first in Illinois. In 1682, Fort St. Louis on the three-fourths acre top of a sandstone promatory that later came

to be known as Starved Rock, was established as the principal French stronghold in Illinois. Both were destroyed, and a new Fort St. Louis and French settlement on Lake Peoria (a wide portion of the Illinois River) was established in 1791. It existed until the War of 1812. A Catholic mission was built at Cahokia in the American Bottoms in 1699, and Kaskaskia was established four years later. Prairie du Rocher, in present Monroe County, was settled in 1723 and nearby Fort de Chartres was begun soon afterward (Figures 4-4 and 4-5). Vincennes, Indiana, established as a fur trading post in 1680 with a fort and mission added in 1702, and St. Genevieve, across the river from Kaskaskia in Missouri, were part of this series of early French forts and settlements in the Midwest, and by their position, immediately across the Wabash and the Mississippi respectively, had some impact upon early development in Illinois.

A few of the early French may have been itinerant fur traders and trappers, but generally they lived in compact villages, whether by habit, or for protection, or both. The French villages were representative of ones in France and differed notably from settlements

FIGURE 4-5. Inside the wall of Fort de Chartres, today only the foundations remain of the barracks buildings. In the right center of the photo is the fort's powder house. (Photograph by R.E. Nelson.)

in the English colonies. The Catholic Church was the center of village life and the priest was the most influential citizen. One-story houses, usually with wide verandas, lined narrow streets. Except for shopkeepers or more wealthy traders, each man farmed a narrow strip of land aligned at right angles to the riverbank and grazed his animals in a common pasture. In general, the villagers have been described as happy, carefree people who enjoyed dancing and card playing to a degree unthought of in early New England. The original French settlements were a dependency of Canada, but in 1717 they were placed under the government of Louisiana. This political alignment to the south, although generally weak and lax, partially helps to explain the *de facto* slavery that existed in Illinois until the Civil War.

By the Treaty of Paris in 1763, title to the Illinois country passed from France to England, but it was a full two years before Great Britain actually occupied the territory. English military presence was never substantial, and their civil government was minimal. The French settlements continued a rather placid existence in the heart of official British territory, and their customs, traditions, and practices remained as outposts of French culture under a British flag.

The colony of Virginia claimed this Illinois country under its colonial charter granted by the King of England. Patrick Henry, as the first governor of the new Commonwealth of Virginia, commissioned George Rogers Clark to attack Kaskaskia. Clark, then a lieutenant colonel in the militia, had organized the defense against the Indians in Kentucky, which was then a county of Virginia. Clark assembled some 175 men and trained them on an island near the falls of the Ohio River, the present site of Louisville. From this location they floated down the Ohio in keelboats to Fort Massac, a former French fort that was abandoned by the British. From there Clark

FIGURE 4-6. The home of Pierre Menard, an early political leader and businessman, is one of the best remaining examples of French architecture from the eighteenth century in southwestern Illinois. (Photograph by R.E. Nelson.)

marched first northward, then westward to capture Kaskaskia from its landward side without a shot on July 4, 1778. The other French settlements in Illinois, as well as British Fort Sackville at Vincennes, then came into possession of the Virginia forces without a struggle. On December 9, 1778, Virginia created the county of Illinois, a vast area with indefinite boundaries that extended from the Ohio and Mississippi northward to Canada.

At this time the settlements along the Mississippi had about three-fourths of the non-Indian population of the entire area. Kaskaskia, with 500 white persons and about the same number of slaves, was the largest community. Cahokia's population was about 300 whites and 80 blacks. The other villages were on their way to abandonment or had already ceased to exist. The population of Illinois had actually decreased between 1763 and 1778. French patriotic sentiment had been high, and many families abandoned their homes and moved to St. Genevieve or St. Louis, the latter having become a major French fur trading center, rather than live under English rule. This strong anti-British feeling probably helps to explain the lack of resistance to

Clark's small, impoverished force at Kaskaskia and Vincennes.

The British, however, had recaptured Fort Sackville at Vincennes in the early winter of 1778-79. This left Clark at Kaskaskia in a rather vulnerable military situation. Upon learning that the French settlers at Vincennes would welcome the Americans he embarked upon a bold, almost irrational move of major strategic significance. In the middle of winter (February) and at the time when rivers were flooded, he marched 140 miles across Illinois with 170 Virginia militia and French volunteers to attack and capture Vincennes. The last 60 miles took eight days, much of that time wading in the ice-cold waters of the flooded bottoms of the Little Wabash and its eastern tributaries, and then skirting the lower Embarrass River and crossing the Wabash River to the south of Vincennes. It was truly one of the boldest, and probably one of the most courageous campaigns in early American history.

By the close of the Revolutionary War settlers were pressing beyond the limits of the original colonies, and Illinois was considered a desirable place for settlement. Some of Clark's soldiers returned to their eastern homes, then came back to the area to establish themselves in the timbered woodlands of southern Illinois. Other pioneers came, mainly from the southeastern states. Their Anglo-Saxon ancestors had crossed the Atlantic from the British Isles or Germany to settle in Virginia, the Carolinas, or Georgia. Later generations had spread into Tennessee and Kentucky. Then, after more time had passed, they moved northward and into southern Indiana, southern Illinois, and eastern Missouri. Those from the middle Atlantic States who came down the Ohio had basically the same backgrounds, with their intermediate ancestral stops in western Pennsylvania, Ohio, and possibly Indiana. As early as 1790 several communities had been established in the area of present Randolph, Monroe, and St. Clair counties, on or near the fertile flood plain along the Mississippi and in the general proximity of the older French settlements. This immigration was the real beginning of settlement by English-speaking peoples and the introduction of English culture. The French culture then ceased to be dominant and gradually declined, preserved today only at Prairie du Rocher.

Whereas the French had lived among the Indians and oftentimes associated with them, the English-speaking settlers viewed the Indians as an inferior race and as people who hindered the welfare of the newcomers and handicapped the progress and development of the area. Obviously, this attitude brought the races into conflict. A running warfare continued until the end of the War of 1812 when the pioneers essentially came into control of the areas immediately around their settlements.

In this period, immigrants began to populate other sections of the state. The Ohio River was the principal east-west artery. Settlers came down the Ohio, then up one of its tributaries or up the Mississippi and later the Kaskaskia and Illinois, settling near their place of debarkation. It was not until after 1830 that the Great Lakes route, the second important western water route, was used to any appreciable extent. The location along, or near the streams gave the settlers access to some of their basic needs (water, fuel, and shelter), which were usually lacking or difficult to obtain on many of the interstream areas. Wild game was more abundant, too, along the watercourses than in the prairie regions.

After the War of 1812 Fort Dearborn was rebuilt, Fort Armstrong was erected at Rock Island, and Fort Edwards was constructed where Warsaw stands today. But southern Illinois was the first region in the state to have significant settlements, a trend that continued for several years. Old Kaskaskia served not only as the capital of Illinois Territory but as the first state capital as well. At the time of admission to statehood, 1818, most of the population was in the American Bottoms, centered in some half-dozen growing communities. Gallatin county, with 3,200 settlers, was the most populous center in the eastern part of the state. Shawneetown, first settled in 1806 and resettled in 1809, was the first permanent community in that area and was already the chief river port on the eastern side of Illinois. It was the southern anchor of a strip of discontinuous settlement extending northward along the Wabash for about a hundred miles. Other established settlements were Palestine and Carmi on the Wabash, Golconda on the Ohio downstream from Shawneetown, Equality, and Albion (Figure 4-7). The two occupied areas on the western and eastern sides of Illinois were connected by significant trails with taverns for overnight stops along them, but there were probably less than a half-dozen settlements of any significance within the interior of the then "settled" state of Illinois (really southern Illinois). Carlyle, on the Kaskaskia, probably was the best known among them. Lawrenceville, Fairfield, and Vienna were new in 1818. Salem was founded in 1823 as the halfway station on the Vincennes-St. Louis stagecoach route.

Central and northern Illinois remained essentially a vast wilderness. As late as 1821 there were a few huts downstream from Peoria, but not a single white habitation between Peoria and Chicago. Permanent occupation was not underway in Galena until 1820, although numerous miners had come upstream for the summer season prior to that time.[3] There was no connection between Kinsey's trading post and Fort Dearborn at Chicago and Galena, nor was there any meaningful association between the Chicago area and either Kaskaskia or Shawneetown, the principal settlements in the southwestern and southeastern parts of the state, respectively. Alton was founded in 1814 and Edwardsville in 1815, a northern extension of the earlier settlements along the Mississippi. But there was no Quincy, Decatur, Champaign-Urbana, or Springfield (the first log cabin in Sangamon County was built in 1817). It was known that there was a Grand Prairie extending southward and westward from Lake Michigan to midstate, but descriptions of it were both vague and inaccurate. Illinois in 1818 was really southern Illinois. Shawneetown was the closest approach to a commercial center, primarily because of the importance of the salt springs twelve miles inland. In 1809 it was described as having more business activity than any other place west of Pittsburgh.

EARLY POLITICAL GEOGRAPHY

At the time of admission to statehood Illinois was characterized by many small, pioneer settlements, almost all of them south of a line drawn from Alton via Carlyle to Palestine, and almost all of them along a river or at least a small stream. In 1821 Illinois had only 22 counties, but this number had increased to 56 by 1830;[4] however, the latter number is misleading, as many of those in the central and more northern parts of the state were paper counties, merely organized political units with very few people and no meaningful government.[5] Interestingly, in 1816 Crawford County extended northward to the limits of Illinois Territory, and included the tiny Chicago settlement. As new counties were created and county boundaries changed, Chicago was successively included in Clark, Pike, Fulton, Peoria, and Vermilion counties from 1819 until 1831 when Cook County was formed. But in 1830, although the stars and stripes waved over the new Fort Dearborn, only about 50 people lived in the village that was destined to become "Hog Butcher of the World," the nation's great freight handler, and the site of the world's busiest airport.

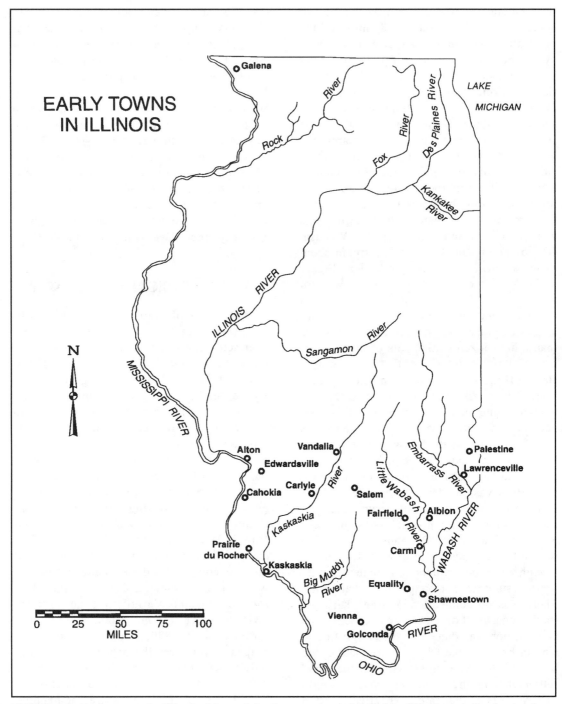

FIGURE 4-7. Early towns in Illinois. Most of the first towns established in Illinois were in the southern one-third of the state.

Except for a fortuitous series of circumstances involving geography, slavery, trade, taxes, politics, and historical accident, Chicago would never have been in Illinois at all. The Ordinance of 1787, which established the Northwest Territory, included the stated intention of ultimately creating two states north of a line "drawn through the southerly bend or extreme (southern end) of Lake Michigan" and three states south of that line (Figure 4-9).

When Ohio filed for statehood in 1803, it prevailed upon Congress to establish its northern boundary five miles north of the 1787 line in order that the mouth of the Maumee River would be included within the state. When Indiana became a state in 1816, it tried to claim territory 25 miles north of the 1787 line, but settled for ten miles which gave

it a shoreline on Lake Michigan from Hammond-Whiting to Michigan City. Two years later the initial proposal for the state of Illinois indicated a northern boundary to correspond in latitude with that of Indiana. This would have given the new state possession of the mouth of the Calumet River, but everything north of that line would have eventually become a part of Wisconsin.

Shortly afterward, Illinois' farsighted territorial representative, Nathaniel Pope, proposed an amendment extending Illinois 51 miles farther north to 42^0 30'. Whatever the stated reason or reasons, the realistic purpose was to give Illinois access to Lake Michigan through both the Chicago and Calumet Rivers. (It is not assumed that Nathaniel Pope ever envisioned a mighty metropolis develop-

FIGURE 4-8. In addition to its prominence as an early lead-mining center, Galena was the home of Civil War hero and President U.S. Grant. Grant's restored mansion, pictured here, is now one of Galena's tourist attractions. (Illinois Geographical Society)

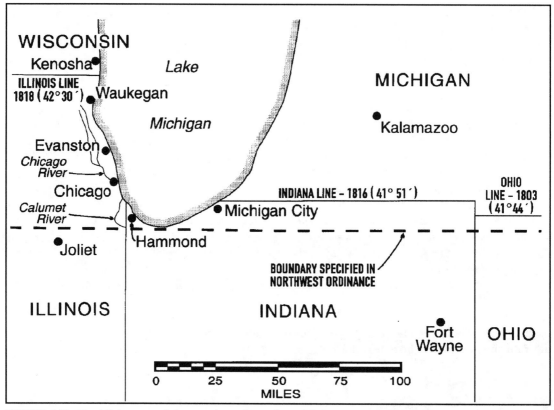

FIGURE 4-9. Establishment of the northern boundary. The sequence of boundaries leading up to the selection of latitude 42° 30′ as the eventual northern boundary of Illinois is indicated on the map.

ing between and around these two small streams with sandbars and mud flats blocking easy entrance into an infrequently used lake.)

Pope's amendment had little opposition. A precedent for violating the 1787 line had already been established by Ohio and Indiana. The states of the South were unconcerned about the boundary, but wanted Illinois admitted quickly in order to attract migration from the North and hopefully give them a free hand to establish slavery in Missouri. Legislators from the eastern states were convinced by Pope that, without good access to Lake Michigan, Illinois would be linked with the South and might well become a slave state. With his proposed access, he argued, Illinois would be oriented to the East and would become "the keystone to the perpetuity of the nation." Probably, too, the proponents of an envisioned Illinois and Michigan canal, not authorized until 1822, wanted to keep the planned canal enterprise within a single state. In any case, the ultimate loser, Wisconsin, had no one to protest on behalf of what was soon to become the Territory of Wisconsin. (In 1842, a mass meeting was held at Oregon City, now Oregon, and the fourteen counties north of the line designated in the Ordinance of 1787 tried to secede from Illinois, either to form a new territory or to be added to Wisconsin, which was then about to become a state. Taxes were believed to have been one motive for this "rebellion."

Obviously the history and politics of Illinois would have been very different without its northernmost fourteen counties, which now represent more than half of the state's population. For example, Illinois would not have gone Republican in 1856 without the vote of these fourteen counties. And, without a Republican administration in Illinois, Abraham Lincoln would not have been nominated for president in 1860. Hence, it may be said that by putting the small settlement of Chicago in Illinois in 1818, Congress did contribute to the perpetuity of the Union in the 1860s, as Nathaniel Pope had so glibly argued almost half a century earlier.

EARLY LIFE-STYLE

Settling the woodlands of southern Illinois was a task of considerable effort. The land usually was cleared of all trees below some definite size, normally those less than fifteen inches in diameter. The larger trees were deadened; that is, they were girdled deeply enough to cause them to die. Occasionally all were cut so that the land was completely cleared. In either case the logs and brush were gathered into heaps for burning. A "logrolling," like house-raisings and barn raisings, was a communal occasion, and in some respects a social gathering as well. The practice of logrolling continued until well past the midcentury, barn raisings sporadically until the World War I years.

Early farmers fenced their cultivated stump fields and turned their livestock loose to forage in the nearby unfenced, wooded areas. Livestock kept in fenced pastures were easy to identify; however, range animals were not, as herds oftentimes combined or stray animals left one herd to join another. Consequently, their owners adopted a system of brands or marks. Early Illinois farmers sometimes used branding irons, a system of identification more commonplace on the western

plains at a later period in history. More often, however, they cropped a portion of the ear, or placed notches or holes in the ear of the animal. By the position of the marks on the top or bottom of the ear, or on the right or left ear, and by the combination of them, a great number of individual identifications were possible. So long as livestock were kept in this semiwild, unsupervised manner, these marks or brands were used and recognized as a legal means of claiming animals or proving ownership. However, it was required that the identifying mark or brand be registered with the county clerk where the owner resided. When stock was traded or sold, the new owner was responsible for changing the marks to indicate the change of ownership. Furthermore, the remarking was done in the presence of two creditable witnesses.

The early farm family was a more-or-less self-sufficient unit. Much of their summer and fall work was in preparation for the winter months. Hay and corn were harvested and stored in some fashion so that the essential animals could be kept through the cold season. Surplus animals were sold to the local slaughter house or livestock buyer sometime during the autumn, at least before Christmas. In addition, some animals were butchered on the farm to supply meat for the family during the winter and even into the early summer months. Most families butchered at least one beef animal and often several hogs, usually sometime between Thanksgiving and Christmas.

Hog butchering became an almost ritualistic procedure and was an all-day affair for the family, usually with the help of a couple of neighbors or a few close relatives from nearby farms. Actually, the work extended far longer than a single day. Preliminary arrangements required the gathering of a quantity of good, dry wood to heat the water for scalding the carcasses, and the placement of the wooden scalding barrel in a firm position and leaned or tilted at the proper angle for convenient use on "butchering day." Iron kettles, usually of 30 or 50 gallon size, were assembled at the site, whether from a summer storage place on the farm or borrowed from a neighbor if the family did not possess its own. The kettles were placed on rocks or, more likely, suspended at the proper height from a small log or heavy pole supported by two forked posts set firmly in the ground. Near dusk the night before, the hogs to be butchered were separated from the other animals and placed in a small temporary pen near the butchering site. The kettles were filled with water from the well, stream, or pond either the night before or very early on butchering day, as the fires were lighted at daybreak so the water would be boiling and the work could begin when the help arrived.

The process itself required shooting the animals with a rifle at appropriate intervals or stunning them with a heavy hammer or an axe and immediate "sticking" (cutting of the jugular vein with a long sharp knife); scalding and scraping to remove the hair; cutting the carcass into the hams, shoulders, sidemeat, and other cuts; making of sausage and possibly head cheese from the leaner trimming; and finally rendering of lard from the trimmed fat, the latter process using the same kettles that had contained the hot water earlier in the day. The major work was done in the morning hours. Making sausage and rendering lard were afternoon work, as was the storage of meat in the "smokehouse" or other appropriate places inaccessible to dogs, cats, and rodents. The curing of the better cuts took place at a more leisurely pace a week or more later. Normally, the hams and shoulders were salted or sugar-cured and probably smoked; on the other hand, some of the side pieces were pickled, partly to provide variety of flavor, but more importantly to preserve these cuts for use during the spring and into the summer months.

From settlement until the early 1900s it was the custom of most farm families to preserve and store food for the winter months and to gather wild products when in season. Dried apples and apple sauce were common winter foods. Autumn cider became winter vinegar. Some families dried corn, green beans, and small fruits. Later home canning became popular. Farm families made sauerkraut, having planted late cabbage solely for kraut. They probably owned at least one five- or eight-gallon, glazed, earthen jar solely for this purpose. Also, wild berries, especially blackberries, were gathered and canned for winter pies and cobblers. Hickory nuts and black walnuts, gathered in October, provided the ingredients for nut-cracking on long winter evenings. Persimmons for pudding, wild crab apples and wild plums for preserves, dandelions, plantains, and sheep's sorrel for early spring "greens," and dewberries, elder berries, and wild gooseberries were gathered from the woodlots or fence rows in season. Sassafras roots were dug in March, preferably from the red sassafras, for a delicious, rosy, springtime tea with a delightful odor. Broken twigs of the spicewood bush were used to make a similar drink, but this practice was far less widespread.

After frost, but before the first really hard freeze, the frugal farmer (more likely the farmer's wife) moved the apples, potatoes, turnips, cabbages, celery, pumpkin, and other vegetables into the fruit cellar. If the family did not have a fruit cellar, and most early farm homes did not have cellars, they made an earthen storage hole. This was really a conical or elongated mound on a well-drained portion of the garden area. First a layer of straw, grass, or leaves was placed on the ground. The products to be stored were arranged in a heap or ridge and a thick layer of straw then thrown over and alongside them. The top and sides were then covered with earth eight inches to a foot in thickness. Usually a small drainage ditch was dug around the mound. Boards or planks oftentimes were laid on top or leaned against these mounds as additional protection against the elements. This method of storage would normally keep the product dry and cool and would prevent freezing. It was a rather crude but reasonably effective method of keeping root vegetables and hardy fruits for use throughout the winter months.

The iron kettle used at butchering time was also used in the fall for making apple butter and, usually in the spring, for making soap. The soap was made from animal fat, most often meat fryings, that had been saved during the winter months. Before commercial lye became available, the household collected an alkaline solution for soap-making by pouring rainwater over wood ashes, usually in a V-shaped, wooden trough tilted at a low angle to facilitate slow movement of the water through the ashes.

Prior to the Civil War, there was little money in circulation in the more rural sections of the state. Many of the small town shopkeepers sold goods on a year's credit and were paid in crops or livestock during the summer and fall. In a few cases the more affluent merchants in towns along the rivers fattened the livestock on their own farms, operated a seasonal packing house, and shipped pork, beef, and grain to New Orleans by flatboat. On occasion, too, the fattened livestock were shipped by flatboat to downriver markets.

DEVELOPMENT OF
TRANSPORTATION

Illinois is a crossroads state, with a major hub at Chicago on Lake Michigan and a secondary center at the St. Louis crossing of the Mississippi. The development of the state's modern transportation system is the result of many factors, both physical and cultural, and

their diverse interrelationships. Relatively flat terrain and easy grades facilitated railway and highway construction. A variety of resources with great potential for trade encouraged investment in transportation. Furthermore, a cultural and economic system based upon exchange necessitated an efficient means of transportation and communication. In little more than a century and a half there developed a vast net of roads, rails, and pipes; a tremendous number of vehicles in all their specialized forms; special route structures; and a variety of specialized terminal services and activities. Facilities evolved from the tavern, wayside inn, stagecoach stop, and obscure railway station to such specialized phenomena as the large grain elevator, piggyback loading dock, and unit train; the ubiquous highway service station and truck stop; the modern motel and fast-food outlet; the pipeline booster-station and oil terminal; the modern airport; and a host of others.

Waterways and Water Commerce

Initially, the movement of people and products was along the waterways (Figure 4-10). The Indian canoe, flatboat, keelboat, and steamboat each played a role in the early commerce of the state. Several flatboat loads of coal were shipped from the banks of the Big Muddy River in Jackson County to New Orleans in 1810. But, in general, the flatboat trade was the movement of agricultural products to downriver markets. Flatboats were built and used on the Saline, Kaskaskia, Embarrass, and Little Wabash rivers in southern Illinois; on the Sangamon and Vermilion rivers in the central section; and probably on several other streams as well. However, both the flatboat and keelboat were replaced by steam traffic, the sternwheeler, on the principal rivers during the early years of statehood.

The first steamboat trip down the Ohio and Mississippi to New Orleans occurred in 1811. By 1820, some 70 steamboats were using

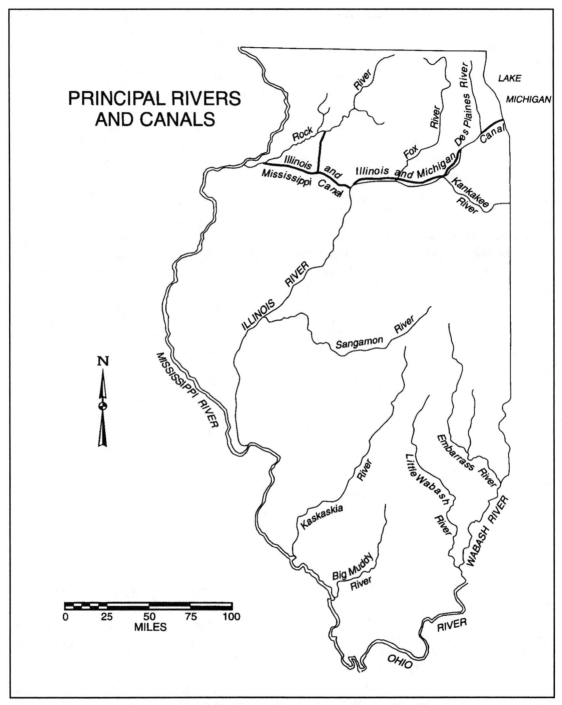

FIGURE 4-10. Principal rivers and canals. The excellent system of internal and bordering waterways has been a major asset for the development of Illinois.

these rivers. The first steamboat upstream to St. Louis was in 1817, and up the Illinois to Peoria in 1820. Shortly thereafter, steam traffic became commonplace on the major rivers and a few of the lesser ones as well. For years Galena was the most important port north of St. Louis. Later, barge traffic became the dominant feature of commercial water traffic in the state and along its borders.

During the first few decades of the nineteenth century, Illinois faced south in its commercial relations. In fact, over half the state still faced in that direction at the middle of the century,[6] and many river communities continued to do so until the outbreak of the Civil War. In 1845, Shawneetown still had varied trade relations with the South, as did some of the towns along the Mississippi. Palestine, Mt. Carmel, Golconda, Cairo, Quincy, and others sent grain downstream, but they probably lacked the variety of trade that characterized Shawneetown, Vincennes, Terre Haute, and St. Louis. Some corn, wheat, and flour moved southward to foreign markets, especially after 1845, with wheat and flour in much greater demand than corn. Coffee and other tropical products were transported north by river until the early 1860s.

Flatboat traffic on the Ohio and Mississippi had been surpassed by steam traffic in 1830, and on the Illinois soon after that date, but it continued to form an important part of river commerce for many years. Steamboats never penetrated the Wabash or Kaskaskia to any appreciable extent, so produce from those valleys had to be transferred to a steamboat or floated on to its destination. As late as 1846, flatboat arrivals at New Orleans numbered 2,792, over 600 of them coming from Indiana and Illinois. Ten years later the number from these two states was reduced to 148, only 12 of them from Illinois.

Early French explorers reportedly suggested the possibility of a canal from the Great Lakes to the Illinois River, and thence a connection via the Mississippi to the Gulf of Mexico. In 1822 the national government granted the state of Illinois a right-of-way for such a canal. Officially named the Illinois and Michigan Canal, construction was begun in 1836, but the nationwide depression following the Panic of 1837 delayed its completion until 1848. Although the canal provided safe and cheap transportation, its slowness and other factors caused a general decline in its importance. Particularly significant was its inability to compete successfully with the railroads that were built soon after the completion of the canal.

For a short period, however, the canal was truly a major factor in the economic development of the state. In fact, so great was the canal's role in developing northern Illinois that, of all man-made waterways in North America, only the Erie Canal surpassed it in importance.[7] Grain growing along the Illinois and upper Mississippi rivers could be shipped to market by way of the canal and Great Lakes at less cost than by way of New Orleans. Canal boats exchanged cargoes with river steamboats at La Salle. Peru also prospered, as did Joliet and Lockport. The long, narrow, rounded boats used on the canal carried passengers along with the freight of the season, but supplied neither food nor bedding. On packet boats, which were introduced later, food service was provided. Especially during the deep mud of spring thaws and heavy rains, these improved craft were more reliable and more comfortable than stagecoaches, which had no competition during winter months when the canal was inoperable. Nonetheless, after the early period of prosperity, there was a gradual decrease in the amount of traffic. The canal showed a profit until 1879, and tonnage declined rather consistently after the peak year of 1881.

In 1890, the Chicago Sanitary District undertook construction of a canal, primarily for sanitation purposes rather than transportation, between Chicago and Lockport, where the new canal joined the western segment of the old Illinois-Michigan Canal and the lower Des Plaines River. Opened in 1900, this work provided a waterway with a minimum depth of nine feet for navigation from the South Branch of the Chicago River to Lockport, but from Lockport to Utica there was no adequate means of water transportation. Consequently, a legislative act in 1919 provided a 20 million dollar bond issue for construction of an eight foot channel between the two towns. This proved insufficient to finish the waterway, and in 1930 the federal government completed the project and extended the improvement along the Illinois River to its confluence with the Mississippi. (The lower Illinois had been canalized, beginning in 1870). The improved Illinois Waterway was opened in 1933. The Sag Channel, a 24-mile southern arm of the Chicago Sanitary and Ship Canal, was added to connect with the Calumet River and Lake Calumet in 1922. It was subsequently widened and deepened and other improvements have been made during recent years.

The federal government also began work, in 1870, on the Illinois and Mississippi Canal from Hennepin to the Mississippi near Rock Island. It was little used by 1950 and abandoned by 1970. Today it is being preserved as the Hennepin Canal Historic District.

The Illinois Waterway, fully opened in 1935, carried 1.7 million tons during the remainder of that year. In 1950, the figure was 12 million tons, and more than 24 million in 1960. The opening of the St. Lawrence Seaway in 1958 greatly stimulated the expansion

FIGURE 4-11. A barge tow on the Illinois River near Beardstown. Grain, coal, and petroleum comprise the bulk of the cargo moved by barge. (Photograph by Allen Englebright.)

of the deep water port of Chicago and contributed markedly to increased use of the Illinois Waterway. Today's use of the Waterway consists primarily of barge traffic, with some private pleasure craft. The barges vary in size from 800 to 3,000 tons. They are of various designs to carry dry bulk, liquid bulk, and general cargo. Special barges are used for petroleum products, cement, and alcohol. The largest towboats are 165 by 35 feet and handle a tow of eight large barges (Figure 4-11). With the state's continuing need for a dependable and economical means of transportation, an increase in traffic on the Illinois Waterway appears to be assured through the remainder of this century.

Early Railroad Development

In 1850 transportation was still tedious over most of the state. The roads were generally inadequate and, except for the Illinois River and the canal connecting it with Lake Michigan, the really usable waterways were along the margins of the growing state. In all cases, traffic was slow. But the basic framework of a great system of railroads was quickly established between 1850 and 1860 when track mileage in the state was increased from less than 100 miles to almost 2,800 miles (Figure 4-12). After 1860 the basic framework was filled in with branch lines, competing lines, and shortcut (cut off) lines, until there were more than 12,000 miles of railroad in 1929 and only one of the 102 counties (Calhoun) was without rail service. More recently, the abandonment of unprofitable lines, which began about 1920, and the consolidation of routes has outstripped new construction. Present mileage is approximately 10,000 miles. Illinois owes much of its development into a great agricultural and industrial complex to the extensive railroad facilities within the state.

The state's first railroad, in Morgan County between Meredosia and Morgan City,

was 12 miles in length and began service in 1839. A year later it was extended to Jacksonville and, despite financial difficulties, to Springfield by 1842. The trip from Jacksonville to Springfield, 33.5 miles, took two hours and eight minutes.[8] This route later became a part of the Wabash System.

The first railroad in northern Illinois was the Galena and Chicago Union, the beginning of the present Northwestern System. Although originally chartered in 1836, work was discontinued during the Panic of 1837 and not renewed for ten years. By the end of 1848 it was completed to the Des Plaines River and brought a load of wheat into downtown Chicago from its terminus ten miles away. Two years later it reached Elgin (via Elmhurst and Wheaton) with a branch south from Turner's Junction (West Chicago) to Batavia and Aurora. It was extended to Belvidere in 1852, to Freeport in 1853, and finally to the Mississippi in 1855. In 1856 it became the first railroad in the West to use telegraph (between Chicago and Freeport). Also in 1856, it purchased two coal burning locomotives, presumably the first ones used in Illinois. The Chicago and Galena put into use the standard T-shaped iron rails in 1851. Earlier rails were made of wood, capped with thin strips of iron, the usual practice on frontier lines.

In 1850 Congress passed legislation of truly major significance in the history of Illinois transportation, an act providing for a grant of public lands to the state of Illinois to aid in the construction of a central railroad— the Illinois Central, later the Illinois Central Gulf. This railroad, the third in Illinois, was to be built through the interior of the state from a point at or near the junction of the Ohio and Mississippi rivers to the western terminus of the Illinois and Michigan Canal at Peru or La Salle, then to the extreme northwestern corner of the state (Galena). Another segment was to be built to Chicago from a site near the present city of Centralia. Interestingly, early

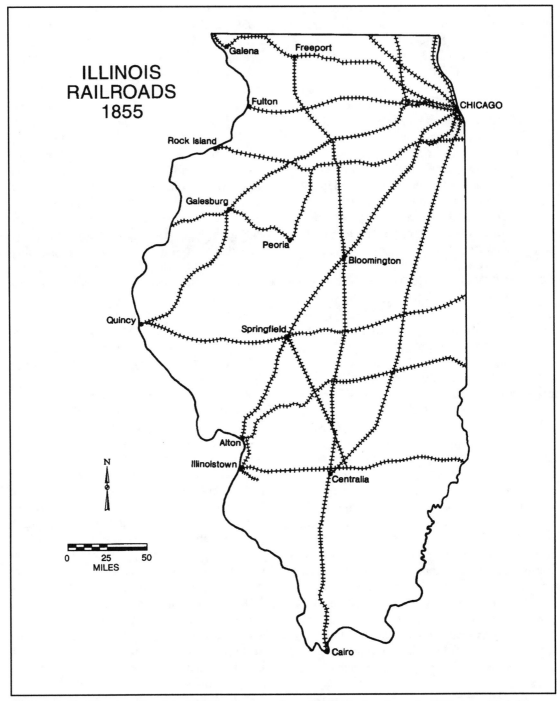

FIGURE 4-12. Illinois railroads, 1855. The construction of track during the early 1850s provided the state with a basic network of rail lines. Note that Chicago was already an important rail hub in 1855.

maps indicate the Centralia-La Salle-Galena segment as the main line and the Centralia-Chicago route as a branch line. The first train from Calumet (Kensington) reached downtown Chicago in 1852 by way of a wooden trestle. The main line was finished in 1855 and the Chicago branch a year later. With more than 700 miles of track at that time, the Illinois Central Railroad was the longest on the American continent.[9] In fact, at that date it was the longest in the world.

Without question the Illinois Central was of paramount importance to the development of the state. First and foremost, it connected the older southern part of the state with the other center of early development (Galena) and with the newer, fast-growing Chicago region. The railroad, especially the line to Chicago, also traversed extensive prairies that were largely unsettled in 1855. Early travellers along the route were surprised to see so small a portion of the land under cultivation and reported virtually no population except along the larger streams and timber areas, which were generally found together. Probably the construction of the Illinois Central was more important than any other single act in opening up the prairie land of east-central Illinois.

Completion of this large mileage of track by a single company within a five-year period during the middle of the nineteenth century was in itself a major achievement, a larger public works project than the digging of the Erie Canal. It was accomplished by simultaneous construction at several points along each line and, at a time of labor shortage, by the recruitment of more than 10,000 men, mostly at New York, New Orleans, and other distant places. Large numbers of Europeans came to America during this period and many of them, especially the Irish, helped to build the Illinois Central Railroad. Many of the Irish workers were recruited directly in Ireland. The practice of assigning an engineer to construct a particular section of track resulted in the identification of that segment of the railroad by the name of the engineer and oftentimes the station on that portion of the track was given his name. For example, Paxton, Gilman, Rantoul, Mattoon, Seidel, and Effingham were stations named for a railroad official or the construction engineer in charge of building the railroad in that particular area.

Although railroad construction continued for more than half a century, the period from 1850 to 1860 was most important for railroad building in Illinois. Among the major lines that had their beginnings in this period was the Chicago and Rock Island, which dispatched its first train to Joliet in 1852, and on to Morris, Ottawa, and La Salle in 1853. It was the first to bridge the Mississippi (in 1856). Both the Chicago and Alton (later a part of the Gulf, Mobile, and Alabama) and the Chicago, Burlington, and Quincy were put into operation in the 1850s. Through the Indiana gateway to northern Illinois came two of America's great railroads: from Michigan City in 1852 was a line that later became a part of the New York Central, and from Fort Wayne in 1859 was the forerunner of the Pennsylvania Railroad into Chicago. The Chicago and Milwaukee line, with train service beginning in 1856, helped develop the North Shore and encouraged the growth of Evanston and Lake Forest as college communities. Into Freeport from the north (Beloit) came the Milwaukee in 1859, but its tracks into Chicago were not built until 1872. The Chicago and Great Western was begun in 1854, but was not completed across Illinois until 1887. By 1857, Chicago was the terminus of eleven rail lines, and their direct connections brought to the city immense quantities of pork, beef, and grain. In general, railroad development lagged in southern Illinois, but a line from Vincennes to St. Louis, later a part of the St. Louis-Cincinnati Division of the Baltimore and Ohio, was completed in 1857. The Terre Haute and Alton was finished a year earlier.

The Illinois portion of the Atlantic and Mississippi from Terre Haute to Illinoistown (East St. Louis), temporarily blocked by Alton interests, also was completed in the mid-1850s. By 1857, 48 railroad projects were either completed or under construction entirely within the state or into Illinois from adjoining states. Almost overnight Illinois became the keystone of the American railway system, and Chicago emerged as the world's greatest railway center with eleven roads ending in the city.

Geography made it inevitable that Chicago would become a transportation center. Aggressive and farsighted business leaders made the most of natural advantages, as the coming of the railroads helped to make Chicago a great city as well as the greatest railroad center in the world. By 1852 Chicago had rail connections with the East Coast, and in 1869 service was inaugurated with the West Coast. Today some 30-odd lines radiate from the Chicago area, and there are about 8,000 miles of railway trackage in the Chicago terminal district.

Trails, Roads, and Highways

Early American trails tended to follow the buffalo trace or the Indian footpath, which may have been one and the same. The old "Saint Louis Trace," believed to be the first overland trail in Illinois, was originally a buffalo trail for much of the way. Really an extension of the long Wilderness Road from Cumberland Gap, this historic route extended from the falls of the Ohio at Louisville to St. Louis, crossing the Wabash River at Vincennes.[10] This trail, subsequently a stagecoach route, followed an alignment later paralleled by the Baltimore and Ohio Railroad and U.S. Highway 50 in Illinois. The vicinity of old Fort Massac, where the city of Metropolis now stands, was the Ohio River terminus of several paths that extended northward and westward.[11] Other early trails led to salt licks both in southeastern Illinois and a few miles west of Danville.

The Indians used the buffalo trails, but they also had routes of their own making that connected their important villages and centers of population. One such center was in the vicinity of Metropolis, another near the mouth of the Kaskaskia River, and a third on the Big Vermilion River near present-day Danville.[12] Some native trails were well-beaten footpaths. Others, used less frequently, were faintly visible and difficult to follow without an Indian guide.

Sometime after Kaskaskia and Detroit were founded by the French, an overland trail partially connecting the two settlements was blazed across Illinois to the Great Bend of the Wabash River in Indiana just east of Danville. At modern Georgetown, just south of Danville, it crossed an ancient Indian trail leading southward to Vincennes, thus linking the three most important centers of French influence in interior America. A more important route from Kaskaskia (and St. Louis) to Detroit, however, used the Mississippi and Illinois rivers to a point near Joliet, and then an ill-defined connection eastward to the beach ridges skirting the southern end of Lake Michigan (Sauk Trail), from where it extended eastward across Indiana and Michigan. The eastern segments later became more widely known and used as the land route from Chicago to Detroit.

One of the earliest land routes in northern Illinois was Hubbard's Trace, also known as Vincennes Trace. It was basically an extension of the Vincennes-Danville route to the newer Chicago community. The name came from one colorful Chicago pioneer and enterprising Indian trader who set up a series of trading posts across the vast prairie wasteland near the Indian border. This was the route later followed by the four Chicago men on horses who made an unsuccessful attempt to get a loan from the bank of Shawneetown. In 1834 the state legislature designated it a state road, marked with milestones and terminating in "downtown" Chicago where it became State Street. Wagons of produce, much of it from the Valley of the Wabash, came to Chicago along this route.

Other early roads of the Chicago region that have remained important over the years are Green Bay Road, St. Charles Road, Naperville Road (now Ogden Avenue and U.S. Route 54), Joliet Road (Archer Avenue), and the present Indianapolis Boulevard. Both the Joliet and Naperville Roads were extended to Ottawa at the confluence of the Fox and Illinois rivers. This was the northern segment of the first land route from Chicago to St. Louis. As far as Naperville, this route was identical with the southern stage route from Chicago to Galena, opened in 1834. The earliest impetus toward an overland highway between these two communities, as might be expected, came from Galena rather than Chicago. The first load of lead to Chicago, in 1829, crossed the Rock River at the site of modern Dixon and the Fox River at Plainfield.

The National Road, sometimes called the Cumberland Road, was the first large highway project in Illinois. It was begun in Cumberland, Maryland, in 1811, and subsequently extended through Pennsylvania and the West Virginia panhandle to Wheeling. Shortly before 1830 it was proposed to extend the road westward to connect the capitals of the new states of Ohio, Indiana, and Illinois. (The capital of Illinois had been moved from Kaskaskia to Vandalia, a mere hamlet, in 1820.) The road was completed from Vandalia to Terre Haute in 1836, although the Indianapolis-Terre Haute segment was not finished until 1840. Shortly thereafter the road was extended to St. Louis. It was still known as the National Road when the present system of road numbers was estab-

lished and it became U.S. 40, now the route of Interstate 70.

The road system of Illinois evolved slowly. Many of the early trails became earthern roads, some of them later stagecoach routes. The principal routes were then improved and became "gravel roads," (i.e., all-weather roads) although oftentimes cluttered with chuckholes. The paved highway is a twentieth century development, and coincided somewhat with the increased use of the motor vehicle. In 1914, when state appropriations for hard roads from proceeds of automobile license fees was begun, there was less than one mile of concrete road in Cook County outside of Chicago.[13] Most of the paved highway net of the state was constructed in the 1920s, the result of several successive state bond issues for road construction. The expansion of multilanes, begun in the 1930s, was primarily a post-World War II development. The super highway (limited access and divided pavement) did not become common until after 1960.

Historical factors explain much of the road pattern of the state. In the earlier settled areas many existing roads and highways follow early trails or traces. A few of the main highways inherited these routes and are therefore not oriented to the cardinal points of the compass. But over most of the state, especially in the less rugged areas, a rectangular pattern predominates. The roads are oriented to the lines of the township and range land survey, a system for surveying the new lands in the west included in the Northwest Ordinance of 1787. Rural roads are commonly spaced at one-mile intervals, following section lines and insuring access to all rural residences. The paved highways built in the 1920s and 1930s are superimposed on this basic rural net and tend to follow this same general pattern. The Interstate system, mostly built in the 1960s and 1970s, follows state and national highways, but is unrelated to section lines.

ALONG THE ROAD TO GREATNESS

Illinois is woven into the history of America with threads of steel. It took hardy settlers to develop the rolling hills and river bottoms of the southern and western parts of Illinois. It took outcasts and adventurers from the East to see and exploit the possibilities of the lakeside swamplands. It took European money to finance the railroads of Illinois, and European immigrants were needed to build those railroads and work in the factories and shops that made the state a major part of the American Manufacturing Belt. Illinois is a great state because of its central position in the midcontinent, its natural endowments, and the foresight and ability of its people. It is a crossroads state that has attained a major position in agriculture, manufacturing, and transportation.

Over the years Illinois has been noted for leadership, progress, and people of vision. The Morrow plots on the Urbana Campus of the University of Illinois connote leadership in agricultural research for more than a century. John Deere of Grand Detour developed a plow with a steel moldboard that made it possible to turn the matted sod of the Midwestern prairies—the plow that broke the plains and set an agricultural revolution in motion. Cyrus McCormick of Chicago built the reaper to harvest the grain that grew in those fertile prairie soils. Joseph Glidden of DeKalb obtained a patent on barbed wire that permitted fencing the land at moderate cost. Years later, A.E. Staley of Decatur proposed that soybeans be used not only as a forage crop for dairy cattle, but also as an industrial raw material. George Pullman manufactured sleeping cars that made train travel more comfortable; he also built the first completely planned community in Illinois. Elbert Gary of Wheaton became Chairman of the Board of Directors of U.S. Steel and helped to build a city amid the sand dunes of nearby Indiana.

William B. Ogden, Chicago's first mayor, talked suburban farmers into joining him in building the city's first railroad. Twenty-one years later, as president of Union Pacific, he drove the golden spike at Promontory, Utah, completing a rail linkup that united the nation. The Chicago-based Illinois Central Railroad developed the first refrigerator car. Richard Sears, an amiable salesman, in the 1890s founded a company that became "the world's largest store." In the 1920s that company operated one of the first commercial radio stations; in the 1930s it marketed the first prefabricated houses; and in the 1970s it built the world's tallest building, the Sears Tower in Chicago. Aaron Montgomery Ward started a mail order business in 1872, but more importantly he fought a decade of court battles to save Chicago's lakefront from commercial development. William Rainey Harper built a major university on the former site of Chicago's first great fair, The World's Columbian Exposition of 1893. Richard J. Daley built McCormick Place on the site of the second great fair, The Century of Progress of 1933-34. Memorial Day was first celebrated in Carbondale, and the idea for the G.I. Bill of Rights at the end of World War II originated in Salem. Gustavus Swift and Philip Armour, along with others, made Chicago the "hog butcher of the world" for many decades, and most of that time East St. Louis was in second position. In 1871, after a fire that left 100,000 homeless, Joseph Medill of the Chicago Tribune said in a now-famous editorial, "Chicago shall rise again." It certainly did. Frank Lloyd Wright received more honors for his works than any other architect of his time. Daniel Burnham exhorted, "Make no little plans." In 1908 he called for developing the Chicago lakefront as a recreation area, for building Wacker Drive in what had been a dirty, smelly wholesale produce area along the Chicago River, and for major traffic arteries radiating from the central city. A practical idealist and activist, Jane Addams, came to Chicago from Iowa and in 1889 converted the former Hull residence, then used for furniture storage, into a settlement house that became known around the world. These leaders, along with many, many others, helped to make Illinois one of the nation's leading states.

NOTES

1. Stuart Struever and Felicia Antonelli Holton, _Koster: Americans in Search of Their Prehistoric Past_ (Garden City, N.Y.: Anchor Press/ Doubleday, 1979).

2. Robert P. Howard, _Illinois: A History of the Prairie State_ (Grand Rapids, Mich.: William B. Eerdmans Publishing Co., 1972), p. 18.

3. Gerald H. Krausse, "Historic Galena: A Study of Urban Change and Development in a Midwestern Mining Town," _Bulletin of the Illinois Geographical Society_, 13 December 1971, p. 7.

4. Randall Parrish, _Historic Illinois: The Romance of the Earlier Days_ (Chicago: A.C. McClurg & Co., 1905), p. 296.

5. Michael D. Sublett, _Paper Counties: The Illinois Experience, 1825-1867_ (New York: Peter Lang Publishing, Inc., 1990).

6. Henry C. Hubbart, _The Older Middle West: 1840-1880_ (New York: Russell & Russell, 1936), p. 76.

7. Howard, _Illinois_, p. 239.

8. _Illinois Business Review_, 5 (May 1948), p. 1.

9. Ibid.

10. Carlton J. Corliss, _Trails to Rails: A Story of Transportation Progress in Illinois_ (Chicago: Illinois Central Railroad, 1934), p. 3.

11. Ibid.

12. Ibid., p. 4.

13. _Chicago Daily News_, 12 August 1974, p. 4.

5

POPULATION AND SOCIAL GEOGRAPHY

Albert Larson
Siim Sööt
Edwin Thomas
University of Illinois at Chicago

Illinois often is perceived as divided into two fundamental regions: the six-county Chicago metropolitan area and the remaining 96-county "downstate" region. The latter is predominantly agricultural but spotted with an irregular geometry of cities and industrial clusters. Although more extensive than the Chicago metropolitan area, the downstate region accounts for only one-third of the state's population. Nevertheless, it is impossible to subdivide the state into two regions with a better balance of area and population without fragmenting the Chicago area. This initial dichotomy, however, may be a gross oversimplification.

While the Chicago metropolitan area dominates the state's population numbers, an examination of Illinois' population distribution shows three urban concentrations:

1. Metropolitan Chicago with extensions to the west to Rockford and south to Kankakee;

2. Metro East, the Illinois portion of the St. Louis metropolitan area; and

3. an interrupted zone from Champaign-Urbana to the Quad Cities that also includes Peoria, Bloomington-Normal, Springfield, and Decatur (Figure 5-1).

This suggests a Von Thünen-like economic pattern. With reference to goods and services, the centers of distribution at the northeast and southwest ends of the state may be too distant to fully serve the more sparsely-settled and areally-extensive intermediate territory. Thus, a zone of secondary centers trending northwest-southeast, and occupying a spatially-central position between the two larger nodes, has developed. This intervening urban corridor provides a tolerable location for the distribution of goods when the locations of the two primary nodes are beyond economically viable distances.

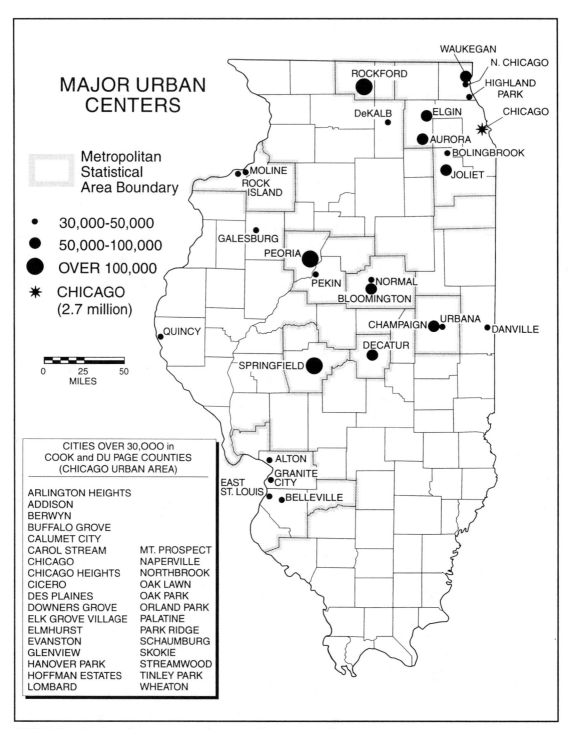

MAJOR URBAN CENTERS

Metropolitan Statistical Area Boundary

- 30,000-50,000
- 50,000-100,000
- OVER 100,000
- ✳ CHICAGO (2.7 million)

WAUKEGAN
N. CHICAGO
ROCKFORD
HIGHLAND PARK
DeKALB
ELGIN
CHICAGO
AURORA
BOLINGBROOK
MOLINE
ROCK ISLAND
JOLIET
GALESBURG
PEORIA
PEKIN
NORMAL
BLOOMINGTON
CHAMPAIGN
URBANA
DANVILLE
QUINCY
DECATUR
SPRINGFIELD
ALTON
GRANITE CITY
EAST ST. LOUIS
BELLEVILLE

0   25   50
MILES

CITIES OVER 30,000 in COOK and DU PAGE COUNTIES (CHICAGO URBAN AREA)

ARLINGTON HEIGHTS
ADDISON
BERWYN
BUFFALO GROVE
CALUMET CITY
CAROL STREAM   MT. PROSPECT
CHICAGO   NAPERVILLE
CHICAGO HEIGHTS   NORTHBROOK
CICERO   OAK LAWN
DES PLAINES   OAK PARK
DOWNERS GROVE   ORLAND PARK
ELK GROVE VILLAGE   PALATINE
ELMHURST   PARK RIDGE
EVANSTON   SCHAUMBURG
GLENVIEW   SKOKIE
HANOVER PARK   STREAMWOOD
HOFFMAN ESTATES   TINLEY PARK
LOMBARD   WHEATON

FIGURE 5-1. Major urban centers and metropolitan statistical areas.

TABLE 5-1. Populations of Illinois and Neighboring States

State	1990 Population (in thousands)	Percent change 1970-80	Percent change 1980-90	Area (mi.2)	Percent urban	Nonmetropolitan population (in thousands)
Illinois	11,431	2.8	0.0	55,930	84.6	1981
Michigan	9,295	4.3	0.4	57,019	70.5	1850
Indiana	5,544	5.7	1.0	36,185	64.9	1748
Kentucky	3,685	13.7	0.7	29,863	51.8	1971
Missouri	5,117	5.1	4.1	69,138	68.7	1730
Iowa	2,777	3.1	-4.7	56,032	60.6	1554
Wisconsin	4,892	6.5	4.0	54,705	65.7	1593

Trends are commonly cyclical, often with numerous positive and negative swings and this is reflected in Illinois' population. During the 1970s, the state's population increased steadily and by approximately 300,000 or 2.8 percent. The 1980s experienced the beginning of an economic slowdown in the Midwest, with population decreasing in Illinois until 1986 when it reached a low of less than 11.4 million. Subsequently an economic recovery spread to the Midwest and the state's population again increased and reached a level in April 1990, when the census was taken, of nearly 5,000 more people than in 1980. Illinois thus escaped the distinction of having lost population from one decennial census to the next. In this regard it is fortunate that the official head-counts are not conducted every five years. In sum, over the interval between 1970 and 1990, the population of the state was generally stable with periods of only modest growth.

In this chapter we will highlight demographic differences within the state by county-level examinations of the principal dimensions of the population. Before we begin the discussion of internal patterns, a short overview of the contrasts between Illinois and its neighboring states is presented. This provides a basis for analysis of Illinois that follows.

COMPARISONS WITH NEIGHBORING STATES

With Chicago dominating its urban structure, Illinois is clearly unique in the Midwest. In contrast to its neighboring states, Illinois has the highest percentage of its population in urban areas (Table 5-1). On the other hand, the nonmetropolitan populations of Illinois and neighboring states are remarkably similar. The Von Thünen model would suggest that the rural population density should decline with increasing distance from the dominant regional market, Chicago. While this is true in Illinois and Iowa, the nonmetropolitan population of Kentucky approximates that of Illinois even though Kentucky has a much smaller area (Table 5-1).

The general population growth in the Midwest also is rather similar from state to state. The decennial data indicate modest but declining population increases over the last 20 years (Table 5-1). Additionally, each of the states had a higher growth rate in the 1970s than in the 1980s; Iowa even registered a loss in the 1980s. None of the states kept pace with the national average, an increase of 9.8 percent in the 1980s.

FIGURE 5-2. Both old and new skyscrapers in Chicago are testimony to the city's continuing dominance of the population and economic structure of Illinois. (Courtesy Illinois Office of Tourism.)

TABLE 5-2. Population Characteristics of Illinois and Neighboring States

State	Median Age	Age 65+	Age <18	Percent Latino	Percent Foreign Born	Percent w/Bachelor Degree	Percent Captia Income	Percent Families Below Poverty Line
Illinois	32.8	14.2	23.7	7.9	8.3	21.0	15,201	9.0
Michigan	32.6	12.7	25.2	2.2	3.8	17.4	14,154	10.2
Indiana	32.8	13.0	25.5	1.8	1.7	15.6	13,149	7.9
Kentucky	33.0	12.9	25.4	0.6	1.2	13.6	11,153	16.0
Missouri	33.5	14.8	24.7	1.2	1.6	17.8	12,989	10.1
Iowa	34.0	15.7	25.5	1.2	1.6	16.9	12,422	8.4
Wisconsin	32.9	14.1	25.1	1.9	2.5	17.7	13,276	7.6

Considering the two decades, 1970-1990, most adjacent states registered a higher rate of population growth than Illinois, despite the latter's balanced economy and greater foreign immigration. Even Michigan, with its problems in the automobile sector, has out performed Illinois in population growth. Kentucky benefitted in the 1970s from movement of industrial capacity to the South and in that decade had the highest population growth of all adjacent states; its rate of growth slowed drastically in the 1980s, however. Wisconsin and Missouri both have been able to register steady growth over the 1970-1990 period, again outpacing Illinois.

The declining population in Iowa also is reflected in other demographic statistics. For example, Iowa has the highest median age and the highest percentage of people over the age of 65 (Table 5-2), indicating that the young in significant numbers have moved from the state. Illinois, conversely, has the lowest percentage of people under 18. The declining birth rates and professional orientation of its population have led Illinois to an age profile that is not high at either end, i.e., neither young nor old.

The professional orientation of the state's population is revealed in two indicators. Among the states under consideration here, in 1990 Illinois had the highest percentage of its population over age 25 with a bachelor's degree and the highest per capita income (Table 5-2). Michigan is second highest in per capita income.

Illinois holds the median position amongst its neighbors in percentage of households below the poverty line. Iowa, Wisconsin, and Indiana have lower poverty rates, but in Missouri, Michigan, and Kentucky they are higher. With 16.0 percent of its families below the poverty line, Kentucky is substantially higher in that regard than Illinois which has only 9.0 percent.

Among midwestern states, Illinois also is distinguished by its immigrant population. In 1990 Illinois had over twice the percentage of foreign born than Michigan, its closet rival, and more than three times the Latino percentage (Table 5-2). Illinois also has 14.2 percent of its population who speak a language other than English. Most neighboring states are closer to 5 percent except Kentucky, which registers the lowest (2.5 percent) of any state in the Union.

Basic Population and Demography

Since the mid-1880s Cook County has had the largest population in Illinois. It surged in population in the 1920s and until the 1970s contained more than one-half of the state's inhabitants. Metropolitan Chicago's population decentralization, which began at the turn of the century, eventually reached beyond Cook County. Currently many of Cook County's close-in suburbs also are declining in population. In the 1980s Cook County's population dropped by approximately 150,000 while adjacent DuPage increased by almost 125,000. In 1990 Cook County accounted for only 45 percent of the state's population. While all of the suburban Chicago counties registered major population gains from 1970-1990, many rural and even some metropolitan counties in the downstate areas lost population (Figure 5-3). Among the metropolitan counties, Kankakee in the northeast and St. Clair and Madison in the southwest had a net loss of inhabitants. Also, Rock Island, Henry, Peoria, and Macon counties in the central to northwest urban corridor all experienced a loss of population between 1970 and 1990.

The age profile of the population also exhibits an irregular pattern. The percentage of Illinois' population over 65 years of age in 1990 was 14.2 percent and growing. On a county basis, however, it varied from a high of 21.7 percent in southern rural Hamilton County to a low of only 8.4 percent in Lake County, in the very northeastern corner of the state. Several counties have values over 19 percent (Figure 5-4). These are predominantly rural counties in western and southeastern Illinois where there are low percentages of people in the child-rearing age categories. The largest proportion of senior citizens, in fact, is in the southeastern part of the state.

The greatest concentration of counties with a low percentage of senior citizens is in the suburban Chicago area. The movement of young families to the urban fringe in search of affordable housing has resulted in a relative decrease in the older segments of the population profile. In addition to Lake County, the following counties have less than 10 percent of their population over 65: DuPage, Kane, Kendall, McHenry, and Will. Outside of the Chicago suburban area, only in Champaign County does the population over 65 account for less than 10 percent of the total.

Counties with small percentages of senior citizens might be expected to have large proportions of their population under 18 years of age, and indeed this is true in Illinois (Figure 5-5). However, there are also a number of counties with high percentages of young people (under 18) that are not particularly low in their percentage of senior citizens. These include non-suburban counties such as Alexander, Pulaski, and Effingham. Alexander and Pulaski are in the far southern portion of the state where family incomes are lower than the state average and the family size tends to be larger than the state average. These counties are characterized by both relatively high birth rates and an elderly population that has been unable or unwilling to migrate elsewhere.

Historic Population Development

The earliest European occupation of what came to be Illinois was in the southwest at Kaskaskia, and early settlers in the American period also gathered in the same general area. Pioneer hunters from the Upland South moved along the watercourses and regarded the open prairie with suspicion. When Illinois became a territory in 1809 and a state in 1818, there was only scant settlement in the central and northern areas (Figure 5-6). Most of the

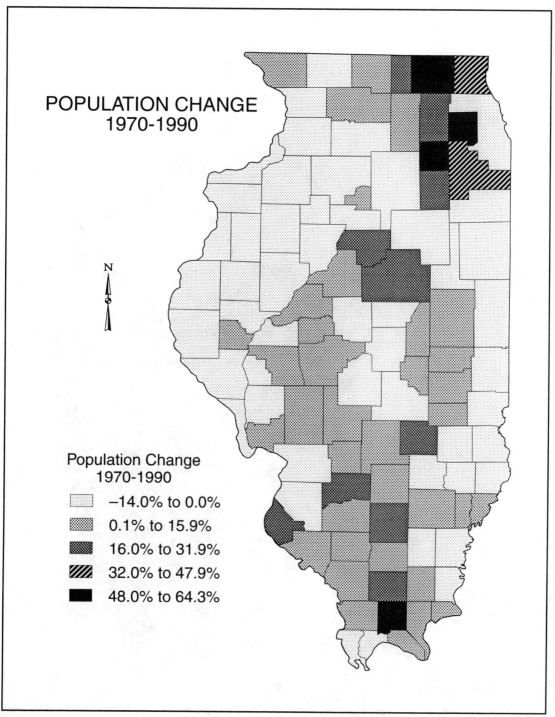

FIGURE 5-3. Population change, 1970-1990.

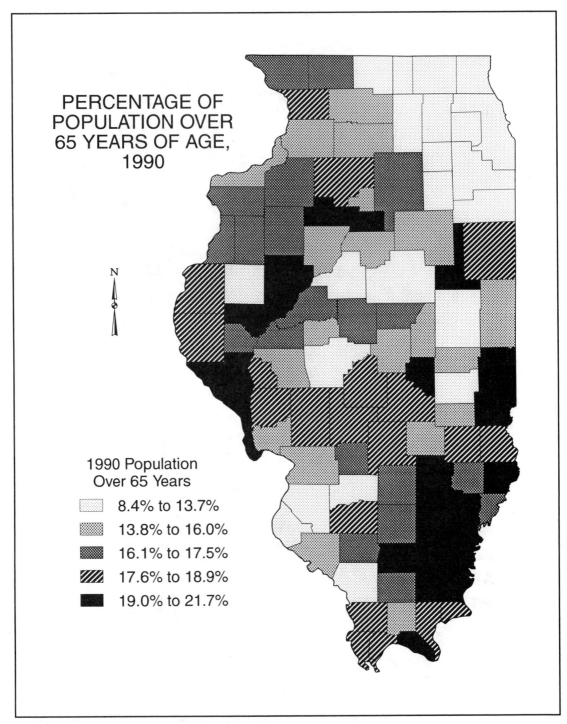

FIGURE 5-4. Population over 65 years of age, 1990.

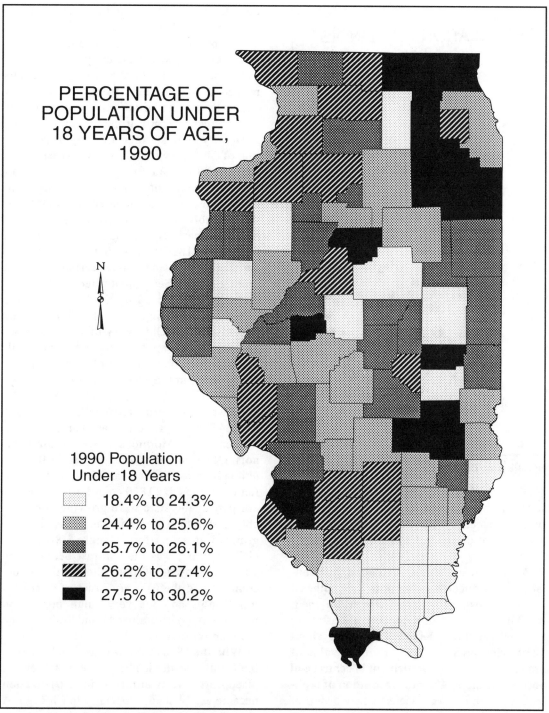

PERCENTAGE OF
POPULATION UNDER
18 YEARS OF AGE,
1990

1990 Population
Under 18 Years

18.4% to 24.3%
24.4% to 25.6%
25.7% to 26.1%
26.2% to 27.4%
27.5% to 30.2%

FIGURE 5-5. Population under 18 years of age, 1990.

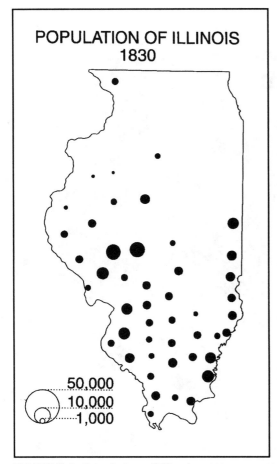

POPULATION OF ILLINOIS
1830

50,000
10,000
1,000

FIGURE 5-6. Population of Illinois, 1830.

dle Atlantic states, reached northern and central Illinois. There was a general land rush helped by the beginning of steam navigation on the Great Lakes in the early 1830s as well as by use of the Erie Canal. Many Easterners moved to the western frontier where they hoped to prosper. Additionally, new settlers came directly from Europe seeking the same prosperity. The strategically important location of Chicago with reference to Great Lakes transportation led to its early growth, although it came into existence in August of 1810 with only 150 inhabitants. Land offices were established in the northern part of Illinois in the early 1830s, and a major land boom followed in 1836. Galena, in the far northwest had been established much earlier as a lead mining center. Settlement after the 1837 depression was spurred by the building of the Illinois and Michigan Canal to connect the Great Lakes and Chicago with the Mississippi system, by the appearance of the first railroads, and by the building of plank roads. In 1833 Alton, in the southwest, was the state's largest city, but increased settlement in the north was fast shifting the population balance. By 1845 the largest city was Nauvoo, settled by a colony of Mormons, in west-central Illinois. Other colonies were planted in Illinois, although most settlers arrived singly or in family groups. By 1850 counties in the northern part of the state were growing considerably faster than those in the south, but the growth was not uniform (Figure 5-8B). Especially striking was the contrast between the sizeable growth of both the northernmost counties and the Military Tract of west-central Illinois and the very small population increase in the ill-drained Grand Prairie of the east-central section.

With the 1850s Illinois ceased to be part of the frontier as the leading edge of settlement disappeared westward. Now industrialization became established, especially in Chicago. A conscious statewide effort to push the growth

state had been ceded by Indians during the first two decades of the century, although the last negotiation for Indian land was not until 1833.

At the time of Illinois' statehood, its population was concentrated in both the southeast and southwest. This pattern was still evident in 1830 with the largest concentrations along the lower Illinois River. Beginning about 1830, widespread realization of the value of prairie soil increased settlement in central and northern Illinois. After the resolution of problems with native Americans, a new group of pioneers, directly from the northern and midof its own towns and cities, while doing little to aid those of adjoining states, allowed Chicago to attain a supreme position. Continued transportation improvements made increased rural settlement feasible and brought new settlers to Illinois who worked as laborers on the construction of new transport facilities. Large numbers of Europeans came to Illinois, many taking up residence in Chicago. Especially numerous in the 1850s were Irish and Germans. By 1860 foreign-born outnumbered native-born. Not all new settlers stayed in Illinois; some moved farther west, but the newcomers always outnumbered those leaving the state. In 1857, eleven railroad main lines radiated from Chicago and served in transporting immigrants to unsettled land and in collecting produce from developing farms. In the 1850s and 1860s, Chicago's growth was supported not only by immigration but also by its use of materials from the Great Lakes area for industrialization. In 1849, 40 percent of the state was still public land, but within six years almost all the remaining public domain passed to private owners. Realizing that woodlots were not a necessity, farmers settled increasingly on the open prairie, especially with the aid of new implements. As the railroads had come into ownership of public land, they also entered into the business of selling land. Even those east-central counties of the ill-drained Grand Prairie began to grow with the construction of the Illinois Central Railroad. Many towns

FIGURE 5-7. The restored home of Mormon Prophet Joseph Smith in Nauvoo on the bank of the Mississippi River. Although it was the largest city in Illinois in 1845, Nauvoo soon was abandoned by the colony of Mormons following the murder of Smith and the destruction of their temple. (Photograph by R.E. Nelson.)

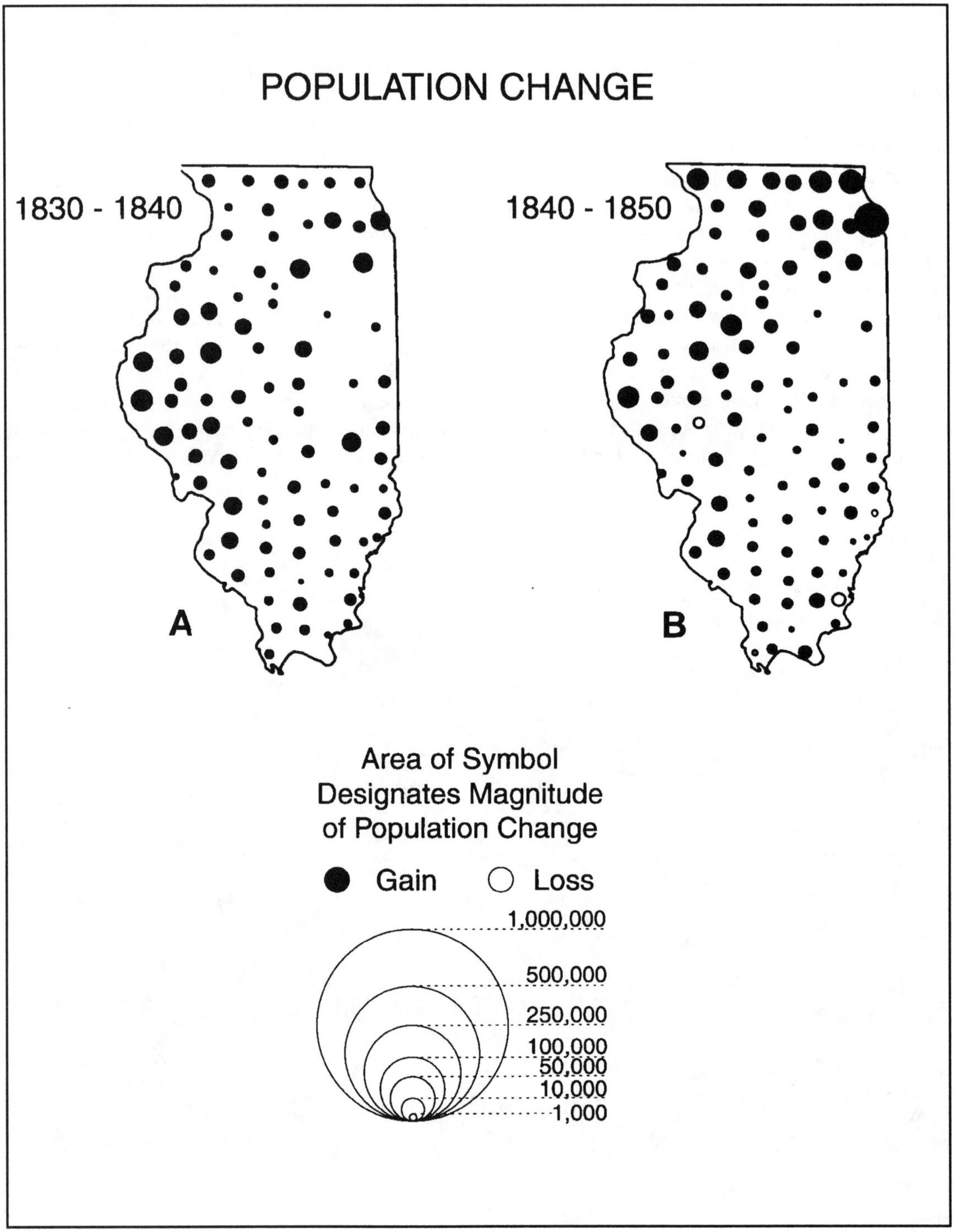

FIGURE 5-8. Population change, 1830-1970.

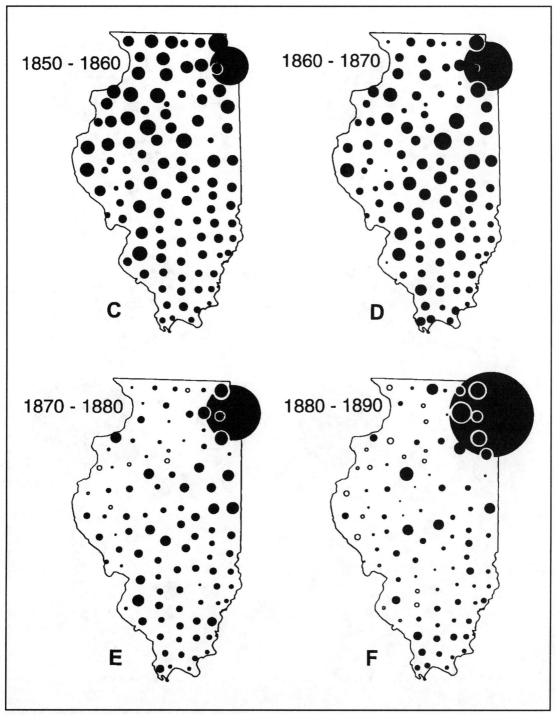

FIGURE 5-8. Population change, 1830-1970.

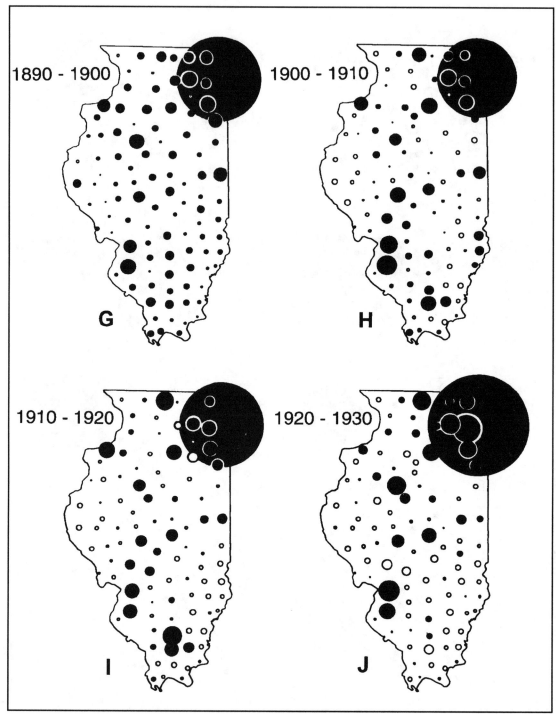

FIGURE 5-8. Population change, 1830-1970.

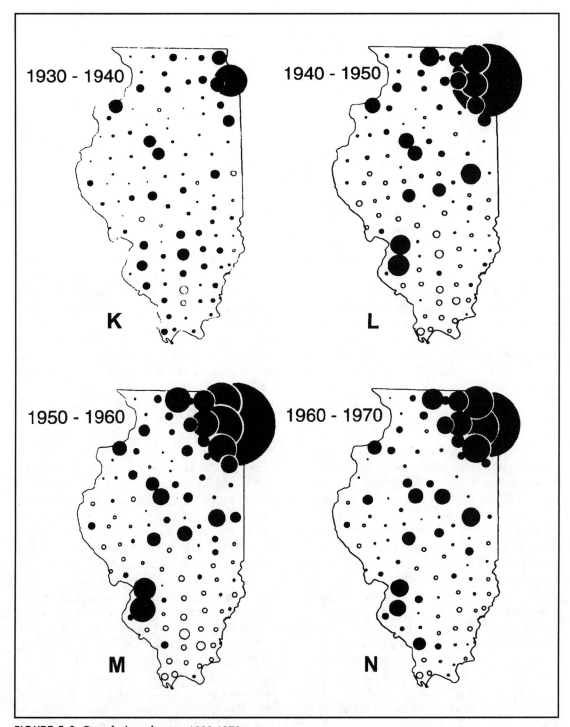

FIGURE 5-8. Population change, 1830-1970.

began as stations on rail lines, and those that prospered grew at the expense of those not so located. The railroads also put long-distance drovers out of business, but concurrently made Chicago the nation's leading livestock market. Chicago bankers who supported the market development also assisted in financing the development of new farm implements as well as new farms. In addition, Chicago established the Board of Trade, which standardized the grain market and constituted the country's leading lumber market. All of this activity provided many jobs in Chicago, which led to large population growth especially in the 1850s. In addition, many small-town production plants moved to Chicago, contributing to its concentrations of trade and industry.

The somewhat uniform growth of Illinois counties in the 1850s and 1860s changed with the beginning of the 1870s (Figure 5-8E). While the growth of rural counties declined from the previous decade, Cook County more than doubled in population, and by 1880 Chicago was the nation's fourth largest city. Following the Civil War, a rural depression led to migration both to newer lands in the West and, even more, to Chicago.

The period beginning in 1880 was characterized by heavy immigration to Chicago, the state's primary industrial center. Increasingly, the new arrivals came from northern, then eastern and southern Europe. Chicago's industrial power grew with production of Minnesota's iron ore (accessible by lake transportation) and nearby supplies of coal. The rail net brought agricultural produce for processing by the city's growing labor supply drawn from its farming hinterland as well as from Europe. Mechanization on the state's farms released population from the land, often in the richest agricultural counties. Thus, while most of Illinois' downstate counties grew, 31 lost population in the 1880s. The state's population changed from 69.4 percent rural in 1880 to 55.1 percent rural in 1890

(Figure 5-8F). In addition to Chicago, some downstate cities also grew in commercial and industrial importance; several of these were mining centers.

General prosperity returned to the nation beginning about 1900, but even the previous decade was one of economic gain for Chicago and the state. Whereas other more totally agrarian states suffered depressed times, the economic diversity in Illinois coupled with its fortuitous location allowed for economic advancement. By 1900 Illinois' rural population had dropped to 45.7 percent. In the face of the depressed conditions of the early 1890s, Chicago held the Columbian Exposition which, among other things, publicized the city's technological and cultural achievements. But the general prosperity and urban growth of the first decade of the twentieth century were accompanied by population loss in 56 of the 102 Illinois counties. Again, technological improvements in agriculture released farm labor. At the same time, downstate urban and coal mining centers experienced population growth, though considerably more modest than that of Chicago.

Heavy immigrant flow from eastern and southern Europe continued for the first 30 years of the twentieth century. Beginning about 1915, migrants from the American South added to the Illinois population mixture. Labor shortages during World War I led to the northward movement of southerners, many of them African-Americans. This movement steadily increased as economic opportunities in Illinois constituted a strong pull. A compilation in 1914 showed important industry in 35 cities of downstate Illinois, while the African-American population of Chicago more than doubled between 1910 and 1920 (Figure 5-8I). To meet the war effort, very high agricultural and industrial production was needed. This production was realized even though many rural counties continued to lose population. These losses, plus Chicago's continued large gains, led to the sectionalism that had been building before 1910 and is still in evidence today. Downstate feelings, manifesting themselves in Illinois' politics, grew out of a distrust of Chicago and a fear that the city would dominate the rest of the state.

Overseas immigration was virtually stopped after the war, but the northward movement of African-Americans from the South continued to increase. With continued industrial growth, especially in Chicago, jobs were available in the North. Chicago's African-American population rose to 4.1 percent in 1920 and 6.9 percent in 1930. They were squeezed into a definite zone on the south side of the city, one of the factors keeping racial tensions alive.

The decline and demise of small towns in Illinois continued as the means of transport were improved. With the arrival of the auto and better roads, a new dimension was added, but results were much the same. The auto enabled farmers and villagers· to travel to larger places where a more extensive array of goods and services was available. The continuing loss of business in small towns hastened their decline. In large urban centers, especially Chicago, the inner city became blighted. Auto transport, just as the streetcar before it, enabled the working population to commute from suburban residential locations. The sizable suburban growth is suggested in Figure 5-8I. At the same time, 63 rural counties lost population in the 1920s.

As a consequence of the Great Depression, the northeastward migration of Illinois' center of population reversed itself for the only time between 1930 and 1940. Likewise, Figure 5-8K reflects a reversal of earlier trends. The vast urban growth of the previous decade virtually ceased. In addition, many downstate counties that had suffered a population loss in the 1920s now were the recipients of modest gains, the result of a general "back-to-the-farm" movement characteristic

of the Depression. In contrast to the previous decade, only 26 counties experienced population losses, and these were generally very small. Illinois' population growth, which had been 17.7 percent in the previous decade, was only 3.5 percent in the period 1930-1940. Rural growth actually outpaced urban, 4.6 percent to 3.1 percent. Unemployment reached approximately 1.5 million during the Depression, and 277,000 families received relief. Immigration was restricted and Chicago lost some of its economic attractiveness. For some people who lost their city jobs, a return to relatives or friends in rural locations meant at least subsistence and avoiding the degrading aspects of the dole.

Although better times began to emerge in the late 1930s, prosperity returned to Illinois when World War II stimulated the economy (Figure 5-8L). Domestic war effort requirements found many willing workers. While one million of the state's citizens took an active part in the war, federal spending provided local employment for many of those who remained on the home front. In some cases labor shortages resulted and the number of African-Americans migrating northward once again helped fill the need. An estimated 100,000 Black Americans moved during World War II to Chicago, where their employment increased from 4.9 percent to 11.7 percent in five years. A similar increase was also experienced in the Metro East urban agglomeration in southwest Illinois tributary to St. Louis. Though the number of farms decreased, improved techniques helped develop better farms. Levee construction and farm drainage districts helped bring 88.3 percent of Illinois under cultivation, the largest proportion of tilled land for any state. Illinois farms produced record crops. In 1945, 13 counties in north-central Illinois were listed among the leading 100 American counties in value of farm products. That this great production was accompanied by increased mechanization is evident in Figure 5-8L. Many counties, especially in west-central and southeast Illinois, lost population in the 1940-1950 decade. These same regions continue to be the most rural parts of the state today.

The return of peace after 1945 found Illinois more firmly established than before as a manufacturing state, reversing the back-to-the-farm movement. While certain Chicago industries lost importance, others readily absorbed the skilled and unskilled workers who remained. By 1950, Illinois was fourth in the nation in population, but a strong third in manufacturing. The massive growth of Chicago and its suburbs, and to a lesser extent the growth in downstate manufacturing centers, is readily apparent in Figure 5-8M. Jobs in the Chicago area continued to exercise a powerful attraction. The new immigrants were from the American South, Europe, and Latin America. As Chicago's suburbs became the recipients of those who were seeking the good life away from the more crowded and pluralistic city, the new arrivals took up the inner city housing of former residents. In the 1950s, Cook County had more people than the other 101 counties combined. The City of Chicago actually has been losing population since the 1950s, but the city's suburbs have gained greatly in the same period (a gain of 77 percent in the decade of the 1950s). Elsewhere, growth continued in counties that contained medium to large urban populations and where industrial activity shared in the prosperous American economy.

Those counties that did not share in the urban-oriented job market lost further numbers as the trend toward greater agricultural mechanization continued. Once again, the counties experiencing greatest loss were in the southeast and west-central parts of the state. Beginning in the late 1880s a county growth pattern emerged that still holds. Three major zones of growth are:

1. The northeast, expanding outward from Chicago
2. A northwest-southeast axis in central Illinois
3. The Metro East area in southwestern Illinois

Areas of population loss have been in the aforementioned west-central and southeastern parts of the state. Separating the northeast and central sections of continued growth is an area of population stability. An examination of Figures 5-8I through 5-8N shows that only a few notable exceptions altered the dominant pattern.

By 1970, Illinois ranked fifth in population, having been leapfrogged by California and Texas. The state's population increased 10.2 percent in the 1960s, becoming 82.9 percent urban and 13.6 percent non-White in 1970. The rush to suburban fringes in the urban northeast has continued since the 1960s, and by 1990 Chicago had more than 200 suburban municipalities. The inner city's inhabitants continue to be African-Americans and other recent arrivals, especially from Latin America and Asia. Chicago's African-American population rose from 10 percent in 1950 and 14 percent in 1960 to 32.7 percent in 1970. Spanish-speaking newcomers constituted 10 to 15 percent of the city's people. Appalachian Whites and Native Americans were among other minorities in the nation's (then) second largest city. Meanwhile, 40 percent of the population of the nation's second largest county lived outside the city of Chicago. The growth went beyond the county, and for the first time since 1920, a majority of Illinoisans in 1970 no longer lived in Cook County. However, those counties showing the greatest increases outside Cook were also in the northeast. Of the state's 19 municipalities over 50,000, seven were in suburban Cook County and four others were in adjacent counties. Downstate, the largest growth in the 1960s was again in these counties with sizable cities. Almost one-half of the state's counties (49) again lost population, mostly in areas that had been losing for some time. Changes in the 1960s also were associated with the new superhighway system. Whole new suburbs and housing subdivisions were created along the transport arteries, enabling workers to commute greater distances to their jobs. Downstate villages declined, hotels were replaced by motels, and city centers decayed as outlying shopping centers flourished. Nevertheless, in 1970 Illinois' advantageous location along with its strong balance of agriculture and industry seemingly assured the state's future economic viability.

CONTEMPORARY PATTERNS

Urban Versus Rural Population

The population of Illinois in the 1990s can be apportioned among five areas:

1. Metropolitan Chicago
2. Other metropolitan
3. All other urban
4. Rural non-farm
5. Farm.

Metropolitan Chicago accounts for approximately two-thirds of the state's population (7 of 11 million) and encompasses six counties, soon to expand to seven by the addition of Kendall County. The Chicago area dominates most population statistics and therefore the following data are commonly reported in relative terms, percentages and medians. There are nine other metropolitan areas with central cities of at least 50,000 residents. Together these ten metropolitan areas (including Metropolitan Chicago) account for 23 of Illinois' 102 counties and 82.7 percent of the state's population (Figure 5-1). There are another 220,000 people who live in urban areas outside the metropolitan counties. These include communities of at least 2,500 residents

but less than 50,000 (larger urban areas would be designated metropolitan). Many of these places can be described as small and medium-sized cities that are scattered throughout the state. The final two categories, both rural, are designated as rural non-farm and farm. Many in the former category live in small communities with populations under 2,500, thereby not qualifying as urban. There also are large numbers of non-farmers who reside outside community boundaries. The last category, the farm population, has been declining in numbers for several decades but continues to be very important in Illinois' economy.

Population growth was experienced throughout most of the state in the decade of the 1970s. Only 15 of the state's 102 counties did not increase. However, that trend reversed in the 1980s, when only 20 counties had a population increase. A variety of factors led to an apparent urban-to-rural movement in the 1970s. This so-called "population turn-around" was short lived, however, as rural areas suffered serious population decreases during the 1980s. By 1990 the farm population in the United States had sunk to less than three percent of the total. Illinois followed the national trend of declining rural population while remaining one of the country's leading farm states.

Largely economic circumstances have led to fewer (although larger) farms, fewer farmers and, consequently, fewer small towns and small town populations. Although desire for a rural-like lifestyle has led some urban dwellers to non-urban areas, they have not become directly engaged in agriculture. These rural non-farm persons have been attracted to the real or perceived amenities of living "closer to nature." But part of this circumstance is one of census definition. The United States census defines anyone living in a place of less than 2,500 as rural. Many Illinoisans living in places of less than 2,500 are not farmers; in fact, they are living in ways that differ little from urban lifestyles. The old dichotomy of city slicker and country hick is largely inappropriate today. But it must be borne in mind as one passes the beautifully planted and cultivated fields of rural Illinois that relatively few farmers are responsible for those fields, and fewer still are successful in the traditional sense of wealth accumulation. The areas of Illinois where the rural population is most dominant continues to be the west-central and southeast (Figure 5-10). These two regions have few cities and, as such, have suffered the most from an economy that no longer favors farming and small service center activities.

Economic and Education Patterns

Economically, Illinois is an enigma. It has for decades ranked in the top dozen states in per capita income, yet recently the state's financial status has been weak. The wealth of the state can be seen in its high per capita income, highest among neighboring states and ranking seventh in the country. College graduation rates also are high, suggesting that Illinois' per capita income will continue to rank among the leading states.

Within the state the distribution of the population with a college degree is rather predictable. The highest concentrations are in urban and in university counties (Figure 5-11). DuPage County ranks highest with 36 percent of its population over 25 years of age having graduated from college. Champaign County is second with 34 percent and Jackson County is third with 29.5 percent. Most of the counties with the lowest percentages are predominantly rural and in the southern one-third of the state.

FIGURE 5-9. The small village of Adair is among hundreds of communities in Illinois that do not meet the census definition for urban status. (Illinois Geographical Society)

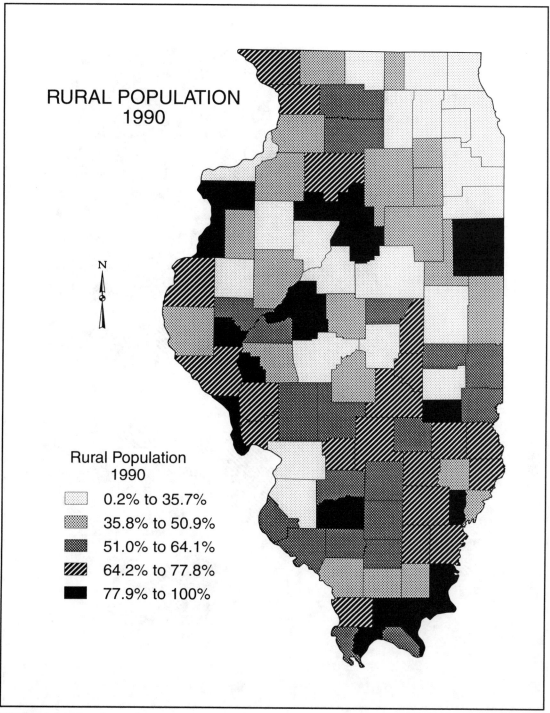

FIGURE 5-10. Rural population, 1990.

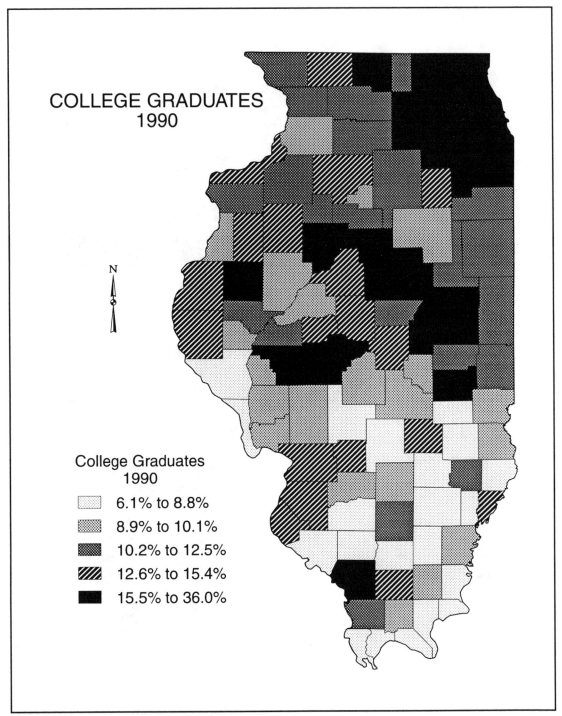

COLLEGE GRADUATES
1990

College Graduates
1990

	6.1% to 8.8%
	8.9% to 10.1%
	10.2% to 12.5%
	12.6% to 15.4%
	15.5% to 36.0%

N

FIGURE 5-11. Percentage of people who are college graduates, 1990.

Similar patterns can be seen in the proportion of the population aged 25 and over with less than nine years of education (Figure 5-12). On this map the north-south gradation is very evident. Percentages less than six are found in Champaign, DuPage, Kendall, Lake, and McHenry Counties, the latter four within Metropolitan Chicago. Conversely, two counties, Gallatin and Hardin, have values over 25 percent. Both of these are on the Ohio River opposite Kentucky. Almost the entire extent of Hardin County and about one-third of Gallatin County are in the Shawnee National Forest. These education data underscore the contrasts between Metropolitan Chicago and the southern portions of the state.

Figure 5-13 shows the pattern of poverty in the state. There is strong correspondence between Figures 5-12 and 5-13 (i.e., between low educational attainment and poverty). On both maps the Ohio River counties have high levels. Illinois has seven counties in which the poverty level is over 20 percent and five are on the Ohio River. Among the six counties on the Ohio River, only Massac County, across the river from Paducah and bisected by Interstate 24, is not on this list. The Interstate highway provides access to employment opportunities in Paducah, and it also provides access to the county from other areas. As a consequence, retail trade in Massac County involves 20 percent of the labor force, a level exceeded by only two other counties in the state. The state's highest poverty levels are in Alexander and Pulaski counties with levels of 32 and 30 percent respectively. While most of this is rural poverty, Cairo, a community with a population of almost 5,000 in Alexander County, also has severe poverty. The continued loss of jobs in mining, coupled with shrinking opportunities in marginal farming operations, are two of the contributing factors to economic problems in southernmost Illinois.

By contrast, Cook County has less than 14 percent of its population classified as below the poverty level. The absolute numbers in Cook County obviously are larger than in the southern counties, but the percentage is less than one-half. The great concentration of low income families in the Chicago inner city projects a powerful visual image of poverty. It has a greater impact than the poverty in southern Illinois, which is much more dispersed over a large territory. Although rural poverty may be less visible, it nevertheless is present and should be assessed and dealt with as thoroughly as that in the inner city.

Not surprisingly, all of the five counties with poverty levels less than five percent are suburban. They are, from lowest to highest, DuPage, Kendall, McHenry, Monroe, and Lake. All but Monroe (suburban St.Louis) are in the Chicago metropolitan area.

Ethnic Patterns

Immigration to the United States increased steadily through the late 1800s and early 1900s until the depression. Since the 1930s, when it hit a low point, immigration has again increased every decade without exception and reached, in the 1980s, the highest level since the 1900-1910 decade. During the twentieth century there has been a dramatic change in the origins of the immigrants to the United States. The largest number of foreign-born U.S. residents now are from Mexico and Central America, followed by Asia and Europe. The origins of immigrants in Illinois are similar.

In Illinois, the pattern of foreign-born is reminiscent of some of the other patterns, with urban and university counties having the highest levels (Figure 5-15). There are six counties with five percent or more:

Cook (14.1)
DuPage (9.1)
Kane (8.2)
Lake (8.0)
Champaign (6.1)
Jackson (5.0)

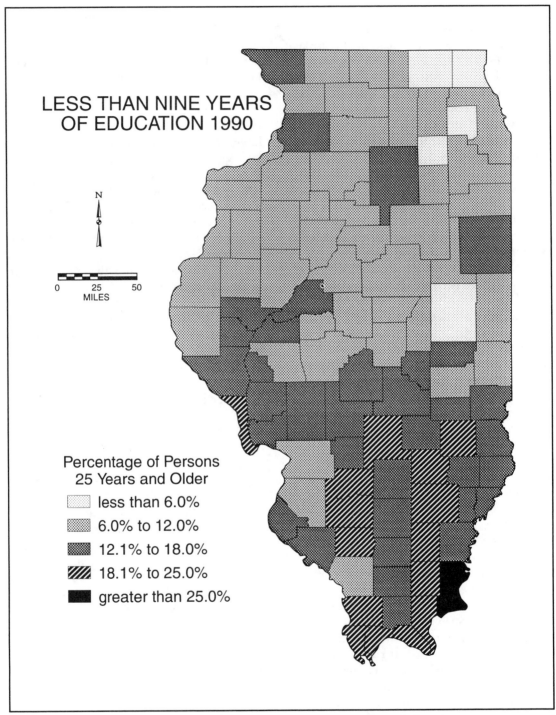

LESS THAN NINE YEARS OF EDUCATION 1990

N

0   25   50
MILES

Percentage of Persons
25 Years and Older

less than 6.0%

6.0% to 12.0%

12.1% to 18.0%

18.1% to 25.0%

greater than 25.0%

FIGURE 5-12. Persons 25 years of age and older with less than nine years of education, 1990.

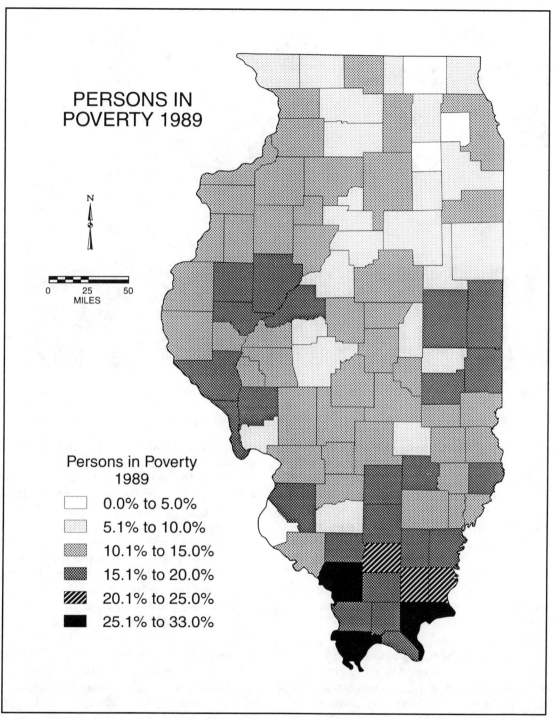

FIGURE 5-13. Persons in poverty, 1989.

FIGURE 5-14. Deteriorated and abandoned commercial buildings in villages and small towns, such as these in the village of Plymouth, reflect the poverty in parts of rural Illinois. (Illinois Geographical Society.)

The first four are in the Metropolitan Chicago area and the latter two are counties with large universities: the University of Illinois at Champaign-Urbana and Southern Illinois University at Carbondale, respectively. American universities continue to attract foreign students in large numbers. The quality of American institutions is widely known in the world, and the presence of foreign students enriches the educational experiences of domestic students. Illinois' institutions of higher education enroll students from throughout the world.

Cook County, with approximately 700,000 foreign-born residents, has more than the other 101 counties in Illinois combined. The majority of the remaining foreign-born are found in suburban Chicago counties. Not all

metropolitan counties in the state have high percentages, however. Peoria, popularly associated with the heartland of the country, has only 2.3 percent of its population foreign-born. Sixty percent of the counties in Illinois have less than one percent.

Some immigration patterns were established decades ago. For example, migrants to the United States often settled in certain urban centers and then informed others from their home region about opportunities in their newly adopted city. This process established distinctive channels of migration and Chicago has been a major destination. Large numbers of Polish immigrants, for example, have settled in Chicago and even today the city continues to attract Poles. Although it no longer holds this status, Chicago has widely been

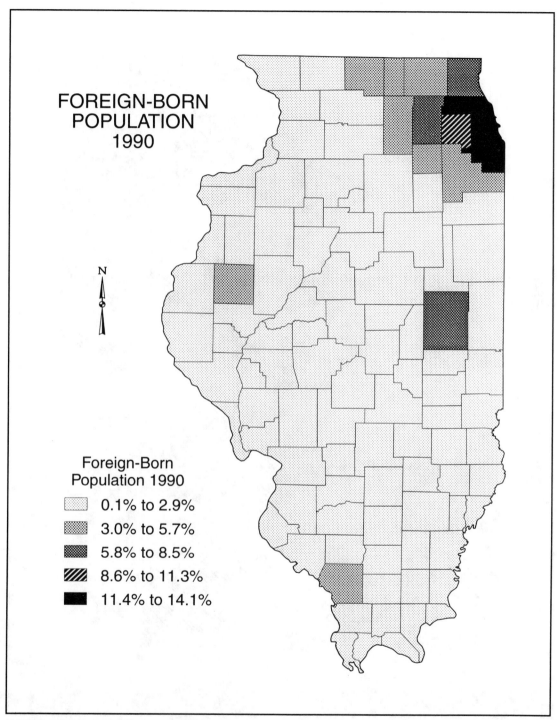

**FOREIGN-BORN
POPULATION
1990**

Foreign-Born
Population 1990

0.1% to 2.9%
3.0% to 5.7%
5.8% to 8.5%
8.6% to 11.3%
11.4% to 14.1%

FIGURE 5-15. Foreign-born population, 1990.

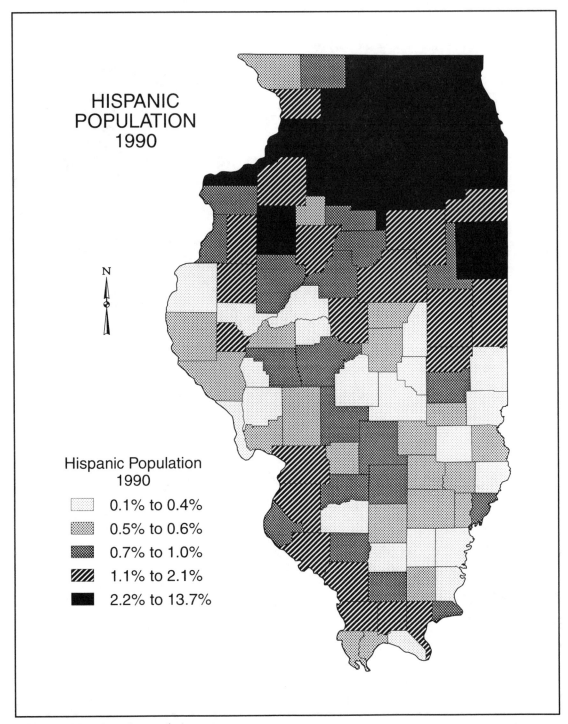

FIGURE 5-16. Hispanic population, 1990.

proclaimed the second largest Polish city, after Warsaw, in the world. This claim tends to linger and may still be heard on the streets of Chicago.

The largest group arriving today is the Latinos, with one in twelve persons in the state having a Hispanic origin. They are clearly concentrated in the northern part of the state (Figure 5-16). This northern cluster of counties includes four metropolitan areas: Chicago, Rockford, Kankakee, and Davenport-Rock Island-Moline. Cook and Kane Counties are just over 13 percent Latino and the only two counties in the state over 10 percent. There are over 900,000 Hispanic people residing in these four metropolitan areas but fewer than 25,000 in the remaining metropolitan areas.

The African-American population of Illinois is much more scattered (Figure 5-17). Although the largest number of African-Americans live in Cook County where they number over one million and make up 25 percent of the county's population, the highest percentages are found in the southern tip of the state. Both Alexander and Pulaski counties are 32 percent African-American. Northward along the Mississippi River, Jackson (Carbondale), and Randolph counties have more than eight percent and St. Clair County (E. St. Louis) even exceeds 27 percent. East St. Louis is widely recognized as a largely African-American community.

The rest of the state's African-American population is scattered mainly in the remaining metropolitan areas. Cities such as Rock Island-Moline, Peoria, Kankakee, Springfield, Decatur, Champaign-Urbana, and Danville have relatively high percentages of African-Americans. The only exception to this pattern is Brown County in rural western Illinois where the Western Illinois Correctional Center is located. Brown County's population in 1990 was only 5,836, but more than nine percent was African-American.

The Asian population of Illinois also is scattered but predominantly in metropolitan and university counties (Figure 5-18). Only in seven counties does the Asian population exceed two percent of the total: DuPage, Champaign, Cook, Jackson, McDonough, Lake, and DeKalb. Four of the seven are university counties that contain the University of Illinois at Champaign-Urbana, Southern Illinois University-Carbondale, Western Illinois University, and Northern Illinois University. The largest Chinese, Korean, and Indian communities in the state are in Cook County where they contribute to the multi-cultural makeup of the population. Many of the new immigrants from these countries live in the city of Chicago, whereas many of the earlier arrivals who now are more established have moved to suburban Chicago. The two highest income counties in the state, DuPage and Lake, also have a relatively high percentage of Asian residents. Although many Asians are professionals and have high incomes, they still rank lower than non-Asians in the seven counties listed above. On the other hand, in many counties where Asians constitute a very small percentage of the population, they have considerably higher incomes than non-Asians.

Health and Quality of Life

The quality of life in Illinois can be characterized by a number of factors, many of which are aspects of personal health and the incidence of crime. In homicides, suicides, and HIV infection, for example, Illinois ranks lower than the national average. Despite the high rates of homicide in some of the disadvantaged neighborhoods of Chicago, Illinois' rate of 9.0 homicides per 100,000 residents is lower than the national average of 9.8. It also is lower than that in Michigan, but higher than the 3.5 level in Wisconsin and 1.9 in Iowa. Suicides in Illinois also are lower than the national level of 12.4 per 100,000 population.

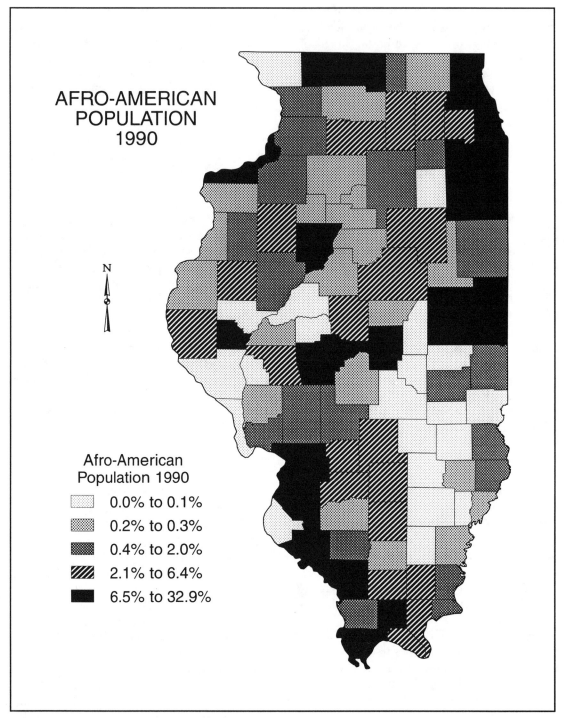

FIGURE 5–17. African-American population, 1990.

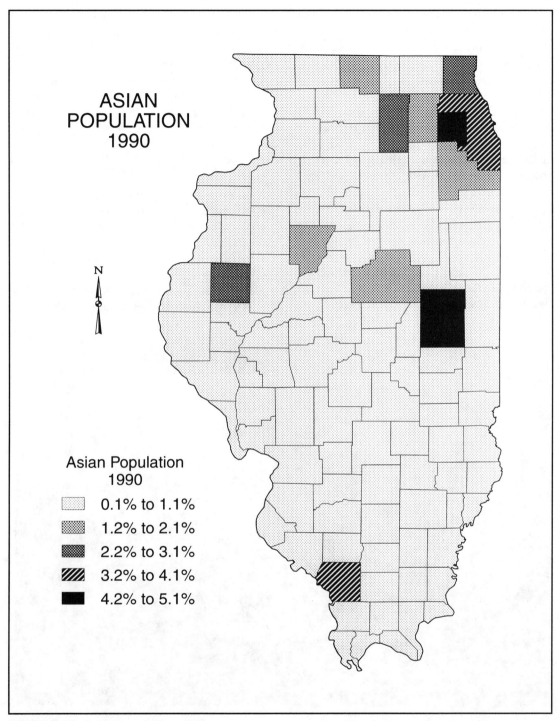

ASIAN
POPULATION
1990

N

Asian Population
1990

0.1% to 1.1%
1.2% to 2.1%
2.2% to 3.1%
3.2% to 4.1%
4.2% to 5.1%

FIGURE 5-18. Asian population, 1990.

HIV infections are most prevalent in large metropolitan areas and constitute a classical example of hierarchical diffusion. That is to say the infections spread from one large metropolitan area to another and then eventually trickle down the metropolitan hierarchy to smaller places. Illinois might be expected to have a relatively high incidence of HIV infection, with about two-thirds of its population in the Chicago metropolitan area. Its level, however, is only 4.0 per 100,000 residents compared to 22.3 in New York and 11.2 in California. (The 1990 national average is 6.8.) In neighboring states around Illinois the levels are even lower with 3.6 in Missouri, 2.0 in Indiana, and only 1.2 in both Iowa and Kentucky. Even in these cases the metropolitan hierarchy is evident; the state levels reflect the size of their respective metropolitan areas.

The Illinois divorce rate also is lower than the national average of 5.0 per 1,000 population in 1985, the most recent year for this statistic. The national divorce rate has been steady for several years, but it is twice the level of the early 1960s. The state level was 4.2 per 1,000 population in 1985, after declining by almost five percent in two years.

SUMMARY

Illinois' present population pattern reflects a long-term movement to the more industrial northeast followed by suburbanization and the development of newer, more modestly sized urban counties located approximately midway between the two largest urban agglomerations, the Chicago metropolitan area and Metro East. It is especially to the largest

FIGURE 5-19. A shopping area for the Korean community in Chicago. (Photograph by Siyoung Park.)

centers and the more cosmopolitan university counties that multi-ethnic migrants have concentrated. Economic well-being in the state is especially reflected by the level of its residents' educational attainment. Illinois' economy overall continues to display the traditionally successful mix of agricultural and industrial activities, although the more rural parts of the state lag in the economic indicators. It should be noted, however, following national trends, the largest percentage of Illinois workers are employed in the tertiary sector of economic activities, i.e., services such as professions, retailing, and wholesaling. Increasingly, the highly mechanized farms of the state require fewer farmers, and that brings about a reduction not only of farms but also a decline in the small towns that traditionally served farms and farmers. Rural poverty, often hard to see, is there nevertheless.

Finally, while it is difficult to see into the future, three main interrelated factors may bring about a movement away from the large metropolitan areas to more rural locations:

1. The desire to flee the more crowded cosmopolitan cities that are perceived as dangerous, unhealthy, and "un-American" in the traditional sense
2. Increasingly, with growing telecommunication technology, it is becoming possible to work at home and thus avoid commuting to work at traditional job sites
3. Continuing dispersal of settlement will require, despite a reduction in the commuting trips of workers, a transport technology permitting the movement of people and goods rapidly over the settled territory

Technologies of this sort continue to be perfected as the twenty-first century approaches. Long distance rapid movement of people and goods, as well as messages, will be increasingly needed. Since at least the early 1970s, questionnaires given to Americans concerning their residential preferences have shown that a majority would prefer to live in a small town or small city, but located near larger places where a greater supply of goods and services are to be found. These factors, in light of the preference of its citizens, may bring about continued decentralization of the state's population.

6

FARMS AND FARMING

Dalias A. Price
Eastern Illinois University

Illinois is one of the leading states in agriculture, ranking fifth in the nation in total farm income after California, Texas, Iowa, and Nebraska. In 1991, Illinois farmers sold over $8.0 billion worth of agricultural products, an increase of nearly $4 billion over the 1975 total. The amount of sales from crops nearly doubled between 1974 and 1993, increasing from $3.7 billion to $6.1 billion. Livestock sales amounted to $2.2 billion in 1993, an increase of about $600 million over the 1975 amount.[1] Obviously agriculture is big business in Illinois and has experienced striking growth in recent years.

Over 80 percent of the state's total land area is in farms, one of the highest percentages in the nation. While Illinois farms have diminished in number during the last few decades, farm acreage has decreased by only five percent. The size of individual farms continues to increase, especially in the cash grain areas where many farmers now cultivate upwards of 2,000 acres. In addition, the value of farmland in the state has increased to a remarkable level. In 1976 the average value per acre reached $858. By 1994 this figure had almost doubled to $1,645, with some of the highest quality cropland selling for over $3,500 per acre. The huge amounts of capital involved in land, taxes, machinery, and production supplies has made small-scale farming economically marginal and has stimulated a growth in size of farm units, especially in the cash grain areas.

On their extensive and fertile lands, Illinois farmers produce a variety of crops and livestock. Corn, traditionally raised as a livestock feed and as a cash crop, long has been the dominant crop in Illinois (Figure 6-1). Among the states, only Iowa competes with Illinois for the national leadership in corn production. The versatile soybean has become a very close contender to corn in acreage harvested and during recent years has occasionally surpassed corn in cash receipts; Illinois ranks first among the states in soybean acreage and production. Wheat represents only about six percent of the state's grain crop, although its production has been stimulated in recent years by growing foreign markets. Illinois also is one of the leading states in pork production. Dairying, while not important enough for Illinois to qualify as America's dairyland, is of significance especially near the large urban markets of St. Louis and Chicago. Beef cattle are less important but still account for some 11 percent of farm income in Illinois (Figure 6-2). The persistence of

FIGURE 6-1. Corn has been the leading crop on most Illinois farms since pioneer times. Hybird varieties with high yielding characteristics are now raised almost exclusively. (Courtesy U.S. Department of Agriculture.)

sheep in the agricultural picture of Illinois is mainly a result of their continued production in the southern and western areas of the state on the more rolling and thus less tillable farms.

Illinois is located in the heart of the most productive agricultural region in the world. The mighty Mississippi, the Illinois, the Ohio, and the Wabash rivers long have provided transportation for produce exported from Illinois farms. In addition, roads and highways criss-cross the state, and Lake Michigan provides ocean transportation to far-flung markets throughout the world. Illinois leads all other states in the exportation of farm commodities to foreign consumers. Nearby major markets, however, are provided by Chicagoland in the northeast and the St. Louis metropolitan area to the southwest.

NATURE'S ENDOWMENTS

The high-ranking position of Illinois in the nation's agricultural economy can be attributed to several factors, among which are its superb natural endowments. Climatically, Illinois is characterized by adequate rainfall the year around, and especially significant is the concentration of rain during the growing season. Average yearly amounts of rainfall range from 46 inches in southernmost Illinois to 34 inches in the northern part of the state. Droughts do occur, but only on rare occasion do they cause severe losses to farmers and then usually in localized areas. Although winters, especially in central and northern Illinois, tend to be cold, most agricultural pursuits are not unduly handicapped by them. Bitter blizzards, so common to the northern

Great Plains, occur only rarely and then in modified form. Heavy snows also are uncommon, yet some snowfall is needed to provide moisture to growing wheat and for early crops in the spring. Yearly snow amounts vary from over 40 inches in the north to about 10 inches at the southern tip.

The growing season varies a great deal across the state since Illinois extends some 387 miles from north to south. The frost-free period ranges from 210 days in southern-most Illinois, sufficient for growing cotton successfully, to about 150 days in the northwestern part of the state. Longer hours of daylight during the summer in the north compensate somewhat for the shorter growing season so that farmers there are able to raise successfully many of the same crops grown in southern Illinois. Occasionally there are problems with early autumn frosts that damage late maturing corn and soybeans, though fortunately not often. In general, climate is favorable to the farmers of Illinois and climatic endow-

ments are many. Farmers have learned to take full advantage of them.

Soils also vary a great deal across the state—east-west as well as north-south. Geologic events of the past have provided soil sources of great value and variation. In Pleistocene times, ice sheets spread great quantities of soil materials over all of the state except the northwest corner and the Ozarks in the south. The last ice age, called the Wisconsinan, spread a highly valuable veneer of soil materials over the northern and central parts of Illinois. Later enrichments were brought about by the development of luxurious prairie grasses that helped to impart great fertility to the soils of most of central and northern Illinois. Some of the most productive soils in the world are found in this prairie region. Farmers have combined science, technology, and hard work to produce crops and raise livestock with great success even in the southern and northwestern areas where natural soil endowments are less generous.

FIGURE 6-2. A northern Illinois beef cattle farm with feed lots, silos, and a large barn. (Photograph by Fred C. Caspall.)

Illinois contributes an inordinately large share of the foodstuffs of the nation from its soils—about eight percent in recent years.

THE EARLY YEARS OF FARMING

When explorers and early settlers arrived in Illinois they found a land ready for the ax and the plow. Conquering the land for crops was a challenging task. Migrating westward, mainly by way of the Ohio River, most of the first pioneer farmers entered southern Illinois, a forested land in which they felt comfortable because they had lived earlier in such an environment. They set to work clearing the timber and planting the land with such crops as corn, wheat, and oats. Since they had to be almost self-sufficient, they also grew vegetables and fruit to preserve for use through the winter. In addition, pigs and cattle were essential elements of every farm in those early years.

As these pioneer farmers expanded their tillable land in southern Illinois by clearing the trees, they began to produce surplus foods for which nearby markets were insufficient. They soon realized that they had to find a way of getting their produce transported to growing cities such as Chicago, St. Louis, Memphis, and even far away New Orleans. Some of the first exports of agricultural commodities consisted of cured hams and other salt meat rafted down the rivers. Wheat and corn, either in the form of grain or as flour and cornmeal, were soon in demand and were eagerly supplied by Illinois farmers to equally eager city consumers.

The Role of Canals

It soon became obvious to residents of Illinois that means of transportation other than rivers were desperately needed, not only for moving farm surpluses to markets but also to enable pioneer settlers to reach interior areas of the state where navigable rivers were non-existent or inadequate. Canals were seen as a partial solution to the transportation problem, since the Erie Canal in New York State, completed in 1825, had proven to be a huge success. Later waves of immigrants to Illinois used that link to the West, but more important to Illinois farmers was the fact that it, together with the Great Lakes, provided a water route to ship their produce to the Atlantic Seaboard. The almost immediate success of the Erie Canal convinced Illinoisans to build a canal linking Chicago with the Illinois River and thus the whole Mississippi system. This canal provided a better means of transportation for the farmers who were venturing out onto the prairies of northern Illinois and for them to ship their produce to new markets. The Illinois-Michigan Canal unfortunately was not completed until 1848, which was almost on the eve of the railroad building period.

Railroad Construction and the Growth of Cities

The Illinois Legislature in its wisdom decided to further open up the interior of Illinois by building a north-south railroad from Freeport to Cairo and from Lake Michigan (Chicago) diagonally to connect with the north-south line. When completed in 1856, the two railroads connected far to the south near Centralia. The Illinois-Michigan Canal was no match for this railroad, the Illinois Central, and for others that quickly followed. By the early 1880s almost every farmer in Illinois had a railroad station within a few miles of his farm. Railroad construction set off an agricultural boom of enormous proportions after the early 1850s. Land from the public domain was purchased by settlers at land offices scattered over the state where it was sold by the federal government usually for $1.25 per acre. In addition, to help defray construction costs, railroads were given vast acreages of land along their rights-of-way that they sold to prospective settlers as quickly as possible.

FIGURE 6-3. Breaking prairie with teams of oxen in Henry County, Illinois, during the mid-nineteenth century, as depicted on a painting by Swedish-American artist Olof Krans. (Courtesy Ron Nelson.)

Coincident with the rapid settling of the land and railroad building during the nineteenth century was the remarkable growth of cities in Illinois. Chicago began significant growth following the opening of the Erie Canal, then experienced rapid development with the building of the Illinois and Michigan Canal and railroads linking the city with farming areas and other urban centers. Along the new railroads, especially at the intersection of tracks and where they crossed rivers, other cities began to mushroom. Almost all people in these burgeoning urban centers were dependent upon the state's farmers for nearly every calorie of food they consumed, and the farmers were eager, willing, and able to supply the cities' food needs.

Agricultural Development and Mechanization

Land values doubled over and over again during the nineteenth century in Illinois, and farmers became very prosperous, especially those who were able to own the land they tilled.[2] An enormous increase in cultivated land was accomplished during the 1880s with relatively primitive tools and teams of horses, even teams of oxen in early days. The invention of an efficient moldboard plow enabled farmers to cut the thickest, toughest grass sod and turn it over to plant corn, wheat, oats, and other crops in the virgin prairie soil.

For a few decades in the early nineteenth century, settlers in Illinois avoided the prairies. This was understandable, because they were accustomed to living in forest clearings.

Forests were valuable assets to pioneer farmers for supplies of nuts, fruit, fuel, and as wood for making such necessities as tools, furniture, houses, and even fencing. Once the prairies were made accessible by railroads, however, settlers began to venture out onto those trackless lands and soon discovered enormous wealth in the black soil beneath the thick, tough sod. Today that very soil provides the backbone of the Illinois farming economy.

As cities continued to grow after the turn of the century, more and more efficient means to provide foodstuffs were acquired by Illinois farmers. Lumbering steam engines, belching great quantities of smoke, were moved from farm to farm to provide stationary power for threshing grain. Steam powered tractors were soon to waddle across fields drawing banks of plowshares, displacing the teams of draft horses that pulled only a single plow or at the most a double shovel affair. When the gasoline tractor made its appearance on the agricultural scene, it gained wide acceptance, especially after World War I, but horse-drawn equipment persisted on Illinois farms through the Great Depression of the 1930s. Today one still can observe the old ways of cultivation and tractorless farms of the Amish religious sect near Arthur, Illinois; it is a fascinating trip back into the past to see horses still providing true "horsepower" for Amish farmers.

Along with gasoline engines for power, of course, came the "gas buggy"—the Model T Ford—which was superior to the horse and buggy almost from the start. Illinois farmers learned they could drive greater distances to larger towns to exchange their produce for items they no longer had the time or inclination to provide for themselves. The difficulty in negotiating a Model T over mud roads in rainy weather stimulated a demand for paved roads. First constructed during the 1920s, brick and concrete roads were and still are called, understandably, "hard roads" by most Illinoisans. These new all-weather roads were a great boon to Illinois farmers, enabling them to market perishable produce (eggs, butter, milk, etc.) more quickly and to haul more easily bulky products from the farm such as grain and animals by truck. The Model T truly liberated farmers from the rural isolation of earlier years.

During the years between World War I and World War II, modern tractors forced horses practically out of existence on Illinois farms. Larger and ever more powerful machines enabled the farmer to cultivate more and more acres with fewer "hired hands." Illinois agriculture experienced a revolution with the acquisition of power tools and equipment by farmers, a trend that continues to the present.

MODERN AGRICULTURE

Today agriculture is a sophisticated, highly technical, thoroughly mechanized component of the Illinois economy. Much evidence of the technical revolution in farming can be observed as one travels in all parts of the state, even the most remote areas. There are enormous diesel tractors, often with air-conditioned cabs and radios; self-propelled harvesters that can pick and shell corn as well as harvest soybeans and wheat; and ungainly sprayer rigs for the application of such agricultural chemicals as herbicides, pesticides, and fertilizers. The fields of high-yielding crops reflect the development of hybrid seeds since the 1930s. Once harvested, grain crops now are artificially dried and stored in enormous bins on the farmstead. The successful modern farmer must possess expertise in mechanics, economics, chemistry, and a variety of other areas of knowledge. But farming as a "way of life" is in jeopardy. Rarely does the modern Illinois farmer consume directly what he produces on his farm; these days he simply buys his milk, butter, bread, and meat at a

FIGURE 6-4. The mechanization of Illinois agriculture is suggested in this photo by the machinery assembled to begin harvesting corn. (Photograph by Scott D. Miner.)

nearby supermarket, just as does his city cousin, even though he may actually produce some of the commodities that he buys back in packaged form.

To illustrate some of the basic characteristics of modern agriculture in Illinois, we will survey three types of farming:

1. Production of cash grain crops
2. Specialty cropping such as truck farming and fruit production
3. Farming operations that center on animals and their products, mainly milk

Cash Grain Farming

With agricultural mechanization and growth in the market demand for foodstuffs during the early twentieth century, Illinois farming experienced several fundamental changes. For several generations, farmers throughout the state had followed the tradition of diversifying their operations by raising hogs, beef cattle, dairy cattle, and other livestock as well as various field crops. Particularly on the flat prairieland of eastern and central Illinois, this traditional general farming was replaced by cash grain farming—the production of grain crops for sale rather than for consumption on the farm where they were grown. The exceptionally high natural fertility of the prairies provided high yields of grains, and the flatness of the land facilitated the use of huge machinery in the fields. Eastern and central Illinois became identified as the Illinois Corn Belt or, more correctly, the Cash Grain Region (Figure 6-5).

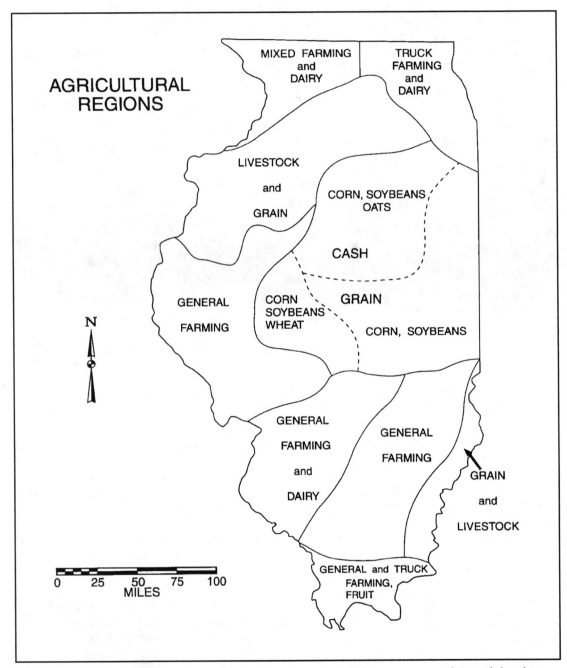

FIGURE 6-5. The agricultural regions of Illinois. This map is based on one prepared several decades ago, but the regions remain essentially unchanged. The Truck Farming and Dairy region in the northeast has been shrinking as increasing amounts of land have been converted to urban uses in the Chicagoland area. The Cash Grain region has been expanding toward the south and the northwest. (Adapted from a map in R.C.. Ross and H.C.M. Case, *Types of Farming in Illinois*, University of Illinois College of Agriculture, 1956.)

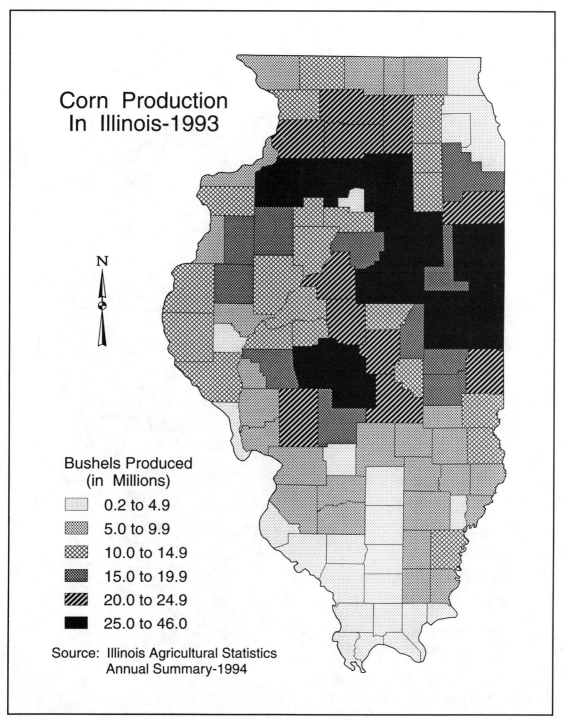

FIGURE 6-6. Corn production in Illinois during 1993.

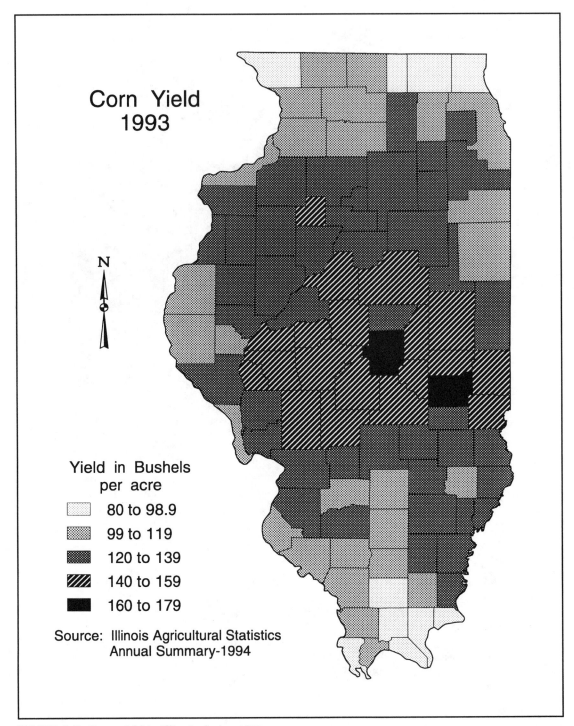

FIGURE 6-7. Corn yield in Illinois during 1993.

Almost from the beginning of settlement, corn was the dominant crop in the prairie areas of eastern and central Illinois. Although new and strange to nineteenth century European immigrants, it soon was adopted by them when they settled in Illinois because the crop was well suited to a pioneer economy. Corn not only gave good yields in a variety of environments, but it also served several functions. It provided feed to farm animals being raised or fattened and served as an important staple for the pioneer farm family in such forms as hominy, cornbread, and "roasting ears." As transportation facilities became more readily available, early farmers in the developing Cash Grain Region discovered that they could sell their corn to the growing market in major cities. In addition to its direct consumption by humans, corn was manufactured into starch and cornflakes, the latter

creating a revolution on the breakfast table of Americans, and used for making syrups and alcoholic beverages. Also, farmers in other parts of the country often encountered shortages of corn and were anxious buyers of some of the surplus Illinois farmers provided. Now corn has become a source of fuel and is replacing gasoline as ethanol.

During recent years in Illinois, corn has been raised on more than 10 million acres of land—an area greater than that devoted to any other crop (Table 6-1). More than 70 percent of the state's farms are involved in the production of the Illinois corn crop. Corn is grown in every county of the state and, as indicated by Figure 6-6, most counties in central and northern Illinois produce more than 10 million bushels annually. Most corn farmers in Illinois now expect a yield of at least 90 bushels per acre and production in excess of 150

FIGURE 6-8. Harvesting a field of wheat. (Illinois Geographical Society.)

bushels per acre in the central part of the state no longer is uncommon.[3] (See Figure 6-7 for yields in 1993.)

Although Illinois does not possess the environmental endowments for raising wheat equal to that of Kansas, wheat figures significantly in Illinois agriculture. To maintain soil tilth and productivity, generations of farmers have rotated wheat with corn and other crops in eastern and southern Illinois. In 1993, wheat was harvested on 38,000 farms in the state and 1.5 million acres were devoted to its production (Table 6-1). A growing demand for food in the world has stimulated an increase in the acreage planted in wheat during recent years. Yields of approximately 50 bushels per acre now are typical of most wheat-growing areas in the state. Although it is grown in combination with corn and soybeans in many counties in central and southern Illinois, wheat is particularly concentrated in an area east of St. Louis (Figure 6-9). This area is involved in dairying and other agricultural pursuits, but wheat is the most important crop here—particularly in Washington County. Nearly all wheat raised in Illinois is winter wheat, a variety that is planted in the fall, has early growth during the winter, and reaches maturity for harvesting in the early summer. Spring wheat is grown on only a very small scale in the northern part of the state.

TABLE 6-1. Grain Acreage in Illinois, 1974-1993 (in thousands of acres)

Grain	1974	1993
Corn	9,400	10,000
Soybeans	8,100	8,700
Wheat	1,600	1,550
Oats	378	90

Oats long have been included in the crop rotation system employed by farmers in Illinois, but they no longer can be regarded as a major crop in the state. Earlier, oats were raised primarily as a feed for horses and served much the same function that diesel fuel and gasoline do for tractors today. With the displacement of horses by tractors, however, the acreage devoted to growing oats has declined drastically. The amount of land planted in oats had decreased to 90,000 acres by 1993 (Table 6-1). Of course, the demand for oats as animal feed has not been eliminated entirely, and the crop still has minor but growing significance as a source of human food for diet conscious people. Most oats raised in Illinois today are concentrated in the cooler northern and northwestern counties. The highest density of the oat crop, however, is in the Amish settlement near Arthur where horses provide power instead of tractors and automobiles.

Soybeans entered the agricultural scene quite late, initially appearing on farms in the 1920s. This remarkable plant subsequently gained wide acceptance among Illinois farmers, and today it rivals corn as a cash grain. The soybean is an absolute wonder. Not only does it produce well on soils of limited fertility, but it also thrives under a variety of climatic conditions. Yields generally range between 25 and 50 bushels per acre in Illinois and average 43 bushels statewide. Some yields have been as high as 65 bushels per acre in recent years. Soybeans serve as a raw material for a variety of industries, including ones involved in making paints, varnishes, plastics, margarine, and cooking oils. In addition, the crop makes an excellent livestock feed, and it is becoming increasingly important as a human food because of its exceptionally high protein content. Some manufacturers recently have developed techniques for converting soybeans into products resembling hamburger, meats, and other foods.

Soybeans are now grown throughout Illinois and the amount of land planted in the crop is rapidly approaching that for corn. The area of most concentrated soybeans produc-

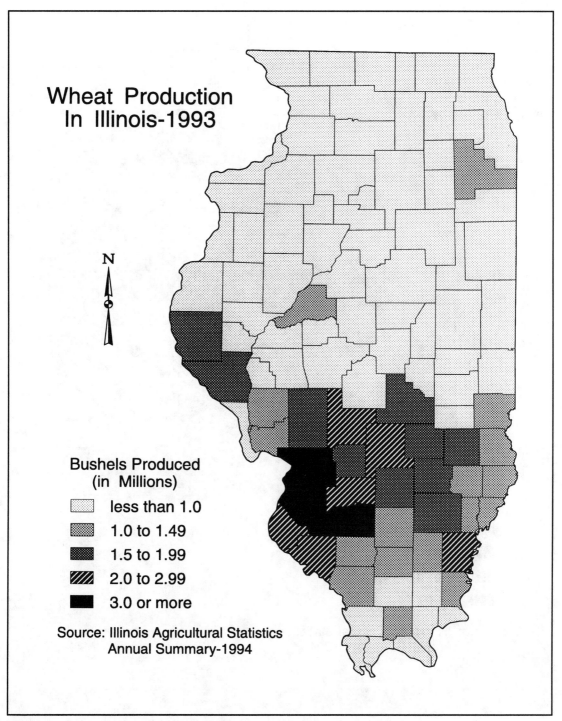

FIGURE 6-9. Wheat production in Illinois during 1993.

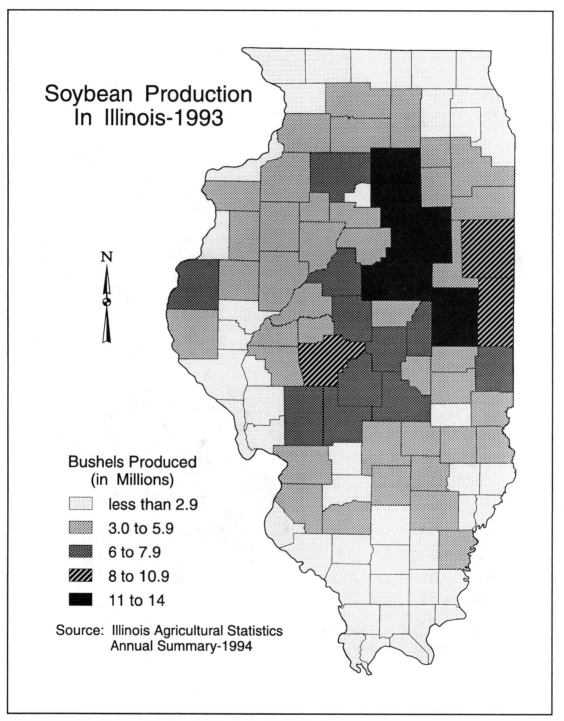

FIGURE 6-10. Soybean production in Illinois during 1993.

tion is the Cash Grain Region of eastern and central Illinois (Figure 6-10). As a result of the widespread adoption of soybeans in the state, cash grain farming is more widely practiced throughout the state than formerly. In 1993 soybeans were harvested on about two-thirds of the state's farms, and the amount of land devoted to the crop had increased to 8.7 million acres (Table 6-1).

Specialty Cropping

There is a general tendency for certain areas to develop specialization in agricultural pursuits involving a particular crop or product. The reasons for this are diverse, but one is the tendency for farmers to imitate the practices of their successful neighbors. If one farmer makes a success at raising strawberries, for example, his neighbors, ever on the alert for getting the most from their farms, follow suit. If economic and especially environmental conditions are suitable, the enterprise becomes implanted, and the area acquires a reputation for that specialty.

Strawberry production, in fact, is a good example of areal specialization in Illinois agriculture. Although the strawberry plant can be raised throughout Illinois, the southern part of the state has a climatic advantage because of its earlier arrival of spring. Therefore, southern Illinois growers can market their produce earlier than upstate producers. The first experiment with refrigerated shipment of a perishable crop in the United States involved strawberries from southern Illinois. A man named Parker Earle, living in Cobden, sent a shipment of strawberries to the Chicago market by rail in 1866. The method of refrigeration was very primitive; in fact, the shipment arrived in Chicago a soggy, rotten mess because railroad workmen forgot to replenish the ice supply in the wooden casks containing the strawberries en route. Not being discouraged by this initial failure, Mr. Earle soon perfected a successful technique of shipping

perishable fruits and vegetables to the Chicago market in refrigerated railroad cars. As a consequence, there became established a major truck farming and orcharding industry in southern Illinois that persists to this day. The refrigerated car soon revolutionized the distributional characteristics of production of perishable agricultural commodities not only in Illinois but throughout the country.

While most Illinois farmers had fruit trees on their farms in the early days of settlement, certain areas with transportational and environmental advantages took the lead, but not until after the railroads provided the necessary "rapid" transportation to city markets. Apple and peach production also illustrates the regional specialization of much of Illinois agriculture. Today the Illinois Ozarks, especially areas around the towns of Anna, Cobden, Villa Ridge, Alto Pass, and Murphysboro, account for the major production of apples and peaches for shipment to the urban areas. Most fruit farmers combine their orchard enterprises with the production of such truck crops as tomatoes, peppers, squash, cucumbers, and beans. This is a happy combination because occasional frosts damage apple and, especially, peach trees. An orchardist can thus weather a bad season by shifting emphasis to truck farming during such years.

Although not as important as the Illinois Ozarks, there are other areas of significant fruit production in the state. One is Calhoun County, that interesting backbone of land squeezed between the Illinois and Mississippi rivers just north of St. Louis, where peaches dominate the fruit production, although apples are significant too. This region has capitalized on its proximity to St. Louis for sales and now engages in what is called "U-pick." The urban dweller enjoys driving out to the fruit farm and picking his or her own fruit; the orchardist and the buyer save labor costs and both are happy. Other fruit and truck farming regions of lesser significance center on Centralia and Salem. In the past there was greater emphasis on specialty farming here, but most farmers have succumbed to the temptation of government price supports and switched to cash grain production, a somewhat more reliable farming enterprise than orcharding.

In the very southernmost part of the state, primarily Alexander and Pulaski counties, cotton is a residual crop of the once more northerly expanded Cotton Belt. Cotton has suffered materially in the face of competition from soybean and corn production. Cotton gins no longer are located here, and it is economically unfeasible to haul this bulky light-weight fiber to Missouri cotton gins. Thus it appears that the future of cotton farming in Illinois is not good.

There is a long list of less important, perhaps more exotic crops produced in Illinois. One is popcorn, a close relative of field corn. Popcorn production is located mainly in the southeastern part of the state, particularly around Ridgeway, which boasts that it is the popcorn capital of the United States. Although Ridgeway's claim is not strictly true, there is considerable local pride in the fact that the area does specialize in commercial popcorn production. Sweet corn is produced in several northern parts of the state. In the Hoopeston area it is a particularly important crop. Most of the corn is processed, canned, or frozen in nearby communities.

Truck farming and the production of additional specialty crops are found in several other parts of Illinois. Surprisingly, horseradish is a specialty crop grown on the mucklands of the American Bottoms in the East St. Louis region. Almost all of the commercial horseradish of the United States is produced by a relatively small number of farmers in this area. Undoubtedly, there are many other parts of the state suitable for horseradish, but local interest and early development of know-how help to explain why there is such a high con-

centration in this one small area. Watermelon and cantaloupe are grown on the sandy soils in scattered areas of the Wabash bottoms near Crossville in White County and in Mason County adjacent to the Illinois River. Pumpkins for commercial canning are raised in Tazewell County in central Illinois and in Ogle and DeKalb counties in the northern part of the state. Pana once was the rose capital of Illinois but no longer. Competition has driven out this enterprise. Greenhouses are used now for bedding plants, flowering plants, etc., thus providing their own climatic environment. Fuel costs are the chief consideration.

All of the specialty crops combined do not account for a very large percentage of the agricultural income in Illinois. Corn and soybeans are far more important, but regional specialization in minor crops adds to the agricultural diversity and variety of land use in the state.

Animal Production

Illinois farmers who specialize in animal production concentrate mainly on hogs, beef cattle, dairy cattle, and sheep, in that order (Table 6-2). Raising hogs has been common since the first pioneers arrived in Illinois. These animals were an important element in the subsistence economy prevalent in the early settlement period, but farmers soon realized that pork in the form of bacon, hams, and lard was in demand by city dwellers. Slaughter houses and butcher shops became common businesses in early towns and villages. As modern transportation facilities developed, however, urban specialization in the meat packing industry evolved. Chicago became the "hog butcher of the world," and other growing cities developed meat packing industries.

Today hogs are raised in most sections of the state, but they are particularly important in the counties north and west of the Illinois River (Figure 6-11). In this subregion of the

Corn Belt, farmers tend to feed some of their corn crop to livestock rather than sell all of it as grain. Henry County, just east of Rock Island, is the leading hog producer of Illinois. In 1993 hogs were raised on 11,500 Illinois farms, or 14 percent of all farms in the state. The inventory of hogs and pigs in Illinois at the end of 1993 was 5.45 million animals. Cash receipts to Illinois farmers from hog enterprises was 14 percent of their total cash income.

TABLE 6-2. Livestock on Illinois Farms, 1974-1993 (Inventory in thousands of head)

Animal	1974	1993
Hogs and Pigs	5,200	5,450
Beef Cattle	873	590
Milk Cows	237	175
Sheep	223	95

Beef cattle, like hogs, require great quantities of corn for feed; however, their distribution is not quite the same as that of hogs. The western and northwestern counties dominate the state in the production of beef cattle and calves (Figure 6-12). Generally beef cattle are both raised and fattened on the same farm in Illinois. While corn is an important element in the feeding of cattle, great amounts of hay and pastureland also are required. This helps explain why the beef and dairy industries are not favored in the Cash Grain Region where the land is so highly productive and expensive that the farmer cannot afford to allow it be used for hay and pasture. The farmer who engages in production of beef cattle has a more diversified enterprise than his counterpart in the Cash Grain Region, even though the product of his efforts is ultimately a fat animal to be shipped to some urban market for processing. At the beginning of 1994 the inventory of beef cattle in Illinois was 590,000 head. The numbers of cattle and hogs on Illinois farms have been declining in recent years, mainly because of changing human diets and fluctuating prices for livestock.

A dairy cow or two was common on every farm in the early days of settlement, but not as a commercial venture. Gradually most Illinois farmers gave up dairying and turned to other specialties, while only a relative few expanded their herds to commercial proportions. In metropolitan areas the demand of urban people for milk products increased through the years. The farmers nearest these markets, of course, had a distinct advantage in that their milk could reach the consumer in short enough time to still be fresh. Those farmers somewhat farther away from the urban markets sold their milk to the producers of cream, cheese, and butter. This pattern now has changed because milk can be shipped hundreds of miles in refrigerated tank trucks. Also, dairying has been affected by technological advances and improved breeds; much more milk now is obtained from one cow than in the past. Consequently, the number of dairy cows and dairy farms has declined. Dairy herds are larger and the dairy farmer has become a highly skilled and trained technician. One farmer alone now can manage a large herd with mechanization at his disposal, including the ingenious milking machine and automatic feeders.

Strong concentrations of dairying in Illinois still remain in the traditional, city milk-shed regions extending from Chicago west and north (eventually into Wisconsin) and from St. Louis east and south—east as far as Effingham County and south as far as Randolph County (Figure 6-5). These are the significant milk producing regions of the state, although there are minor concentrations of dairying near most larger cities such as Peoria, Decatur, Springfield, Danville, and Carbondale. The inventory of milk cows in Illinois at the beginning of 1994 was 175,000 head, down from 237,000 in 1974.

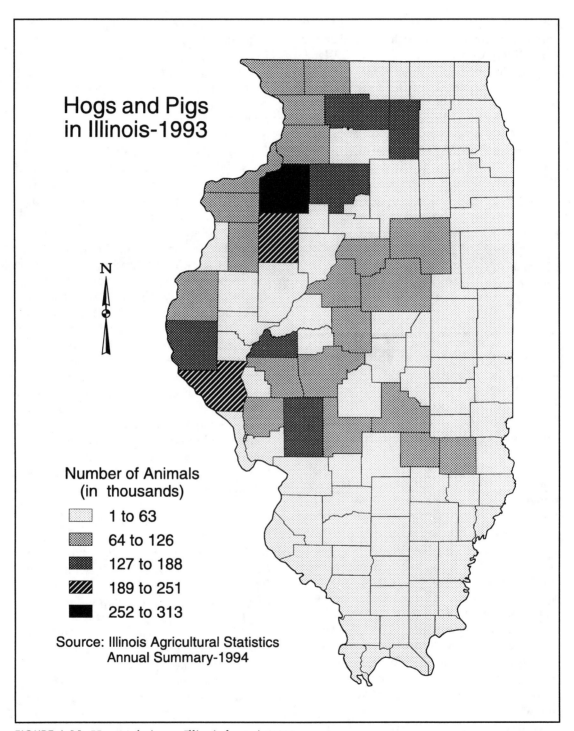

FIGURE 6-11. Hogs and pigs on Illinois farms in 1993.

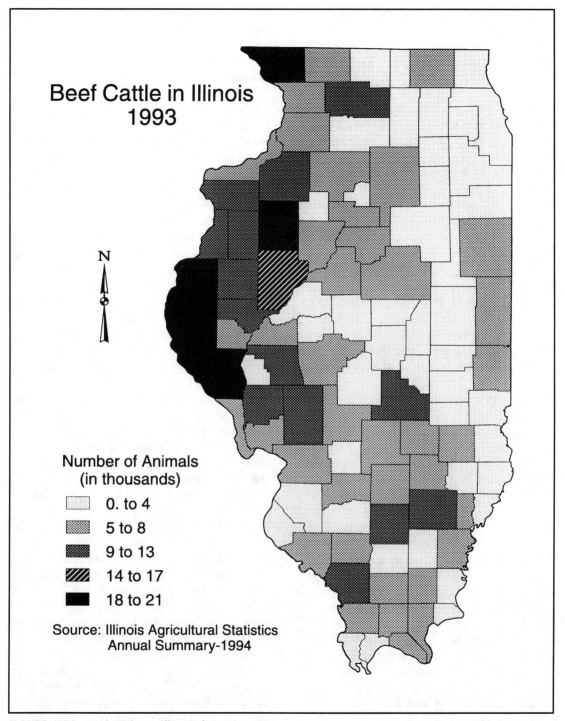

FIGURE 6-12. Beef cattle on Illinois farms in 1993.

Illinois still persists as a sheep producing state, although certainly not to the extent of some other states. Flocks of sheep are grazed in the hillier sections of the state in the Ozarks and in the northwest; however, the areas west of the Illinois River account for a large share of the sheep produced in Illinois. There is, in addition, concentration in McLean County and some adjacent areas in the central part of the state. Sheep are raised for their fleece and production of wool and for meat products. The inventory of sheep in Illinois at the beginning of 1994 was 95,000 animals, a decrease from 223,000 counted in 1974.

Regionalization

From the foregoing descriptions and maps it is apparent that Illinois agriculture varies greatly in character from one part of the state to another. The general regional pattern of agriculture is shown in Figure 6-5.[4] The largest and most important agricultural region in Illinois is the Cash Grain Region, which was singled out for attention earlier; it occupies the prairies of the central and eastern parts of the state. Regions of distinctive agricultural characteristics associated with dairying are in northernmost Illinois and in the southwest near the St. Louis urban complex. The regions west and north of the Illinois River and adjacent to the Wabash River in the southeast are distinctive in their mix of crop production and livestock combinations. South of the Cash Grain Region and extending to the northern edge of the Ozarks is a region mainly devoted to general farming in which no particular single enterprise dominates; however, cash grain farming has gained steadily in importance, especially since science has made it possible to produce corn and soybeans for a profit in this area of limited soil resources. In the hill country and stream valleys of southernmost Illinois, truck farming and fruit production are combined with general farming in a final distinctive agricultural region.

The identification of precise reasons for regional specializations is a difficult task. The more obvious reasons, of course, center on variations in soils, terrain, the length of the growing season, the severity or mildness of winters, and the time of the arrival of spring. We cannot overlook, however, the element of judgement by an individual farmer. An assortment of vague reasons, even including personal preferences, may enter into a farmer's adoption of a certain agricultural enterprise. If his experiment proves to be successful, other farmers in the immediate area are likely to follow the example of the initiator.

There are advantages, of course, in concentrating production of certain agricultural commodities in a particular area. Problems common to all farmers can be solved more readily. Dissemination of information and of results of certain farming experiments is easier. Also, there is the matter of marketing in which cooperative efforts are needed to efficiently market products from the farms. A dairy farmer located 50 miles from any other dairy farmer undoubtedly would suffer serious handicaps; no processing plant would send a truck that distance to pick up only one dairyman's daily supply of milk. Regional specialization just makes good sense in Illinois agriculture.

Another significant regional variation in Illinois agriculture involves the ownership of land and farm tenancy. Less than one-half of the farms in the state are now operated exclusively by the owners themselves. The practice of landlords leasing farmland to tenants is widespread and dates from the early days of settlement in Illinois. Undoubtedly most farmers would prefer to own the land they farm, but this is often not feasible or even possible. Because of the enormous expense involved when a young man chooses agriculture as a career, he is likely to have to rent land for many years before accumulating enough capital to become the owner of a farm. Furthermore, an increasing number of owner-operators must rent additional land in order to make economical use of all the machinery they must acquire. Today about half of the farmland in Illinois is not owned by the farmers who cultivate it.

The origin of farm tenancy in Illinois is associated with the acquisition of large landholdings by speculators during the period of early settlement and the railroad boom. Several people of means acquired vast acreages, particularly in the prairies of central and northern Illinois, by purchases from the federal government as it disposed of the public domain, purchases from the Illinois Central Railroad as it disposed of land granted to it by the government, and other means. The owners of these huge tracts were mostly speculators who were not inclined to settle on the land and cultivate it themselves; therefore, they either hired other people to farm the land or rented it to tenants. Some of these enormous tracts have remained largely intact, a factor which has restricted the amount of land available for purchase to the present time.

Owner-operated farms are concentrated mainly in the southern one-third of the state, in the western Illinois counties bordering the Illinois and Mississippi rivers, and in the tier of counties along the northern border of the state (Figure 6-13). In general, dairy farmers, livestock farmers, and specialty crop farmers have been primarily owner-operators in the past and continue to be so. With the exception of machinery, investments on such farms are of a decidedly permanent nature. A silo or a milking parlor, for example, cannot be moved from one farm to another. Likewise, fruit trees cannot be moved; they are long term investments, take several years to begin producing, and require careful attention in order to produce well. Only owners of farms are usually willing to invest in such permanent facilities and trees. Owner-operators have the opportunity to exercise complete control over their

farming operations. Demands on their time are year around.

The greatest concentration of tenant-operated farms is in the Cash Grain Region of eastern and north-central Illinois (Figure 6-14). Most cash grain farmers have heavy investments in huge tractors, combines, cultivators, fertilizing equipment, etc. Such equipment obviously can be moved about from one unit to another, so when a tenant farmer loses his lease or for some other reason moves to another farm there is little difficulty associated with the transfer. The tenant farmer simply moves his equipment to the new farm being leased and avoids loss on his investment. Another reason for the close association of tenancy with the Cash Grain Region is that most landlords prefer the production of cash grain crops on their land because it involves less risk and provides a more dependable and remunerative income than other types of farming.

FARMING AS A WAY OF LIFE

It has long been common to refer to farming as a "way of life," but this characteristic of agriculture in Illinois is rapidly disappearing. Many high officials in government and business still extoll the virtues of agriculture as an industry and insist that the nation no longer can afford the relatively small farms on which farmers and their families have earned a livelihood generation after generation. We are told that agriculture must become organized like big business and adopt principles of industry such as automation and mass production.

One cannot deny that agriculture has undergone startling changes during the past several decades, for better or worse, and in some areas it is already nearly as mechanized and automated as is other industry. The level of mechanization has reached a point that the farmer himself often is able to do all of the

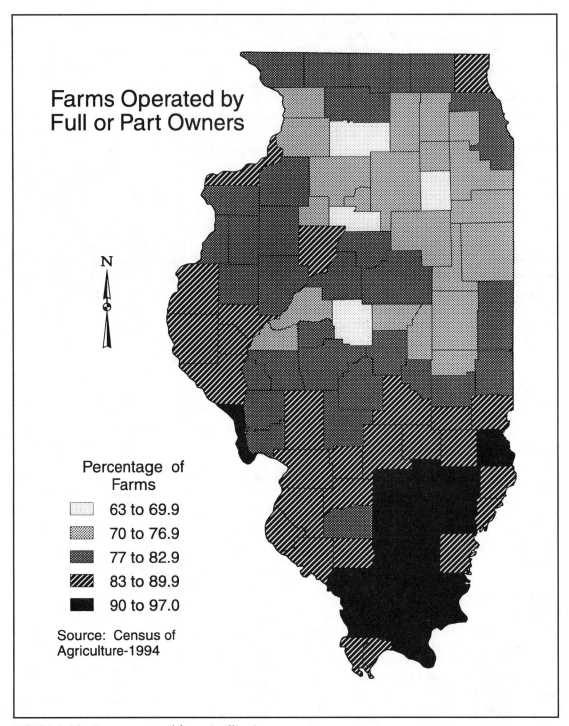

Farms Operated by Full or Part Owners

N

Percentage of Farms

63 to 69.9
70 to 76.9
77 to 82.9
83 to 89.9
90 to 97.0

Source: Census of Agriculture-1994

FIGURE 6-13. Owner-operated farms in Illinois.

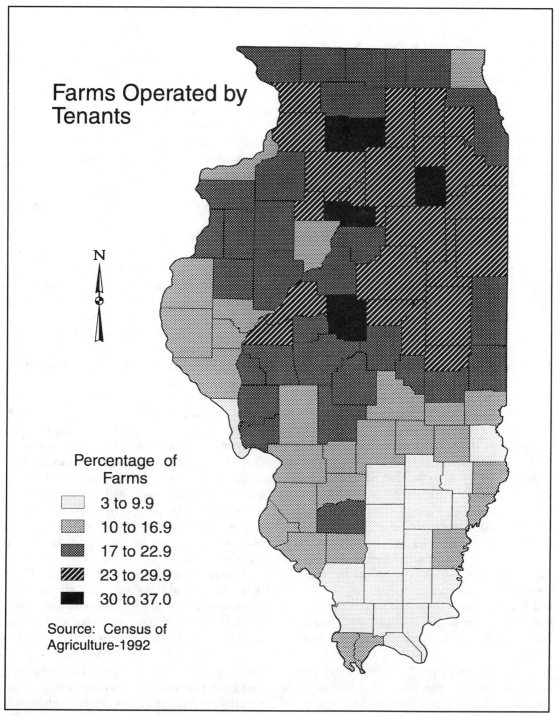

Farms Operated by Tenants

N

Percentage of Farms

	3 to 9.9
	10 to 16.9
	17 to 22.9
	23 to 29.9
	30 to 37.0

Source: Census of Agriculture-1992

FIGURE 6-14. Tenant-operated farms in Illinois.

work necessary to operate the farm. There simply no longer is need for additional "hands," including the farmer's children, on many Illinois farms. The children consequently develop a detachment from the land and are forced to seek employment elsewhere. The family thereby becomes scattered and farming as a way of life deteriorates.

The decision by some farmers to move to town rather than live on the land they work also can have an adverse effect on attitudes toward the land. Especially in the Cash Grain Region there are growing numbers of what Walter Kollmorgen called "sidewalk farmers."[5] These are people who live in town but drive out to their farms when necessary to attend to crop production. There also is an increasing number of farms in Illinois operated by what Kollmorgen called "suitcase farmers."[6] These are people who live at such a distance that they carry a suitcase of some sort when they briefly visit their farm to perform such needed work as plowing, planting, cultivating, and harvesting. Suitcase farmers commonly have another occupation in addition to their farming operation.

In 1974, nearly 30 percent of all farms in Illinois were operated by individuals who did not consider farming to be their principal occupation. In 1993 an increasing number of farmers feel that they are forced to "moonlight" (i.e., work at second jobs) because of the economic uncertainties of farming. Of course, this practice was impossible before farm technology and mechanization reached a high level, but now it is practical for some farmers to work at two jobs. Farming as a way of life may not be eliminated by this practice, but it definitely suffers as farmers devote an increasing amount of attention to their second job.

Some types of farming in Illinois still persist as a way of life. Dairying is the most notable example; it is a year-round operation and requires the farmer's constant attention to his herd. At least twice daily the farmer must attend to the needs of his cows; even with mechanical assistance, he still must work every day. In addition to carrying out the milking, he must keep constant watch on the health of his herd, act as a midwife when calves are born, and make sure that the farm produces as much of the feed for the hungry cattle as possible. The whole family is involved in the many daily duties required to operate a successful dairy farm.

Many farmers are discouraged by the slim profits in dairying, especially when they observe what seems to be a better life enjoyed by their neighbors who engage in cash grain or other types of farming. They often decide to sell their dairy operation and move to a nearby town to find a different job or retire. It has become increasingly difficult to find young men, even the sons of dairymen, to take over when a dairy farmer retires. The younger men prefer jobs that require only 40 hours per week of work time and provide an annual paid vacation.

While dairying is probably the best example of farming as a way of life in Illinois at present, there are others. The raising of hogs and beef cattle and orcharding may not have the demanding daily responsibilities associated with dairying, but they nevertheless require dedication, skill, and almost constant attention on the part of the farmer. Truck farming also is a confining enterprise, although its demands on the farmer are much more seasonal.

A consideration of farming as a way of life in Illinois would not be complete without some attention to the Amish community centered on Arthur in the east-central part of the state.[7] For well over a hundred years a settlement of Amish has thrived here in Douglas, Moultrie, and Coles counties. This religious sect feels that modern conveniences such as electricity, automobiles, telephones, and other forms of mechanization are not in

FIGURE 6-15. A shed is provided for the buggies of Amish farmers while they shop in Arthur, Illinois. In the background are large elevators to serve the area's cash grain farmers. (Photograph by D.A. Price.)

keeping with their religious beliefs. Consequently, a visit to the Amish community is like taking a trip 60 years or more into the past. Common sights are teams of horses pulling plows, cultivators, binders, and wagons through the fields, as the Amish use true horsepower on their farms. Buggies are common sights on the roads. The Amish do not own automobiles.

Amish farms have distinctive characteristics that enable an observer to distinguish them from the farms of their neighbors. With tractors, trucks, and other types of mechanized equipment lacking, the farms are necessarily small—generally no more than 80 to 100 acres in size—but well cared for and intensively cultivated. The land is divided into many fields, each bordered by a tight fence to contain the large number of livestock. To feed the horses and other animals, oats are raised as a major crop much like they were in the early twentieth century on most Illinois farms. After the oats and wheat are harvested, strawstacks dot the feed lots of Amish farms. To increase the income from their small farms, most Amish have hogs, dairy herds, poultry, and even large brooder sheds and egg factories. Large barns and other buildings have been constructed to house animals and to store the hay, feed, and other produce of the farms. Commonly two houses stand off the

corner from each other on an Amish farmstead; the small one is occupied by the grandparents after they retire. The main house is usually quite large to provide room for the many children that are typical of Amish families and to accommodate the crowd of neighbors on the Sunday it is their turn to host the church services.

The Amish way of life has a charm, serenity, and single-minded purpose all too rare in Illinois agriculture today. It would be worthwhile for anyone to travel through the Amish country to watch these industrious people at work, to see them riding in their buggies and wagons on the roads, and to buy some of their produce such as honey, eggs, bread, and other bakery goods. They are a peace-loving folk, dedicated to the simple life.

Although agriculture generally has become mechanized, automated, and oriented toward mass production, there are enough hearty souls in Illinois who fortunately love the land and the job of tilling it to prevent a complete disappearance of farming as a way of life. To the Amish, farming is more than simply a means of making a living; it allows greater independence than other types of employment and represents "the good life." They may grumble and complain, as most of us do on

FIGURE 6-16. Draft horses resting during noon hour after working in the field on an Amish farm near Arthur, Illinois. The smaller horse in the background is used only for pulling a buggy. (Photograph by D.A. Price.)

occasion, but they continue to stay close to the sod. Some young married couples work in cabinet shops, etc. until they can buy their own farm.

AGRICULTURAL TRENDS AND PROSPECTS

While prediction is always hazardous, such variables as weather, technological developments, and market conditions make it particularly risky in the case of agriculture. Nevertheless, many trends in Illinois farming are evident, and from them we, perhaps, can make intelligent guesses as to what the future holds.

The tendency toward larger farms continues, especially in the Cash Grain area, although the rate is somewhat slower than in the past. The average size of individual farms in Illinois increased from 228 acres in 1969 to 368 acres in 1994. A continuing increase in the acreage devoted to cash grain farms throughout the state seems likely, with most notable growth to be expected in the Cash Grain Region where generally large farms already are concentrated (Figure 6-19). Agricultural experts now recommend that cash

FIGURE 6-18. Amish farm houses over 100 years old in the vicinity of Arthur, Illinois. A huge porch connects the grandfather house with the main home. Amish houses provide ample space for church services conducted on a rotation scheme among members of the congregation. (Photograph by D.A. Price.)

grain farmers cultivate at least 1,000 acres in order to earn a reasonable profit, especially during times of high inflation rates. With tractors and other equipment increasing in size and efficiency, the individual farmer undoubtedly will have the capability to plow, plant, and cultivate a greater acreage in the future. As a result of the consolidation of smaller farms into larger ones, there is destruction of fences and old farm buildings, wider spacing of remaining farmsteads, and the development of a more open landscape; the Cash Grain Region, in fact, is looking more like rural Kansas each year. Also related to the growth in farm size is the decline in numbers of Illinois farmers, a trend that can be expected to continue into the foreseeable future.

In comparison to cash grain operations, other types of farms in Illinois generally are smaller in size and can be expected to have little or no increase in their future acreages. Among these, beef farms tend to be the largest and dairy farms, hog farms, and orchard farms are relatively moderate in size. The smallest units commonly are truck farms devoted to the production of vegetables and small fruits. Livestock farmers have turned to the use of

FIGURE 6-17. Threshing oats on an Amish farm in the area of Arthur, Illinois. This scene was common on most farms in Illinois over a half century ago. Power for operating the threshing machine is provided by a tractor. Amish may own tractors to use their power take-off, but not for work in the field. (Photograph by D.A. Price.)

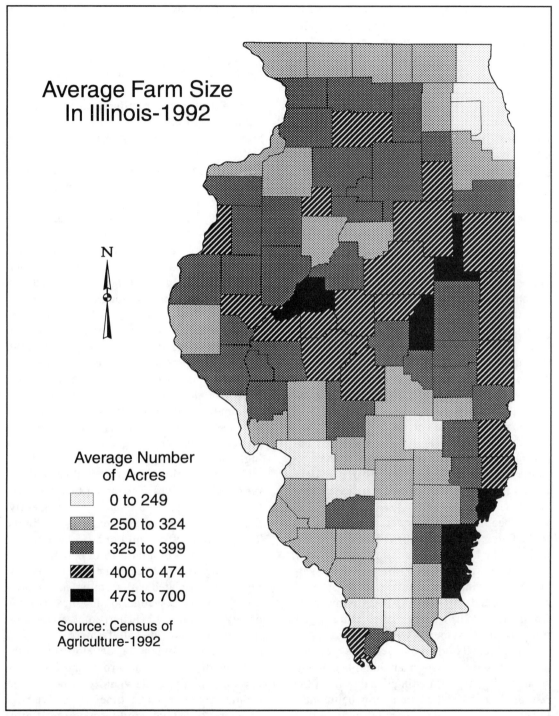

FIGURE 6-19. Average farm size in Illinois.

feed lots and commercial feeds to increase their production as an alternative to the purchase or rental of additional acreage. The process of confined feeding and fattening of livestock undoubtedly will become even more widespread on Illinois farms in the years ahead.

While cash grain farming continues to expand, other types of farming in Illinois are experiencing stagnation or decline. For example, the number of dairy farms is decreasing and the inventory of milk cows dropped from 237,000 in 1974 to 175,000 head in 1994. The per capita demand for milk and milk products is declining and probably will continue to do so because of competition from soft drinks and other beverages and dietary changes. Specialty cropping is another type of farming in the state that is not experiencing significant growth. Illinois orchardists and small fruit producers are suffering keen competition from growers in western states, especially California. Although Illinois fruit is of equal or superior taste, it cannot compete in appearance with the beautiful western fruit that attracts supermarket shoppers.

The technological and chemical revolution already has significantly changed most aspects of agriculture, and its impact may be even greater in the future. The use of chemical fertilizers, herbicides, and pesticides has enabled the farmer to obtain greater crop yields while spending less time operating equipment in the fields; no-till farming is catching on. To improve the health and rate of growth of their animals, livestock farmers have adopted scientific breeding techniques and the use of chemical feed supplements and disease control materials. Growers of soybeans are beginning to benefit from the development of ersatz food products based on that versatile crop; we are promised that soybean hamburger and steak soon will be indistinguishable from the real thing. Artificial bacon is now being marketed, and many other substitute and imitation foodstuffs undoubtedly will soon appear. Fortunately for farmers, most substitute food products must come from the soil. Therefore, the trend toward production of substitute foodstuffs will result in shifts in utilization of Illinois farmland, but the state will continue to be a leader in agricultural output. The development of ethanol from corn promises to stimulate more production of that crop in the state.

An increasingly critical problem in Illinois agriculture, especially for young men starting out in farming, is the burgeoning cost of land, machinery, and other necessities. Land prices have spiraled to the point that the best farmland in the state now sells for more than $3,500 an acre. In addition, the cost of machinery and equipment has skyrocketed to a staggering amount. Although different types of farming have different machinery requirements, some combination of tractors, combines, planters, trucks, milking equipment, orchard equipment, and other implements is required. The investment in machinery for even a modest farming operation can easily amount to many tens of thousands of dollars. Because of the enormous capital requirements, farms are being purchased increasingly by professional people, corporations, and even foreign investors rather than by farmers themselves. The land is then rented to those who actually cultivate it; consequently, farm tenancy continues to expand in Illinois. The young person who hopes to become the owner-operator of a farm faces a difficult and often lengthy struggle. In the face of increasing costs associated with farming, many farmers have taken a second job in a nearby town or city; as indicated earlier, some are becoming suitcase and sidewalk farmers.

A serious problem has developed with the transfer of farmland to urban and industrial uses, especially in the vicinity of larger metropolitan centers. The time has arrived for planners to reserve good farmland in Illinois for agricultural purposes alone, so that it cannot be diverted from the all-important role of producing foodstuffs. Priorities must be established and restrictions imposed on the conversion of farms into industrial sites, parking lots, shopping malls, bedroom suburbs, and super highways.

SUMMARY

Illinois is one of the nation's most important agricultural states; it has played a dominant role in farming for well over a century. The state enjoys many advantages for a wide variety of farming types. Nature bestowed fertile soils and gently rolling terrain over most of the state; in addition, the variety of climatic conditions is favorable to many different kinds of farming endeavors.

Locational factors also are advantageous for agriculture in Illinois. The state is bordered and crossed by navigable rivers that provide inexpensive transportation linkages. Lake Michigan allows the state to be accessible by ocean-going ships for inexpensive export of Illinois farm products to foreign countries. Major east-west railroad and highway routes cross Illinois and give the state an excellent land transportation network. Many nearby midwestern cities, especially Chicago and St. Louis, provide eager markets for Illinois farm products. The enormous growth of the Chicago area alone has stimulated a corresponding increase in agricultural production in Illinois.

Illinois agriculture has experienced striking development since its humble beginnings. Modern farming is increasingly mechanized and automated, and it requires great technical and scientific skills on the part of farmers. With mechanization, particularly in the case of cash grain farming, one man now can operate an entire farm single-handedly. There are, however, some types of farming that are more demanding of labor inputs. Orcharding, dairy-ing, truck farming, and livestock raising are examples of such intensive types of agriculture.

Certain areas in Illinois have developed specialization in a common type of agricultural enterprise. Cash grain farming and dairying represent this tendency, but even small operations producing such crops as horseradish, sweet corn, pumpkin, and popcorn have become concentrated in distinct regions. Regionalization of farming types offers a number of economic advantages.

Agriculture in Illinois has become "big business," and farming is acquiring the characteristics of industry—for better or worse. The trend toward larger-sized farms continues, especially in cash grain farming areas. The growth of farm size has been slower in areas of dairying, orcharding, and truck farming, however. Farmers, particularly young people beginning their career, are finding the spiraling cost of land and machinery a difficult problem. Consequently, the practice of tenancy is growing and an increasing number of farmers are working at second jobs in nearby towns. Farming as a way of life is diminishing. The most interesting exception can be found in the Amish settlement near Arthur, where farming continues to be practiced much like it was many decades ago.

Science and technology have become the handmaidens of Illinois farmers. The production of most agricultural commodities has steadily increased as a result of careful plant and animal selection and the use of chemical fertilizers and weed and pest control products. Technology has made possible the conversion of soybeans into artificial beef, bacon, and other foods. Ethanol derived from corn is gaining favor as a fuel because it causes less pollution and corn, unlike petroleum, is a renewable resource. New scientific discoveries will likely continue to bring about significant changes in Illinois agriculture.

Farming can be expected to continue to make important contributions to the economic well-being of the state and the nation in the future. Urban expansion and other non-productive inroads upon agricultural lands must be carefully controlled or halted, however, if the nation is to be assured of an adequate supply of foodstuffs in the years ahead.

Notes

1. Several changes in Illinois agriculture are briefly summarized in Andrew J. Sofranko, "Illinois Agriculture: The Changing Scene," *Illinois Research* 15 (Fall 1973), pp. 3-5, and Michael Bowling and J.C. Van Es, "Changes Over a Decade in Illinois Agriculture," *Illinois Research* 19 (Spring 1977), pp. 14-15.

2. For a detailed study of nineteenth century agriculture in Illinois and neighboring Iowa, see Allen G. Bogue, *From Prairie to Corn Belt: Farming on the Illinois and Iowa Prairies in the Nineteenth Century* (Chicago: University of Chicago Press, 1963).

3. Yield variations of the major grain crops in Illinois are reviewed in J.H. Herbst, "Twenty-Year Trends in Crop Yields: Corn, Soybeans, Wheat, and Oats," *Illinois Research* 17 (Summer 1975), pp. 8-9.

4. R.C. Ross and H.C.M. Case, *Types of Farming in Illinois,* Agricultural Experiment Station Bulletin 601 (Urbana: University of Illinois College of Agriculture, 1956).

5. Walter M. Kollmorgen and George F. Jenks, "Sidewalk Farming in Toole County, Montana, and Traill County, North Dakota," *Annals of the Association of American Geographers* 48 (December 1958), pp. 375-97.

6. Walter M. Kollmorgen and George F. Jenks, "Suitcase Farming in Sully County, South Dakota," *Annals of the Association of American Geographers* 48 (March 1958), pp. 27-40.

7. See Lois F. Fleming and Dalias A. Price, "The Old Order Amish Community of Arthur, Illinois, Part I," *Bulletin of the Illinois Geographical Society* VI (June 1964), pp. 4-13, and Lois F. Fleming and Dalias A. Price, "The Old Order Amish Community of Arthur, Illinois, Part II," *Bulletin of the Illinois Geographical Society* VII (June 1965), pp. 4-24.

7

THE METRO EAST AREA

Robert L. Koepke
Donald W. Clements
Southern Illinois University at Edwardsville

Metro East (Figure 7-1), the Illinois portion of the St. Louis Metropolitan Statistical Area and the second most populous area of Illinois, is one of many regions where the struggle with the physical environment continues, although it has been surpassed in importance by the challenge of the cultural environment (including human attitudes, social structures, and institutions). The aborigines and early European inhabitants, who were hunters/gatherers and farmers, were required to work directly with the physical environment but had limited ability to manipulate it. In contrast, the twentieth-century urban inhabitants of this five county area generally are not directly linked to nature for their livelihood or comforts but nevertheless possess a great ability to manipulate their physical habitat. This chapter is a study of the Metro East area, its physical attributes, and its people and their relationships with their environment in the past and present.

PHYSICAL SETTING

Surface Features

Metro East is composed of two major and several secondary landform regions.[1] Most of the area of the five counties (Clinton, Jersey, Madison, Monroe, and St. Clair) is a loess-covered plain of predominately gentle slopes. Cut into this surface are several noticeable floodplains, including those of the Kaskaskia River, Silver Creek, Cahokia Creek, and Indian Creek. The second major physical region is called the American Bottoms, a lens-shaped major alluvial valley of the Mississippi River of nearly 175 square miles extending from Alton to Dupo along the western side of Metro East. Adjacent to the eastern and northern edges of the Bottoms is a strongly dissected portion of the plains region.

Though it is everywhere an alluvial valley, the American Bottoms landform region has some very important internal variations which can be grouped into seven secondary physical regions (Figure 7-2).

1. The largest of these is the ridge and swale, an area in the central part of the Bottoms around Granite City and in the southern part around Cahokia, that has an undulating surface of linear ridges, generally composed of sand, and long narrow swales of clay.

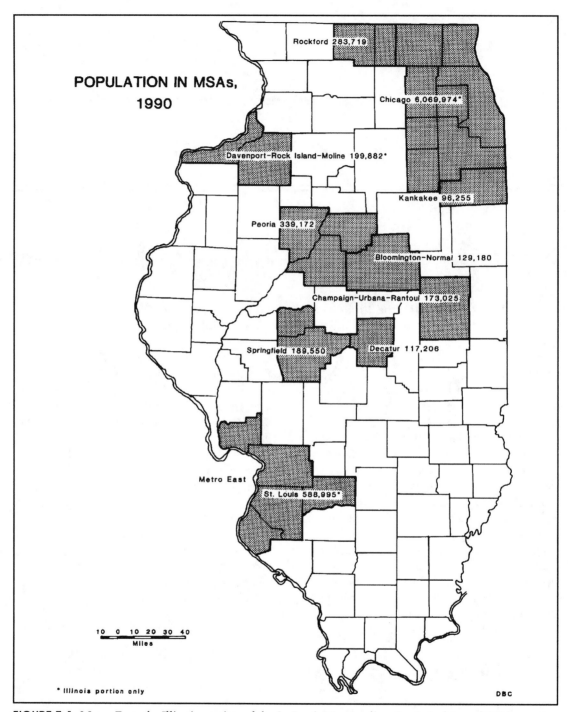

POPULATION IN MSAs, 1990

Rockford 283,719

Chicago 6,069,974*

Davenport–Rock Island–Moline 199,882*

Kankakee 96,255

Peoria 339,172

Bloomington–Normal 129,180

Champaign–Urbana–Rantoul 173,025

Springfield 189,550

Decatur 117,206

Metro East

St. Louis 588,995*

10 0 10 20 30 40
Miles

* Illinois portion only

DBC

FIGURE 7-1. Metro East, the Illinois portion of the St. Louis Metropolitan Statistical Area (MSA) contains nearly 600,000 inhabitants. Only Chicago has a greater MSA population.

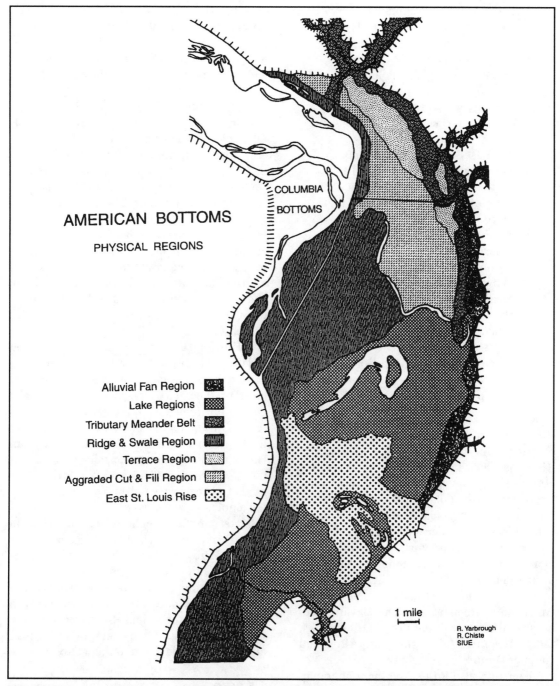

FIGURE 7-2. The American Bottoms, a wide segment of the Mississippi River floodplain in Metro East, has complex internal variations. (From Ronald E. Yarbrough, "The Physiography of Metro East", *Bulletin of the Illinois Geographical Society* XVI [June 1974], p. 18.) Reprinted by permission of the editor of the Bulletin of IGS.

2. The lake region contains oxbow lakes, large flat old lake beds, and meander scars.
3. The aggraded cut and fill region, or the Mitchell flats is an area in the northeastern portion of the Bottoms with very little local relief. The surface material of these "flats" is uniformly clay.
4. The East St. Louis rise, a complex area, is generally higher and flatter than the lake region and the ridge and swale topography that nearly surrounds it.
5. In the far northeast section of the American Bottoms is the terrace region, a high, sandy remnant of an earlier flood plain.
6. An area of alluvial fans, composed of silts washed from the adjacent uplands, is along the base of the bluffs on the eastern and northern fringes of the American Bottoms.
7. A tributary meander belt is adjacent to the bluffs in the north.

Running water is the chief geomorphological agent in the Metro East area. The stream of paramount importance, the Mississippi River, has been flowing through the area for 200 million years. Encountering some relatively soft rocks (shale and coal), the river removed most of them and then continued cutting into the harder materials underneath, creating within the previously deposited sedimentary rocks a valley about as wide but deeper than the present one. Since the softer materials did not exist north or south of the American Bottoms at the same elevation because of the nature of the slightly tilted rocks, the Mississippi Valley north of Alton and south of Dupo is consequently narrower.

Glaciation also influenced the topography of Metro East. It is not clear if the first two Pleistocene glaciers reached the area, but it is known that the leading edge of the third (Illi-

noisan) ice sheet covered Metro East. The observable results of this glacial activity are large boulders on the bedrock surface of the Mississippi Valley; the deposition of unsorted and some sorted material on the uplands, such as the ridges around O'Fallon and Belleville; and the creation of a soil, called Sangamon soil, in the uplands. The most recent continental glacier (Wisconsinan) stopped north of the area approximately 75,000 years ago, but its meltwaters had a marked impact on the present terrain. The valley of the Mississippi, which was a major drainage-way for the heavily laden meltwater, became filled by glacial material to a depth of at least 150 feet. In addition, the uplands to the east and north received fine materials that the wind picked up from the river valley during periods of low water. This material, called loess, which is up to 50 feet thick adjacent to the floodplain, decreases in thickness eastwards.

Following the final disappearance of the glacier, the Mississippi River meandered across the American Bottoms and removed most of the meltwater-deposited materials. Only a few remnants of the earlier floodplain remain, the major one being the terrace region in the northeastern portion of the American Bottoms. The upland streams that drained into the Bottoms and the Kaskaskia River also eroded some of the wind-blown materials, but the amount of loess removed is not known.

Karst landform features are found in the vicinity of Columbia and Waterloo in Monroe County. Here rain has solution eroded the limestones, resulting in a cratered surface with sinkholes that grow trees and brush as farmers crop around them. Beneath the surface are caves with their dripstone formations. Subterranean water in the karst areas often emerges as springs along the bluff adjacent to the Bottoms.

Climate

The climate of Metro East is characterized by continental temperatures and sufficient precipitation to produce a humid environment. The January average temperature is 32°F, while the July average is 79°F. Severe outbreaks of cold, polar air move through the area during many winters and produce minimum temperatures as low as -15°F. These cold waves are generally of short duration, however, and in three or four days more moderate temperatures, usually above freezing, return. As one would expect at this latitude, winters are noticeably milder than in northern Illinois. In fact, the chief method of removing snow in Metro East is to wait a few days for the sun to melt it. The 80-year average annual precipitation is 35.4 inches, with June being the wettest month. Thunderstorms produce much of the summer rain. Although snow may occur on any day from early November through April, most of the winter precipitation is in the form of rain. The average annual snowfall of approximately 17 inches is substantially less than that of upstate Illinois.

EARLY OCCUPANCY

Aborigines

People have been living in the Metro East area for a least 10,000 years, but it was only during the last 1,700 years that their activities had a significant and permanent impact.[2] Paleoindian (possibly), Archaic, and Early and Middle Woodland people lived in the area but left no enduring imprint on the landscape. On the other hand, there is evidence of the effect upon the area of the Late Woodland and especially the Mississippian people. Beginning around 300 A.D. both the bluffs and the American Bottoms were heavily utilized by Late Woodland people. During the Mississippian period, from 900 A.D. to 1250 A.D., the Bottoms contained a number of major settlements, including a community of possibly 40,000 people at Cahokia Mounds (a Mississippian settlement built on the present-day

East St. Louis rise region near the geographical center of the Bottoms). These people probably farmed most of the surrounding arable land. They also cut trees in the nearby bottomland and upland areas to construct enormous defensive palisades. The palisade on the eastern side of Monks Mound (the largest mound in the Cahokia Mounds site) was nearly a mile long and contained about 4,500 logs (Figure 7-3). It was reconstructed at least four times. Archaeological evidence suggests that deforestation of the uplands resulted in flooding and silting of the bottomland, producing difficult living conditions in that area. Beginning around 1250 A.D. the power and influence of the Mississippian "city" of Cahokia declined rapidly. In fact, the decline was so rapid and so complete that when Marquette and Jolliet came into the area approximately 400 years later, they were unaware of the existence of the mounds at Cahokia.

Native peoples who lived in this area left little evidence to explain their life style, social organization, and their sudden disappearance. The reflections of a local professional archaeologist on the possible reasons for the decline of the Mississippian communities offer some insight into the struggles of these people.

The decline of Cahokia may well be the result of many factors. The apparent abandonment of villages and farmsteads in the period after 1050 A.D. and the accompanying concentration of population in the city of Cahokia itself could indicate pressure from hostile Late Woodland groups. The building of palisades at this time might reflect the same kinds of pressures. The resulting concentration of population at Cahokia and the abandonment of supporting farming villages could have overburdened the economic system and led to collapse. Another possibility, however, could have been a minor climatic shift which led to crop failure and thus to the collapse of the economy. A third alternative explanation of the collapse . . . involves the hypothesis that Cahokia was a major center of trade and that its wealth and power are reflective of its virtual monopoly of the mechanisms and raw materials of trade. The hostility of Late Woodland peoples could have disrupted the trade networks and contributed to the decline. Finally, it is quite possible that the decline of Cahokia is the result of the exhaustion of natural resources in the immediate area. The destruction of the natural habitat due to the enormous use of wood seen in the construction of palisades, temples and houses may well have defor-

ested much of the bottoms. Deer would have been chased out of the area due to hunting pressure and the reduction of habitat, and it is even possible that the populations of fish and waterfowl which constituted important supplements to the diet at Cahokia may have been adversely affected. Additionally, the continued use of agricultural fields near the city could have led to the exhaustion of the soils. All of these factors working together might have led to starvation and a mass movement of Mississippian peoples out of the American Bottoms.

FIGURE 7-3. Monks Mound, the largest of the Cahokia Mounds and the largest prehistoric man-made structure north of central Mexico. The dimensions of the mound are 1,000 feet north-south, 800 feet east-west, and 100 feet from the base to the highest point. The volume is 21,690,000 cubic feet. The mound contains four terraces, some of which are visible in this photograph. A portion of the reconstructed palisade (east of the mound, not on it) can be seen in the upper center. (Courtesy of the Illinois Department of Conservation.)

Clearly the present state of knowledge does not allow us to choose between these hypotheses and answers must wait on further research.[3]

European Settlement

The year 1699 marks the beginning of the European settlement period in Metro East. In that year, following the assignment of the Governor of Quebec, the Seminary of Foreign Missions founded a mission to the Tamoroa Indians in the southern portion of the American Bottoms. In addition to serving as a religious center, the mission, called Cahokia (but located approximately ten miles southwest of the Mississippian Cahokia Mounds site), was also an agricultural village and a center for commercial activities. Cahokia became, in time, one in a series of French communities established southward along the Mississippi River through Ft. Chartres and Kaskaskia to Ste. Genevieve, the latter on the west bank of the river. With the demise of some earlier French midwestern settlements, Cahokia became the site of the longest continuous European settlement in the Midwest. Unfortunately, little has been written about the organization and administration of this early French settlement, but the founding of the mission itself is indicative of the importance of the societal element in the establishment and operation of French Cahokia. Even less is known about the struggle of the people with the physical environment, but one can surmise from the establishment of St. Louis in 1764 on a portion of high ground on the west bank of the Mississippi River, overlooking, but not in, the American Bottoms, that settlers were forced to deal with the problem of flooding.

The French who settled Cahokia left a distinctive mark on the present-day landscape of Metro East. The French long lot system of land division (Figure 7-4) has had a perma-nent effect on the shape of agricultural fields in western St. Clair County, the direction of major roads in this portion of Metro East (such as Illinois Route 15), and the orientation of some later settlements, such as East St. Louis, Alorton, and Centreville.[4] At least three buildings from this first European settlement period also remain in the Village of Cahokia; however, they have not received the public attention they deserve.

The impact of the British on Metro East was as limited as their tenure. In 1763, the area was part of the land ceded to the British by the French, and just 15 years later George Rogers Clark occupied the village of Kaskaskia for the United States. The period of American occupation had begun.

AMERICAN OCCUPANCY

It is this last cultural group, the Americans, who are responsible for the current Metro East landscape. They built the cities in which a half million people live today, the railroads and roads that traverse and tie the area together, and a levee system to hold back Mississippi River water during all but excessively high river stages. The Americans also established institutions and produced a social system to manage the area.

Early Period

For the first 75 years of American occupancy, the history of Metro East primarily involved the expansion of agriculture, the establishment of retail/governmental centers, and the beginning of coal mining. Subsistence farming probably was common in the area, but it is likely that many farmers also found a ready market for their agricultural goods in the growing city of St. Louis. In order to transport the commodities of these and other farmers in southwestern Illinois to St. Louis, a ferry across the Mississippi River was established. In 1817, Illinoistown (later renamed

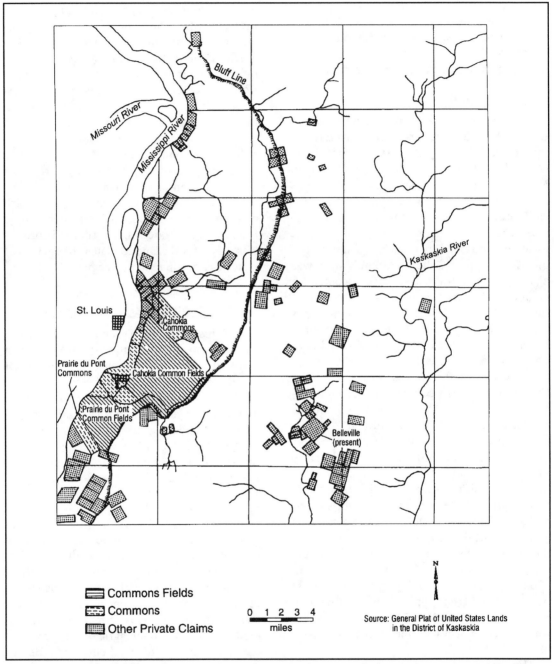

FIGURE 7-4. Private land claims in the vicinity of Cahokia, ca. 1810-1813. The French community of Cahokia is southwest of the Cahokia common fields. The common fields are oriented as shown on the map, but the width of the lots is not drawn to scale. (From William Baker, " Land Claims as Indicators of Settlement in Southwestern Illinois, circa 1809-13," *Bulletin of the Illinois Geographical Society* XVI [June 1974], p. 38.) Reprinted by permission of the editor of the Bulletin of IGS.

East St. Louis) the first permanent settlement at the Illinois end of the ferry, was laid out. It did not experience immediate growth, however, for even as late as 1837 the Reverend John Mason Peck still referred to it as a small village. In the same year a young Corps of Engineers lieutenant by the name of Robert E. Lee completed a plan to improve the harbor of St. Louis, which also was to have a lasting impact on the Illinois side of the river. Basically, Lee's plan involved the building of structures that would narrow the river. Not only did this help to maintain St. Louis as a riverport (rather than Illinoistown, perhaps), but it also aided the development of a large flat expanse of land between today's downtown East St. Louis and the Mississippi riverfront. Also in 1837, a small coal railroad was built across the Bottoms, generally following the relatively high ground found in the East St. Louis rise region. Beginning at the bluffs near Belleville (founded in the early 1800s), the line extended to Illinoistown. In addition to the cities noted, this early period of settlement saw the establishment of Edwardsville in Madison County and Alton on the bluffs at the north end of the American Bottoms.

Railroad Period

The beginning of construction of the Ohio and Mississippi Railroad from Illinoistown in 1852 marked the commencement of the railroad era in Metro East. By 1857, Metro East was connected to Vincennes, Indiana, by the Ohio and Mississippi, and soon thereafter to Baltimore by what was later the B & O/C & O Railroad. Other rail lines were constructed in Metro East, and by 1875 ten railroads traversed the area, most of them focusing on East St. Louis. Because the Mississippi represented an obstacle, these rail lines from the east terminated in Illinois and the railroad cars were ferried across the river. Although the opening of the first bridge (Eads Bridge) across the Mississippi at St. Louis in the

1870s eliminated the need to ferry all of the cars, the freight yards remained on the East St. Louis side of the river. The citizens of the area, and especially of East St. Louis, have been trying for years to originate and implement a plan whereby this now underutilized space of the railyards (much of which was created by the actions of Lt. Lee) can be rejuvenated.

The Urban/Industrial Period

The activities during the early and railroad periods of American occupancy left their mark on the present landscape of Metro East, but most of what exists in the area today is a result of developments over the last century. During this period manufacturing became the paramount economic activity in the area, stimulating not only substantial growth of the pre-1890 cities but the creation of other, completely new, urban centers as well (e.g., Granite City and Wood River). Interurban rail lines and major highways were built to connect these urban complexes. Also, flood protection levees and internal drainage facilities were constructed in the American Bottoms.

The settlement pattern resulting from the urban growth during the American period is dominated by moderate-sized urban communities scattered throughout Madison and St. Clair counties and generally separated from each other by agricultural land (Figure 7-5). These community areas are Alton-Wood River on the north; the Tri-Cities of Granite City, Madison, and Venice in the center; East St. Louis-Cahokia on the south; and Collinsville-Edwardsville and Belleville-Fairview Heights-O'Fallon on the east. The smaller towns, mainly in the primarily rural portion of Metro East, include Mascoutah, Lebanon, Jerseyville, Highland, Columbia, and Waterloo.

Population growth during the urban/industrial period occurred in different portions of Metro East at different times. During the 40

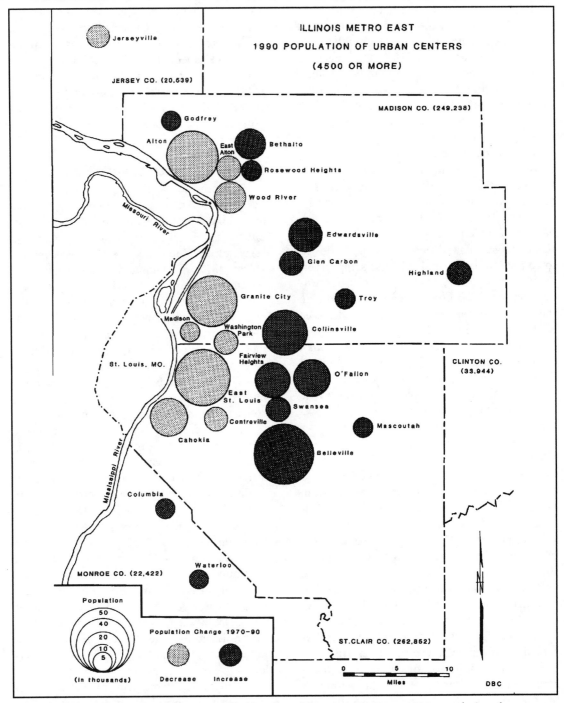

FIGURE 7-5. The largest urban centers of Metro East. Those with 1970 to 1990 population decrease are virtually all on the American Bottoms.

years from 1890 to 1930, most of the population increase took place in the American Bottoms in the communities of East St. Louis-Cahokia; in the Tri-Cities of Granite City, Madison, and Venice; and in the Wood River-Roxana segment of the Alton-Wood River area. During the years between 1930-1970, increases occurred over a much wider area. A large part of the total increase took place in the upland communities of Belleville-Fairview Heights-O'Fallon and Alton-Wood River, with substantial increases occurring in the bottomland centers of East St. Louis-Cahokia and the Tri-Cities. The region experienced an overall population decrease of 12,751, or two percent, from 1970 to 1990. This decrease was realized during the 1970s with a slight upturn during the 1980s. Continued small gains also have been registered in the early 1990s. The Census Bureau reported 588,995 inhabitants in 1990 for the five-county region. During the two decades, the three rural counties actually grew slightly and Madison County virtually retained its 1970 level. However, St. Clair County lost 22,322 or eight percent of its 1970 population. Madison and St. Clair counties both experienced internal population shifts. Places on the Mississippi flood plain such as East St. Louis, Granite City, and Alton, all of which experienced significant losses of manufacturing jobs during the period, lost population. By contrast, communities east of the bluffs experienced growth between 1970 and 1990. The most notable of these were Swansea, Fairview Heights, O'Fallon, Collinsville, Edwardsville-Glen Carbon, Troy, Bethalto, Godfrey, Columbia, and Highland. Other population related characteristics are presented in Figure 7-6.

The scattered urban settlement pattern found in Metro East is not characteristic of the entire St. Louis region or other metropolitan areas similar in size to the St. Louis Metropolitan Area (Figure 7-7). In St. Louis City

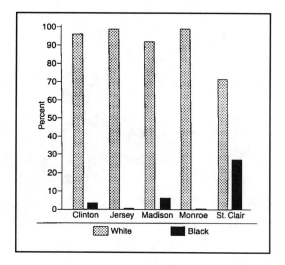

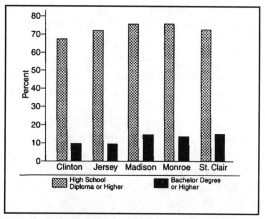

FIGURE 7-6. Population by race (upper) and education attainment (lower) in the Metro East Counties for 1990.

and St. Louis County, Missouri, which are the major elements in the western portion of the St. Louis SMA, most of the people live in one large urban complex. Similarly, in Chicago most of the people are concentrated in a single urban cluster. When the six United States metropolitan areas with populations of between two and three million are studied, it becomes clear that four areas have a primarily compact settlement pattern (Washington D.C., Pittsburgh, Baltimore, and Cleveland), while just

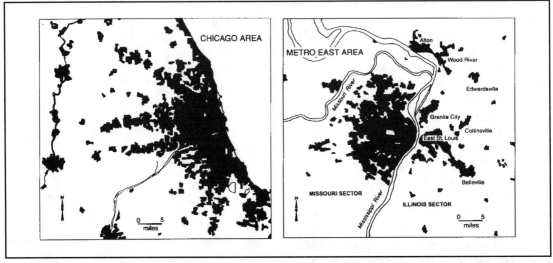

FIGURE 7-7. General urban patterns of the two major Illinois metropolitan areas. The separate urban nuclei in the St. Louis area are relatively widely scattered and mostly on the Illinois side of the Mississippi River.

two, Boston and St. Louis, have a large number of separate suburban centers along with a compact segment. The Boston area exhibits a "shotgun" pattern, in which the separate population concentrations are found scattered around the urban core. By contrast, the separate urban nuclei of the St. Louis area are found primarily in Metro East Illinois.

Life in these moderate-sized Metro East communities is similar to that in nearly any American community of comparable size. Generally, comfortable older residences, some from the 1890-1920 era, and new structures in the small subdivisions on the edge of the community are characteristic of the Metro East urban centers. Some of the commerce revolves around the ever-present community central business district, which contains the banks, savings and loans, and numerous specialty retail shops. Elsewhere, discount stores, unplanned shopping areas, and regional shopping centers in Fairview Heights and Alton all compete for the shoppers' dollars. Children in the towns of Metro East attend schools of local community school districts and participate in strongly-supported high school sports. Many people work in industrial and commercial establishments in their own communities, although commuting to jobs in other parts of Metro East, as well as to the Missouri portion of the urban area, is common.

A strong feeling of community identity prevails within the cities of Metro East. People from Alton think of themselves as Altonians, Belleville residents as Bellevillites, and those from Granite City as Granite Citians. The schools and central business districts, acting as focal points of individual and community thoughts and activities, help to develop and maintain this sense of local identity. The residents of these cities are kept informed about their towns by locally-oriented newspapers such as the *Edwardsville Intelligencer, Alton Telegraph,* and *Belleville News Democrat* which help maintain the cohesiveness of the communities. This sense of belonging also comes from the proud and, in several instances, extensive histories of the communities in Metro East. Cahokia, of course, is the oldest permanent European settlement in the Midwest, predating St. Louis by more than 50 years. Alton was an early rival of St. Louis, and Belleville and Edwardsville were developing communities when most of St. Louis was still farmland. The continuity of life— some would call it "roots"—is part of the life of residents of Metro East.

To most people today, however, the "good life" means more than just living in a small town. It also includes access to professional sports, major recreational facilities, first-class medical services, and the variety of goods offered in large urban shopping centers. Normally, the availability of such urban services is dependent upon living in a large urban complex. Most people must choose between the advantages of living in either a small town or a metropolis. The residents of Metro East, however, can enjoy the benefits of both. The assets of the major urban area of the region in St. Louis City and St. Louis County, Missouri, are accessible to most area residents in less than 30 minutes (Figure 7-8). Because of the short distance and good road transportation, much of it interstate highways, Metro East residents are able to enjoy professional baseball in Busch Stadium and hockey in the newly renovated Kiel Arena. Professional football in 1995 returned to the area in a newly-built stadium near the new downtown

St. Louis convention center. Shopping in the large department and the speciality stores of downtown St. Louis centers, such as Union Station and St. Louis Center, or in the major outlying shopping centers, such as River Roads, Northland, Galleria, Northwest, Jamestown, and South County, is relatively convenient for most residents of Metro East. Forest Park, with its zoo, art museum, and Municipal Opera, is handy, as are the medical facilities near that area. Barnes Hospital, Marlinckrodt Institute of Radiology, and Washington University Medical School are ranked highly at the national level. Stately Powel Symphony Hall and the refurbished art deco Fox Theater, both on St. Louis' Grand Avenue, provide excellent and varied entertainment.

In most instances, the scattered, small-town character of the Metro East population nodes is a significant, positive feature. Such a settlement pattern, however, also causes some problems, the major one being a weak-to-non-existent identification with the entire Metro East area. Partly because they too seldom come in contact with people from other parts of Metro East, most residents do not feel that they are part of an area larger than their home-town. They may be part of the same physical region, but they are not part of the same "mental" region. With such an orientation, the initiation and completion of projects necessary for the betterment of the entire Metro East area is difficult at best. People are interested in local needs, not area-wide ones, if they are interested at all. The political clout of the nearly 600,000 people also suffers at state and federal levels because of this lack of unity. Too few people in the area realize that allegiance to an area is in addition to, and not in place of, allegiance to a town.

THE METRO EAST ECONOMY

Incomes received by Metro East households are generally less than those of the

FIGURE 7-8. View of St. Louis from the Illinois side of the Mississippi River. (Courtesy SIUE News Service Bill Brinson.)

Chicago area but are considerably higher than those in several other parts of the state. Also, considerable differences are found in income levels between the counties and between the urban centers of Metro East. The urbanized counties, Madison and St. Clair, had 1989 median household incomes of $29,861 and $26,813 respectively, but these statistics do not indicate the great range of incomes found therein (Figure 7-9). Many communities east of the bluffs have high income families and are often characterized by expensive homes with attractive landscaping. This contrasts with the often deteriorating condition of the cities down on the American Bottoms. Such places often have more than their share of crime, have a substantial number of their inhabitants with low-educational attainment, and some have a large proportion of minorities in their population. Over recent decades these cities have lost large numbers of manufacturing jobs, and have experienced a commensurate loss of workers and their families. The people remaining are frequently unemployed or underemployed, recipients of public assistance payments and, too often, have broken families. Within the last few years, however, these cities have succeeded in making small gains in manufacturing jobs and capital investments in plants, infrastructure, and public facilities. Moreover, some jobs are being created in health care and other service activities. Even riverboat casinos now operate on the Mississippi River at Alton and East St. Louis. Nevertheless, the immediate future appears to favor higher rates of growth in population and income for the bluff cities.

In 1992 Madison and St. Clair Counties had about 20 percent (242,763) of the total St. Louis MSA labor force, but 8.3 percent of these were unemployed. (The labor force consists of workers who reside in an area but do not necessarily work there; in Metro East many commute to Missouri for employment.) Private sector employment or work force, in

the two counties during 1992 was 131,719, which represented an increase of 17,784 or 15.6 percent from the year 1981. (The work force consists of those employed in an area, but they do not necessarily reside there. Some workers commute from Missouri or other Illinois counties to jobs in Madison and St. Clair counties.) This increase can be attributed to significant gains in the service sector, plus some increases in wholesale and retail trade and in finance/insurance/real estate. Those employed in Madison and St. Clair counties in 1992 were mainly engaged in the activities of service, retail, and manufacturing industries (Figure 7-10). Collectively these three industries accounted for 77.6 percent of the work force in these two counties. Public sector employment also is a strong contributor to the area's economy. The public sector includes military bases, prisons, state universities, community colleges, public hospitals and schools, and numerous federal, state, county, and municipal service facilities. Three of the six largest 1992 employers in Metro East were in this category: first, Scott Air Force Base (12,400); fifth, Southern Illinois University at Edwardsville (2,000); and sixth, East St. Louis Community Unit School District Number 189 (1,958).

Primary Activities

Although most of the five-county Metro East area is farmland, only about one percent of the work force is engaged in agricultural activities. The small towns and villages of the area often contain grain elevators, evidence of local reliance on agriculture. The main crops are winter wheat, corn, soybeans, grain sorghum, and alfalfa. The area also is a significant producer of hogs, cattle, and milk. An interesting specialty crop is horseradish, centered near Collinsville. A significant portion of many farms is wooded, mainly in stream-dissected valleys and along bluffs. Oaks, hickories, and other deciduous hardwoods are

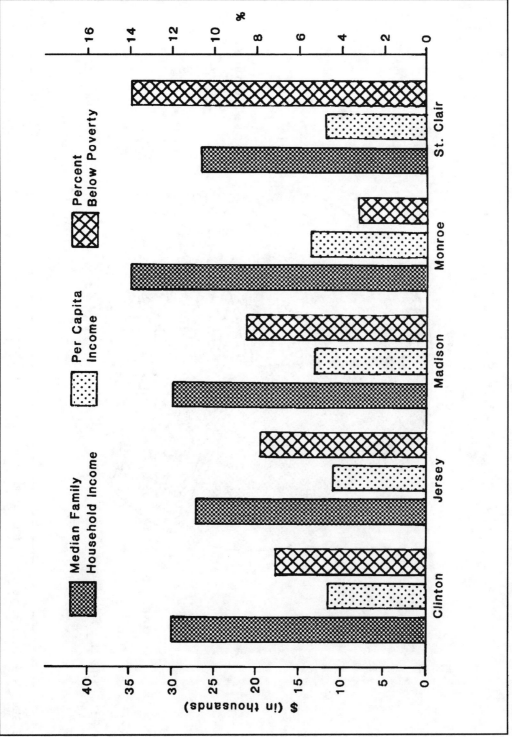

FIGURE 7-9. Income characteristics of Metro East Counties for 1989.

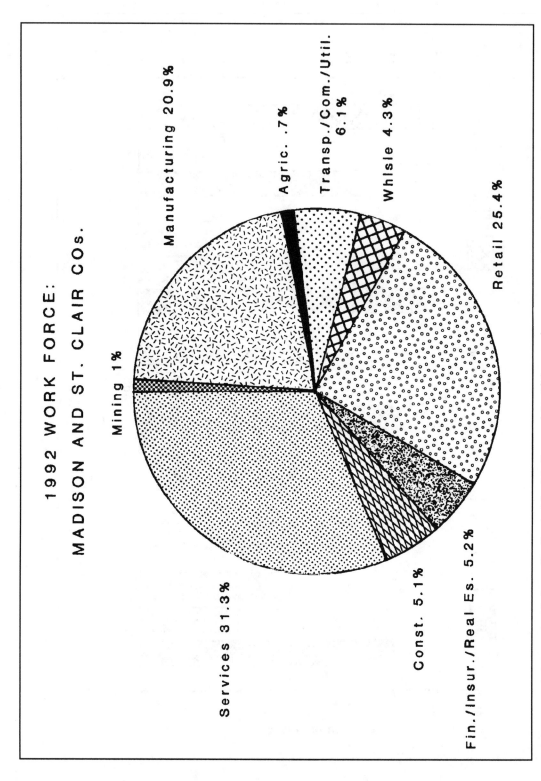

1992 WORK FORCE: MADISON AND ST. CLAIR COs.

Manufacturing 20.9%

Agric. .7%

Transp./Com./Util. 6.1%

Whlsle 4.3%

Retail 25.4%

Mining 1%

Services 31.3%

Const. 5.1%

Fin./Insur./Real Es. 5.2%

FIGURE 7-10. The 1992 work force distribution for Madison and St. Clair counties.

of little economic value but are a pleasant aspect of the landscape and provide sanctuary for wildlife.

Mineral production in the region exists on a small scale. Pumps and small clusters of storage and brine separator tanks are visual evidence of oil production. However, its significance is very limited. Area coal mining has had a long history and is somewhat more important, accounting for most of the region's workforce in mining (one percent). Current operating mines are in the southern portions of the region. Productivity is high in these efficient mines and most production is sold to electric power producers. However, coals produced have high sulfur content, and their use is limited by the costs resulting from abatement of sulfur-compound smokestack emissions. Also, mined-out areas under communities such as Glen Carbon, Maryville, and Belleville have created subsidence problems, mainly damage to roads and buildings on the surface above.

Manufacturing

One of the most important economic functions of Metro East is manufacturing, which, as was pointed out earlier, has played a key role in the growth and establishment of the area's cities.[5] Manufacturing plants employing hundreds of people and occupying extensive sites are prominent features of the landscape in and around these cities. The manufacturing in Metro East is best characterized by the word concentrated—i.e., concentrated in location within the Madison-St. Clair bicounty area, in a few types of industries, in a few large plants, and in the time period during which the plants were established.

The Bureau of the Census last provided information for manufacturing in Madison and St. Clair counties in 1987 and, for that year, indicated a total of 411 establishments, with 27,200 employees earning $759 million

and producing value added by manufacturing (VAM) of $1,866 million. Of these 411 establishments, 287 (70 percent) were concentrated in only ten industry groups. Employment was especially concentrated in primary metal (Figure 7-11) and fabricated metal products, petroleum refining (Figure 7-12), chemicals, and paper products. By 1990, 39 percent of all manufacturing employment was concentrated in only four firms: Olin of East Alton (4,000), Granite City Steel (3,500), Shell Oil of Wood River (2,088), and Laclede Steel of Alton (1,800). Obviously, this is a precarious position as the demise of even one firm can be disastrous in terms of lost jobs, lost orders for supplies, reduction of the local tax base, and less money available for purchases in local stores. However, an increase in orders and overall growth in production and employment can have a very positive economic outcome for the community. In the early 1990s some area manufacturing firms had made, or were considering, considerable capital investments in plant facilities, an indication of some optimism on the part of their officials. It should be noted that increases in manufacturing output do not necessarily equate with commensurately greater manufacturing employment. Increasingly, manufacturing is more automated and efficient, more capital and less labor intensive.

The period 1890-1919 has been designated by Robert Harper in his study *Metro East: Heavy Industry in the St. Louis Metropolitan Area* as the "golden era" for manufacturing. Harper notes that in addition to being the period when many of the area's large plants were established, this was the time when the employment growth in Metro East was nearly equal to that of the larger St. Louis, Missouri, area. Between 1890 and 1919, the employment in manufacturing in Metro East grew by 20,972, while the increase in St. Louis was 25,247. During these 30 years the location of manufacturing was localized in and around

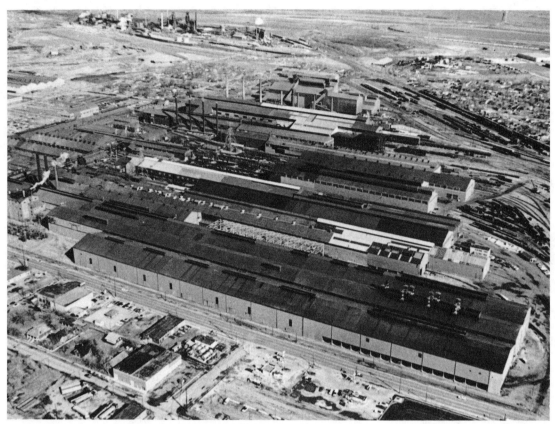

FIGURE 7-11. The Granite City Steel works of the National Steel Corporation (looking to the Southeast). The computerized rolling mill is in the long building in the foreground, the basic oxygen furnace is in the tall building in the center, and the two blast furnaces and nearby coke ovens are near the top of the photograph. Granite City Steel is the only integrated iron and steel plant in the St. Louis Metropolitan Area and the only one in downstate Illinois. (Photograph by T. Mike Fletcher. Reprinted by permission of National Steel Corporation.)

the cities of Alton, East St. Louis, and Belleville. By 1919 it was well established in the Tri-Cities (Granite City, Madison, and Venice) and on the floodplain terrace in Wood River-Roxana.

Our knowledge of the reasons for the development of manufacturing in these two counties is not complete, but what is known is mainly the result of work by Lewis Thomas and Robert Harper. Both conclude that manufacturing in Metro East is a result of low-cost, flood-protected land; the large number of rail-roads, which made possible the rapid receipt of material and shipment of products; the availability of cooling and process water from either the river or the floodplain aquifer; and the considerably lower freight costs on coal, an important industrial material around the turn of the century, in comparison to the costs in the City of St. Louis. Because the Mississippi River is on the west side of its valley in most of the St. Louis area, the floodplain and its associated extensive level sites are nearly all in Illinois. According to Thomas, land costs in 1927 in the American Bottoms, such as in the Tri-Cities area, ranged from 5 to 35 cents per square foot, while those in the Mill Creek Valley in central St. Louis were one to five dollars per square foot. The cost of delivering one ton of coal to East St. Louis in 1913, Thomas notes, was 32 cents, while it was 52 cents to transport a like amount to St. Louis. The difference was in part because the interstate transportation of coal from Illinois into Missouri came under the jurisdiction of the Interstate Commerce Commission, but the Illinois Public Service Commission set the intrastate rates from the Illinois mines to East St. Louis.

Since the 1950s, Metro East has lost several of its large manufacturing plants, most of them in one of the area's oldest centers of manufacturing, East St. Louis. The now underutilized buildings and sites once occupied by Alcoa, Armour, American Zinc, Emerson Electric, and Mobil Oil that ring East St. Louis are reminders of the growth and decline

which are part of all aspects of life, including manufacturing. With changing conditions, some plants survive at their older sites, while others meet the challenge by changing their method of production or their location.

The people of Metro East are beginning to cope with this loss of manufacturing and especially with the type and locational pattern of manufacturing they wish to have in the future. Many in the area are now thinking that light industry in addition to heavy manufacturing would be desirable. Such new manufacturing, which will not be spatially restricted to the floodplain, must produce "contemporary" products and do so very competitively. Moreover, such plants will likely require a highly skilled and adaptable work force. It also is realized that, in the future, manufacturing will likely contribute relatively less to the area's overall economy and that the service sector will continue to grow. The service sector is a broad category ranging from low-paying jobs requiring few

FIGURE 7-12. The Wood River refinery of the Shell Oil Company, with a capacity of 300,000 barrels of petroleum per day, is one of the two largest of this corporation. Originated in 1918, the refinery today covers 2,000 acres in the northern end of the American Bottoms. (Courtesy of Shell Oil Company.)

FIGURE 7-13, An aerial view of Southern Illinois University at Edwardsville in 1993. The view is to the south-southeast. (Courtesy SIUE News Service Bill Brinson.)

skills to professions requiring many years of education but yielding high levels of remuneration. Achieving the latter may be accomplished by attending one of the region's several institutions of higher learning (Figure 7-13).

TRANSPORTATION

Historically, water transportation has been an important feature of the area's economy and St. Louis maintains its premier position in tonnage handled when compared with other U.S. inland waterway ports. The Tri-City Regional Port at Granite City is the largest of three Metro East port districts; in 1992 it handled approximately 2,000 barge loads or 2.7 million tons valued at $650 million (Figure 7-14). This district has a foreign trade zone that may become more important as NAFTA (North American Free Trade Association) accelerates trade with Canada and Mexico. Currently, Mississippi-Missouri-Ohio River barges carry mainly grains, petroleum products, and coal. Generally, grains are

moved to New Orleans where they are loaded onto ocean-going vessels, while petroleum products from the Gulf region are conveyed northward. Recent completion of the new Melvin Price Locks and Dam system at Alton relieves the congestion associated with the earlier Lock and Dam Number 26. This is the largest Corps of Engineers single-site project and took over a decade to complete

FIGURE 7-14. An aerial view of the Tri-City Regional Port District at Granite City in 1993. (Courtesy of Tri-City Regional Port District.)

at a cost of approximately a billion dollars (Figure 7-15).

The St. Louis area is the third-ranked rail center in the nation. Madison and St. Clair counties have access to 13 trunk line railroads that operate 28 lines radiating to destinations throughout North America. Eight of the region's rail systems have internodal facilities, including four piggyback terminals. Thus, the rail system is effectively integrated with the area's highways and ports. An electric light-rail passenger system (Metro Link) connecting Scott Air Force Base and East St. Louis to St. Louis and other Missouri metropolitan communities as far west as Lambert Airport opened in 1993. Its future expansion is contingent on attracting sufficient passengers and continued government subsidies.

Metro East is served by over 200 miles of interstate highways including I-70, I-55, and I-64. These are linked by I-255 and I-270, which provide the circumferential system for

FIGURE 7-15. The Melvin Price Lock and Dam complex during the flood of 1993. The new Clark Bridge at Alton is in the background. (Courtesy of the U.S. Army Corps of Engineers.)

FIGURE 7-16. The Clark Bridge at Alton. (Photograph by Donald W. Clements, March 1994.)

the entire St. Louis area. In addition, these highways link with I-44 in Missouri. Sadly, the interstate system does not effectively link the Metro East communities with each other. This is accomplished by essentially single-lane highways such as Illinois 157, 143, 111, and 159. Unfortunately, these roads often carry through-traffic into the crowded central business districts of communities such as Belleville, Collinsville, and Edwardsville.

Numerous bridges connect Metro East with Missouri. Most accommodate automobiles and trucks, but some permit rail traffic. The majority of bridges are at or near downtown St. Louis, although one is south of Dupo, another north of Granite City, and another at Alton. The old Clark Bridge at Alton was replaced in 1994 by a new structure with an exciting fan suspension design (Figure 7-16).

The region's major air transportation facility is Lambert St. Louis International Airport, located northwest of St. Louis. However, Scott Air Force Base, east of Belleville, is currently being converted to joint military/civilian use. It will provide relief for Lambert as its traffic increases to saturation in the near future. Two major regional commercial and general aviation airports are located in Metro East: St. Louis Regional at Bethalto and Parks Airport at Cahokia, which is near downtown St. Louis.

DRAINAGE AND FLOOD CONTROL

Control of the water that has formed and influenced the American Bottoms portion of Metro East has been necessary in order to make the area habitable and a satisfactory location for homes and businesses. Before its significant modifications by humans, the Mississippi River floodplain in Metro East was a wet land. Around the turn of the century the Bottoms contained several lakes including Cahokia Lake, Smith Lake, Pittsburg Lake, Spring Lake, Horseshoe Lake, Indian Lake, and Grassy Lake. The American Bottoms also received the overflow of the Mississippi River itself, such as the 1844 record flood during which steamboats traveled directly to the bluffs, according to local histories. The Bottoms also received the water of "yazoo" streams, many with large tributary areas. Cahokia Creek, for example, has a tributary area of 290 square miles, most of it on the uplands. In addition, water falling on the Bottoms itself often ponded because of the slight gradients and numerous low areas, some without outlets, that were produced by the complex actions of the Mississippi River in the formation of the surface of the Bottoms.

It was not until 1907, in response to the flood of 1903, that the East Side Levee and Sanitary District (now the Metro East Sanitary District) was formed to provide a major levee system for protection from the river, diversion of upland streams, and drainage ditches for interior flood control. Whereas some levee districts had been formed earlier and some levees had been constructed locally, the East Side Levee and Sanitary District was the first to plan for flooding in nearly the entire Bottoms and to construct facilities to protect and to drain the Bottoms.

One of the first accomplishments of the district was the construction of the Mississippi River levee during the period 1911-15. This system consists of a riverfront levee along the east bank of the Mississippi River and flank levees on the north along the Cahokia Diversion Channel and on the south along the Prairie du Pont floodway. The flood control act of 1936 authorized the U.S. Army Corps of Engineers to strengthen these levees. Today most of the American Bottoms is protected with a 200 year levee (i.e., the probability of it being overtopped is .005 during any one year).

The East Side Levee and Sanitary District also tackled the water problem produced by the upland tributary streams that empty into the floodplain. In fact, in 1910 the first action of the district was the construction of a diversion across the floodplain to the Mississippi River. One of the previously noted flank levees was constructed to prevent Mississippi River water backing up into this diversion channel. In 1917 the district began a similar project on the south end of the Bottoms by building a floodway with flank levees for Prairie du Pont Creek. A year earlier, in 1916, improvements consisting largely of straightening and deepening were begun on the portion of Cahokia Creek still remaining within the Bottoms. In 1920 construction began on Canal No. 1, which was to go along the base of the bluffs from Prairie du Pont floodway and cut off all the "yazoo" streams before they traveled through much of the Bottoms. The canal was never completed and now receives the water from just one upland stream, Powermill Creek, and a small area around the canal.

Hence, within two decades after the flood of 1903, the East Side Levee and Sanitary District had a system to protect the American Bottoms from the Mississippi River and, to some extent, from the major upland tributary streams. Additional laterals to the major

ditches have since been constructed, such as Goose Lake Ditch; some streams have been relocated, such as the portion of Cahokia Creek through East St. Louis; and, as has been noted, the levees have been raised. But by and large the flood protection and drainage system established by the district during its initial years is the system that exists today.

Because of the accomplishments of the early engineers, the East Side Levee and Sanitary District, and efforts of the Army Corps of Engineers, some sections of this system have been quite effective. The protection offered by the levees has so far been complete, with the levees never having been overtopped, even during the great floods of 1973 and 1993. The Cahokia Diversion Channel and the Prairie du Pont floodway also have worked effectively since their installation. By the 1980s the Bluewaters area in the southern part of the American Bottoms was effectively drained and protected from flooding. All of the floodplain is now protected except from the most catastrophic floods.

The Future

Illinois Metro East has an intriguing present and a proud past, but the future should be more important to the citizens, institutions, and researchers of the area. Of course, it is necessary to understand the past and the present, for they are the foundations upon which the future will be built; but the people of today can only study, not influence, the activities and patterns in either of these time periods. It is only the future over which people have some degree of control.

Unfortunately, the future is a period about which little information is available. Researchers in Metro East, as elsewhere, have not pushed enough of their investigations into what tomorrow may be or should be like. Local citizens have generally been so busy with the present (or so disinterested) that the future has been almost ignored. The governmental sector also is oriented towards handling today's problems and leaving tomorrow's needs for tomorrow.

If Metro East is to be a good place to live in the future, the local people must make some major commitments. One is to dedicate themselves to work toward a better tomorrow. Without this dedication there will be little attention given to the future and without this attention the future will be left to take care of itself. As noted earlier, the settlement pattern, the quantity and type of employment, the transportation network, and the solution of environmental problems such as drainage are features of the area over which people have some control, if they will simply "put their hands on the plow." A second commitment is to work toward answering the following series of questions about themselves and their area: What is the area like? How did it get this way? What can it be in the future and what do the people of the area want it to be? The third commitment is to develop a system or procedure whereby people can effectively transform this area from its present to its ideal condition.

While the building of a better tomorrow is not a high priority item in the area today, the seed of "what about tomorrow?" is beginning to sprout. Growth, development, and planning are not major topics of citizen discussions, but the local press is pointing out more and more the potential, need, and problems of growth and the need for management. The professional planners are gathering information on the present character of the area and its future potential.

In addition to recognizing the need for dealing with tomorrow today and the consideration of a series of questions about what tomorrow can and should be like, the people of Metro East need to come to grips with the system whereby the area can be changed. So far, this important element has received the least amount of attention by both laymen and professionals. How much of the management of an area should be at local, regional, state, and federal levels? How does the need and desire of local people to manage their own activities mesh with the advent of regional, state, and perhaps even federal planning and the appearance on the scene of the federal Environmental Protection Agency? Who should make the decisions at these various levels—citizens, politicians, elected public officials, appointed officials, local technocrats, or regional or state technocrats? In other words, what form should an important element of today's cultural environment take?

The struggle of people in Illinois Metro East to make a home for themselves goes on. For previous inhabitants a large part of the struggle was with the physical environment. Today people seem reasonably well equipped to manipulate the physical environment, but less well equipped to struggle effectively with their cultural environment. Yet the character and patterns existing in this part of Illinois when the 300th anniversary of the founding of the United States is celebrated will depend to a considerable degree upon the outcome of the people's attempts (if any) to form their own future through their cultural resources.

NOTES

1. For a more detailed description of the landforms of Metro East, see Ronald E. Yarbrough, "The Physiography of Metro East." *Bulletin of the Illinois Geographical Society* XVI (June 1974), pp. 12–28.

2. See Harry B. Kircher. "The Sequent Occupance of Metro East." Bulletin of the *Illinois Geographical Society* XVI (June 1974), pp. 3–11 for an account of the progression of human groups (Indian, French, English, and American) that have occupied the area.

3. Sidney Denny, "Cultural Elements Archaeology." Section XIV, Environmental Inventory Report Part A (St. Louis: U.S. Army Corps of Engineers, St. Louis District, August 1973), pp. XIV–3, XIV–4.

4. William Baker, "Land Claims as Indicators of Settlement in Southwestern Illinois, Circa 1809–13." *Bulletin of the Illinois Geographical Society* XVI (June 1974), pp. 32–34.

5. Robert Koepke, "Manufacturing in Metro East." Bulletin of the *Illinois Geographical Society* XVI (June 1974), pp. 43–55.

6. Robert A. Harper, *Metro East: Heavy Industry in the St. Louis Metropolitan Area* (Carbondale, Ill.: Southern Illinois University, 1958).

8

THE CHICAGO METROPOLITAN AREA[1]

Irving Cutler
Chicago State University

Less than two centuries ago the Chicago area was a wilderness of flat, poorly drained land blanketed with prairie grass, clusters of trees, and foul-smelling marshes. Occasionally Indians would pass through the area in quest of game. Today, with a population of 2,783,726, Chicago ranks third among the cities in the United States. Moreover, the population of its burgeoning suburban area is now much greater than that of the city itself. The six counties of Northeastern Illinois that generally constitute the Chicago metropolitan area, although occupying only 6.6 percent of the area of Illinois, contain 7,261,176 people or 63.5 percent of the state's 11,430,602.

Despite the area's inauspicious setting, the essentials for its rapid growth were present when the first settlers arrived. These essentials included:

1. Location near the geographic center of the vast and fertile plains between the Appalachian Mountains to the east and the Rocky Mountains to the west (Figure 8-1). Chicago's situation enabled it to become the center of the most productive agricultural hinterland in the world. The flat terrain permitted easy access to this rich tributary empire and allowed the city itself to expand unimpeded.

2. Conveniently located and economically accessible important natural resources—the forests of the north, the iron ore of Minnesota and Wisconsin, the coal of Illinois and nearby states, and an unlimited supply of fresh water.

3. Location at the southwesternmost tip of the world's greatest lake system. This made possible exceptionally low transportation costs and a great range of domestic and overseas connections. In addition, Chicago's location is at a natural point of convergence for land traffic between the East and Northwest that had to find its way around the southern tip of Lake Michigan. Long before the coming of the white man, numerous Indian trails joined at Chicago.

4. The short natural waterways of Chicago, which eventually were modified to provide the only all-water connecting link between the

Great Lakes-St. Lawrence Seaway and the rich Mississippi Valley.

THE NATURAL SETTING

The topography of the Chicago region is the result of millions of years of geologic action (Figure 8-2). The limestone bedrock that underlies the area and is occasionally visible in quarries, road cuts, and river channels, is largely the compressed remains of the limey skeletons and shells of countless creatures that lived in the tropical sea that covered the mid-continent millions of years ago. This limestone provides basic building material and also solidly anchors Chicago's giant skyscrapers.

The glaciers that once covered the Chicago area ground down elevations, polished rough surfaces, gouged and deepened such areas as the basin of Lake Michigan, and left behind over the limestone bedrock a covering of glacial drift averaging between 50 and 60 feet in depth. In the Chicago region, the last glacier receded about 13,500 years ago, having sculptured the basic landscape surface. Chicago now occupies a lake plain that is hemmed in by the Valparaiso Moraine, a ridge of glacial drift parallel to Lake Michigan. This crescent-shaped moraine averages 15 miles in width, is from less than 100 to over 500 feet above the level of the lake, and is significant especially in regard to the drainage pattern. It stretches from southeastern Wisconsin to southwestern Michigan; its inner edge is followed approximately by the Tri-State Tollway some dozen or so miles from Lake Michigan.

The northern part of the Valparaiso Moraine is characterized by rounded hills and undrained depressions. In Lake County, Illinois, and crossing into Wisconsin many of these depressions are occupied by about 100

FIGURE 8-1. A major asset of Chicago is its excellent location. (Based on a map from Chicago Association of Commerce and Industry.) (Courtesy City of Chicago.)

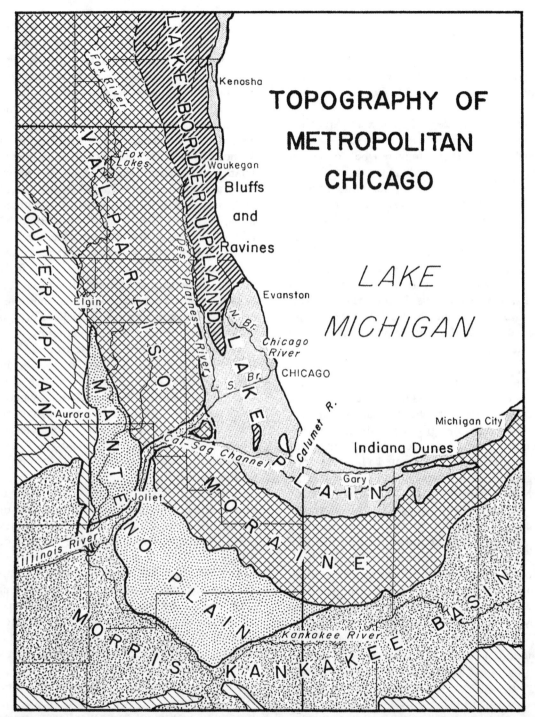

FIGURE 8-2. Topography of Metropolitan Chicago. From OPEN LAND IN URBAN ILLINOIS by Rutherford H. Platt, 1971, Northern Illinois University Press. Copyright 1971 by Northern Illinois University Press. Used by permission of the publisher.

small lakes and ponds. This inland lake region is an important recreational and residential area, with sizable settlements around such larger lakes as Fox Lake, Pistakee Lake, Round Lake, Long Lake, Grays Lake, and Lake Zurich.

On the lake side of the northern part of the Valparaiso Moraine is the much smaller Lake Border Upland, an elongated belt of nearly north-south ridges with a width of five to fifteen miles. The main segment extends northward from about Des Plaines and Winnetka, with a narrow extension south into the Lake Plain as far as Oak Park. Some ridges rise to about 200 feet above the lake level and are interspersed by gentle sags occupied by small streams and an occasional marsh, such as the Skokie Lagoons. Lakeward of the Valparaiso Upland and the Lake Border Upland spreads the flat Lake Plain on which Chicago is situated.

The Lake Plain

Since water drainage to the north was blocked by ice when the last glacier retreated, glacier meltwater filled the depression between the receding ice front and the Valparaiso Moraine. This created a lake that at its highest elevation rose about 60 feet above the present surface of Lake Michigan. This enlarged Lake Michigan, geologically known as Lake Chicago, covered all of the present city of Chicago. The boundary line of the lake reached from about what is now Winnetka through the present communities of Maywood, La Grange, and Homewood, crossing the state line at Dyer and continuing eastward beyond Chesterton, Indiana.

The accumulated water receded in stages, finding its way into the Illinois-Mississippi River drainage system by enlarging two outlets through the Valparaiso Moraine drainage divide. One of these outlets, which now holds the Calumet Sag Channel, was through the Sag Valley south and southwest of the city; the other, to the southwest, sometimes known as the Chicago Portage, contained first the Illinois-Michigan Canal and later the Chicago Sanitary and Ship Canal as well as other important transportation arteries.

The lake bottom of Lake Chicago left the Chicago area remarkably flat—a lake plain—except for a few small islands that had existed in the lake such as Mount Forest Island, Blue Island, and Stony Island, and some spits, sandbars and crescent-shaped beach ridges that emerged as the water receded in three distinct stages. These ridges stand about 60 feet, 40 feet, and 20 feet higher than the present approximately 580 foot height above sea level of Lake Michigan. Driving away from the lake on an east-west street, such as Devon Avenue or 111th Street, will take one over each of the beach ridges of Lake Chicago within a distance of ten to fifteen miles.

Often being the best-drained ground in an otherwise marshy area, some of the sandy spits, bars, and beaches of the Chicago area became Indian trails, and some are now parts of main roads. This good drainage also made these areas attractive locations for cemeteries and golf courses. Both Graceland and Rosehill spits bear the names of large cemeteries on them.

Three small lakes near the Chicago-Hammond state boundary are isolated remnants of the glacial Lake Chicago. In recent decades these lakes have declined in size because of marginal filling and drainage alterations. Lake George, on the Indiana side, has virtually disappeared; Wolf Lake is a recreational area; and Lake Calumet has been developed as the major port of Chicago. Beach ridges separate the three lakes from Lake Michigan. A series of such ridges has hampered drainage in the Calumet district.

The basic topography of the Chicago area resulted from the superimposition on the limestone bedrock of an uneven layer of glacial drift and of Lake Chicago's deposits.

Since the Ice Age, a number of limited topographic changes have occurred. Through weathering, wind and water deposition, and vegetative growth, the present soils have been formed on the surface of the deposits from the glacial period. The soil is generally of rich, fine quality, very productive agriculturally except where there are major drainage problems or where extensive sand deposits have accumulated, such as along parts of the shore at the southern end of Lake Michigan.

Another noteworthy postglacial change has been the development of the very scenic bluffs and ravines along the lake between Winnetka and Waukegan by shoreline erosion. Some of the bluffs are almost 100 feet high, and many of the more than twenty major ravines extend a mile or so inland. The effect of lakeshore erosion in this area is dramatically illustrated by the following report:

In 1845 and for about ten years following there was a village located in the southeast corner of what is now the Fort Sheridan grounds. This village was known as St. Johns. The chief industry was brick making, the yards employing as many as eighty men. ...North of the clay pit remnants of a foundation and of an orchard are at the very margin of the lake cliff. Reports differ as to the amount of land that has been cut away at this point, but all agree that it was more than 100 feet. Some old settlers insist that 300 to 400 feet have been removed, and that the cliff and even overhanging are reported by some to have been in the yard to the west of the westernmost house in the village. If this is true, the entire side of the village of St. Johns is east of the present shore line.[2]

Today even these meager traces of the village have vanished, prey to the attacking waves and currents.

Rivers and Drainage

The Chicago River is the outstanding topographic feature of the rather featureless lake plain that contains the city of Chicago. Though short and sluggish (the important South Branch is only about half-a-dozen miles in length), the river has been a major factor in the city's establishment and growth. It was the early connecting route between the East and the commercial wealth of the Middle Prairie. Early Chicago centered upon the river and in the period of its greatest use the river handled huge cargoes of grain, lumber, and manufactured goods.

The river's main channel with its two branches forms a *Y*, with the junction near the Merchandise Mart. This configuration has by tradition divided Chicago into three broad sections—the North, South, and West sides. The North Branch originates in Lake County, Illinois, as three small streams flowing southward in the sags of the moraines of the Lake Border Upland. The streams join in northern Cook County and flow southeastward toward the junction with the main channel. The South Branch was usually navigable only as far west as the present Leavitt Avenue (2200 West). Often, however, during spring high water it was possible to push canoes across the marshy divide all the way to the Des Plaines River (near 49th Street and Harlem Avenue) by using seasonal Mud Lake that bridged most of the six-mile portage between the two rivers.

The Chicago River has been greatly modified since it was navigated by the early explorers. It has been straightened and widened. The sandbar that blocked its mouth and caused it to bend southward and flow into the lake opposite the foot of Madison Street has been removed. An artificial island, Goose Island, was created and a sharp bend bypassed by the construction of the mile-long North Branch Canal. Most important, the portage was eliminated, and the South Branch of the river was connected with the Illinois-Mississippi River waterway system, first by the Illinois and Michigan Canal, later by the Chicago Sanitary and Ship Canal. The flow of the Chicago River was reversed to flow into these connecting waterways.

Unfortunately, the river was largely walled off by industrial development and its aquatic life drastically curtailed. However, in recent years various environment-conscious groups have promoted limited improvements and have dedicated themselves to the rebirth of the river as a multipurpose, clean, pleasant, life-supportive waterway that could serve as a vibrant artery through the heart of the city as well as a controlled transportation and industrial corridor.

At the southeastern end of the city is Chicago's other important river—the Calumet. Unlike the Chicago River, the Calumet played an insignificant role in Chicago's early history. Later, however, it was one of the most industrialized rivers in the world, "the Ruhr of America," although declining in recent years. The early settlers found the Calumet to be an erratic, meandering stream that had formed an elongated loop parallel to the lake. Later inhabitants altered this river also to suit their purposes. The Calumet River was reversed to flow away from Lake Michigan into the Calumet Sag Channel near Riverdale and thus drain the Lake Plain of southern Chicago and northwestern Indiana westward into the Chicago Sanitary and Ship Canal, and so eventually into the Illinois-Mississippi system.

The reversal of the Chicago and Calumet rivers altered the unusual drainage pattern of the Chicago region. In creating moraines parallel to Lake Michigan, glacial action also created a divide parallel to the lake and relatively close to it. Water on one side of the divide flows into the St. Lawrence River system; on the other side, into the Gulf of Mexico. The divide is less than four miles from the lake in the Waukegan, Illinois area, and at its farthest point, south of Hammond, Indiana, only about 20 miles from the lake.

A few short rivers on the eastern side of the divide such as the Chicago and Calumet broke through sandbars to reach Lake Michigan, but the major rivers, such as the Fox, Des Plaines, and Kankakee, never penetrated the moraines. They flow into the Mississippi Basin. In places, the divide is less than 100 feet above Lake Michigan and at its lowest point, the Chicago outlet of Summit, Illinois, it is a barely discernible 15 feet above the lake.

The land is flat and consequently the rivers are generally sluggish and drainage poor. Furthermore, layers of impermeable clay left by the glaciers hampered the drainage of surface waters, created a high water table, and helped make early Chicago a virtual sea of mud for at least part of the year. To get adequate gradient for storm and sanitary sewers, the Metropolitan Water Reclamation District of Greater Chicago has had to provide more than a dozen pumping stations to enhance the flow to the extensive drainage canal system.

HISTORICAL DEVELOPMENT

In 1673, Louis Jolliet and Pere Jacques Marquette, while returning to Canada after exploring the Mississippi Valley, became the first whites to pass through the Chicago area. In the ensuing decades, French explorers, trappers, and fur traders occasionally traversed the area mainly because of its excellent geographical location and short portage between the Chicago and Des Plaines River. Indians, chiefly Potawatomi, traded furs and sometimes camped in the area they called "Checagou," evidently referring to the wild onion smell that permeated the air.

The area was ruled by the British from 1763 until their evacuation was finally procured by the Jay Treaty of 1794. During the late 1770s, Jean Baptiste Point du Sable established what was probably Chicago's first permanent dwelling on the north bank of the Chicago River near the present Tribune Tower (Figure 8-3). From this base, du Sable carried on trade with the Indians for about two decades.

In 1803, the United States Army erected Fort Dearborn at an elevated point in the bend near the mouth of the Chicago River, to secure the area and protect the important waterway linkage, which became even more important with the Louisiana Purchase the same year. Despite the fort, settlement in the area remained very sparse, due mainly to the hostility of the Indians who were angered by the continued takeover of their lands. This hostility was brutally manifested in the War of 1812 when the Indians ambushed and killed 53 settlers and soldiers and burned the fort to the ground. The fort was reestablished in 1816. A few settlers, tradesmen, and agents were attracted to its vicinity, but large scale settlement did not begin until the conclusion of the Black Hawk War in 1832. The treaty with the Indians provided for their relocation west of the Mississippi River in return for certain payments in cash and goods.

In 1832, the Indians assembled in Chicago for their final payments. Also present were a motley collection of wayfarers—horse dealers and horse stealers, peddlers, grog sellers, and "rogues of every description, white, black, brown, and red—half-breeds, quarter-breeds, and men of no breed at all." By ruse, whiskey, and thievery, they managed to separate the Indians from a good part of their money and goods. About 800 Indians joined in a last defiant dance of farewell before crossing the bridge over the South Branch of the Chicago River and heading westward until Chicago saw them no more.

Town and City

No longer impeded by fear of the Indians, and aided by improved transportation, such as the Erie Canal-Great Lakes route, the trickle of newcomers to the little military and trading outpost grew into a stream. People were drawn to Chicago by cheap land, jobs, and a speculative fervor stimulated by plans for a canal that would connect the Mississippi River with Lake Michigan and terminate in Chicago.

Incorporated in 1833, the new town covered a 3/8 square mile area centered upon the main channel of the Chicago River. In the same year, it was described by the Scottish traveler Patrick Shirreff:

> Chicago consists of about 150 wood houses, placed irregularly on both sides of the river, over which there is a bridge. This is already a place of considerable trade, supplying salt, tea, coffee, sugar and clothing to a large tract of country to the north and west; and when connected with the navigable point of the river Illinois, by a canal or railway, cannot fail of rising to importance. Almost every person I met regarded Chicago as the germ of an immense city, and speculators have already bought up, at high prices, all the building-ground in the neighborhood.[3]

Chicago's rapid growth can be shown in a number of ways. In 1833 only four lake steamers entered its harbor; by 1836 the number had increased to 450. A parcel of land at South Water and Clark streets purchased for $100 in 1832 was sold for $15,000 in 1835. And by 1837, when Chicago was incorporated as a city, its population exceeded 4,000 and it encompassed some ten square miles.

The author John Lewis Peyton portrayed the burgeoning Chicago of 1848 as follows:

> ...The city is situated on both sides of the Chicago River, a sluggish, slimy stream, too lazy to clean itself, and on both sides of its north and south branches, upon a level piece of ground, half dry and half wet, resembling

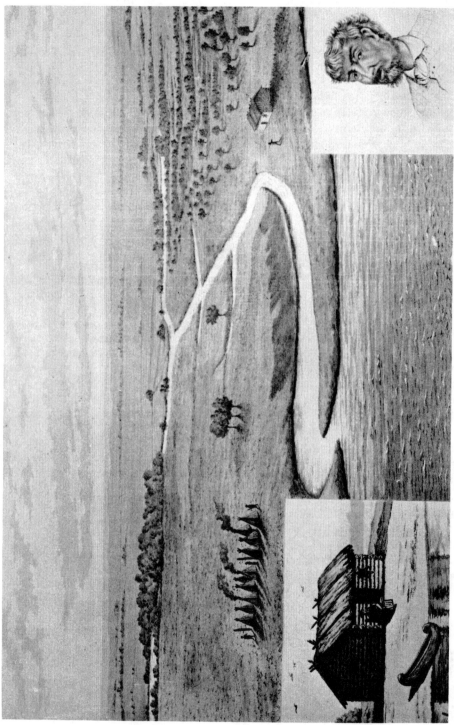

FIGURE 8-3. Looking westward at the site of Chicago, 1779. North of the bend in the Chicago River is the cabin of Jean Baptiste Point du Sable, the first permanent settler (shown in insets). The sand bar at the river's mouth caused the river to bend southward and flow for almost a half mile along approximately what is now Michigan Avenue until entering the lake at about the present Madison Street. In the distant center are the forks of the river where the South Branch and North Branch of the Chicago River join to form the main channel. (Courtesy of the Chicago Historical Society.)

a salt marsh, and contained a population of 20,000. There was no pavement, no macadamized streets, no drainage, and the three thousand houses in which the people lived, were almost entirely small timber buildings, painted white, and this white much defaced by mud....

...Chicago was already becoming a place of considerable importance for manufacturers. Steam mills were busy in every part of the city preparing lumber for buildings which were contracted to be erected by the thousand the next season. Large establishments were engaged in manufacturing agricultural implements of every description for the farmers who flocked to the country every spring. A single establishment, that of McCormick employed several hundred hands, and during each season completed from fifteen hundred to two thousand grain-reapers and grass mowers. Blacksmith, wagon and coachmaker's shops were busy preparing for a spring demand, which with all their energy, they could not supply. Brickmakers had discovered on the lake shore near the city and a short distance in the interior, excellent beds of clay, and were manufacturing, even at this time, millions of brick by a patent process, which the frost did not hinder, or delay. Hundreds of workmen were also engaged in quarrying stone and marble on the banks of the projected canal; and the Illinois Central Railway employed large bodies of men in driving piles, and constructing a track and depot on the beach. Real estate agents were mapping out the surrounding territory for ten and fifteen miles in the interior, giving fancy names to the future avenues, streets, squares, and parks. A brisk traffic existed in the sale of corner lots, and men with nothing but their wits, had been known to succeed in a single season in making a fortune—sometimes, certainly, it was only on paper.[4]

By 1850, Chicago's population had grown to about 30,000 and its future role as a great transportation and industrial center was al-ready clearly evident. The 97-mile-long Illinois and Michigan Canal opened in 1848, connecting the Great Lakes with the Mississippi Valley. Shortly thereafter, a period of vigorous railroad building brought railroad tracks to Chicago from almost every direction. By 1855, Chicago was already the focus of ten trunk lines.

Chicago's location and its excellent transportation connections with the rich agricultural hinterland helped forge strong bonds of interdependence between the city and the farmers of the Midwest. The farmers funneled their produce to Chicago; the city provided stockyards, food processing, and grain elevators, as well as ships and trains to deliver the farm commodities eastward. And from Chicago the farmers were shipped clothing, processed foods, household items, lumber, and farm equipment. Much of the farm equipment was manufactured by the McCormick Reaper factory, which had been established in 1847 on the north branch of the Chicago River at the site of the former du Sable cabin. Cyrus McCormick was among the first of a long line of commercial and industrial entrepreneurs who, together with their employees, were to help make Chicago "Hog Butcher for the World, Tool Maker, Stacker of Wheat, Player with Railroads, and the Nation's Freight Handler."

On the whole, political and economic dominance was held initially by men from the eastern United States. With remarkable combinations of thrift, shrewdness, and drive, they acknowledged no barriers to the successful expansion of a wide range of enterprises. In their climbing to the top, they often had little regard for others, but perhaps they did what had to be done to raise a city out of a swamp. Among these pioneer leaders was William B. Ogden of New York, who was elected the city's first mayor in 1837. One of the earliest of many Chicagoans to promote railroad building, he later became the first president of the Union Pacific. Potter Palmer, who arrived in 1852, made a fortune in dry goods and cotton speculation and added to his wealth by developing State Street. In 1867, Marshall Field, who came from Massachusetts, became a part owner in the firm that later bore his name. Two farm youths from the East, Philip D. Armour and Gustavus Swift, helped make Chicago the meat packer of the nation. A new era in railroad travel began in 1864 when George Pullman invented the sleeping car. Later his shops for building passenger and freight cars spread over 3,500 acres near Lake Calumet. Julius Rosenwald, a native of Springfield, Illinois, learned the clothing business, went to work for Sears, Roebuck and Company, and eventually became its president and board chairman as well as one of the nation's great philanthropists. Sears' major competitor was founded shortly after the Chicago Fire of 1871 by A. Montgomery Ward who had lost everything in the conflagration but $65 and the clothes he wore. He later earned the nickname "watchdog of the lakefront" for his long but successful struggle to save the Grant Park area from being commercialized.

Mudhole of the Prairies

The land survey provisions of the Federal Ordinance of 1785 imparted to early Chicago a basic, functional pattern of land subdivision and roads. Using a rectangular grid system, the survey divided the land into square-mile sections. The section lines were a mile apart and ran either north-south or east-west. They became main traffic thoroughfares, major routes of public transportation, and ribbons of commercial development that ultimately became over-extensive and inefficient. Major shopping areas often developed at the intersections of these major arterials. Superimposed on the rectilinear street system were a number of diagonal streets, some of which began as Indian trails.

Mud was the main problem of the early Chicago streets. In 1848 Chicago

> could boast of no sewers nor were there any sidewalks except a few planks here and there, nor paved streets. The streets were merely graded to the middle, like country roads, and in bad weather were impassable. A mud hole deeper than usual would be marked by signboards with the significant notice thereon, 'No bottom here, the shortest road to China.'[5]

The difficulty arose because Chicago was flat and low, being only about two feet above the river level. Moreover, the sewage that did drain off into the Chicago River flowed into Lake Michigan, the city's source of drinking water. The resulting epidemics of cholera, typhoid, and other diseases were not finally curtailed until the flow of the river was reversed in 1900. Before that, Chicago tried to lift itself from the quagmire by raising the elevation of the city. Fill—sometimes

FIGURE 8-4. State Street in the late 1860s after it had already replaced Lake Street as the main commercial thoroughfare. Chicago's first mass transportation line—the horse-drawn street railway car—was established on State Street in 1859. View is looking south from Lake Street. (Courtesy of the Chicago Historical Society.)

FIGURE 8-5. The forks of the Chicago River looking north from Randolph Street before the fire of 1871. The North Branch flows from the top left and the South Branch from the lower right to form the main stem of the river. Two of the major activities along the river are evident here: lumberyards and grain elevators. The open bridge on the right is at Lake Street. In the right background are Illinois and Michigan Canal barges tied up near Wolf Point on the north side of the main stem of the river. (Courtesy of the Chicago Historical Society.)

obtained from the dredgings of the water-ways—was used to raise street levels. For a time, however, Chicago exhibited a confusing pattern of disjointed sidewalks, some up and some down, depending on whether the owner had raised his property or not. And even today, especially in the inner city, one can find yards and homes below the raised street level, some with stairs leading down to the first floor.

The Fire and Its Aftermath

Despite such problems, Chicago continued to grow and prosper. It was buoyed by the opening of the Illinois and Michigan Canal, the coming of the railroads, the development of substantial industry (fueled somewhat by the Civil War), the further settling of its rich hinterland, and the accelerating influx of newcomers. Between 1850 and 1870 the population increased tenfold, from 30,000 to about 300,000.

The disastrous Chicago Fire of 1871, however, cast a temporary pall on the city. Chicago in 1871 was a city of wood—wooden houses, wooden roofs, even wooden sidewalks (Figure 8-5). After an unusually long period of drought, the city became tinder dry. The stage was set for the fire that broke out on October in Mrs. O'Leary's barn (at the site of the present Chicago Fire Academy). Fanned by a southwest wind, the fire spread rapidly. When it finally subsided two days later, it had thoroughly gutted about four square miles of the city including the entire downtown area (Figure 8-6). The fire took more than 250 lives, destroyed some 17,000 buildings, and left almost 100,000 people homeless (Figure 8-7).

The determined Chicagoans, who had already created a city on marshland, immedi-ately turned to the task of rebuilding it. By 1875, little evidence of the catastrophe remained. The fire accelerated the movement of residential homes from the central business district. The new buildings in the downtown area were larger and higher, conforming to a new city ordinance that outlawed wooden buildings there. In other parts of the city thousands of homes were going up, many of brownstone and brick. The rebuilding activities attracted thousands of laborers and numerous architects to Chicago. Many of them helped construct the first skyscrapers using the innovative steel skeleton, elevators, and the floating foundation.

Despite the fire, depressions, and such sporadic violent labor strife as the Haymarket Riot, Chicago continued to grow rapidly. Besides its commercial and transportation importance as a major handler of grain, cattle, and lumber, the city was becoming

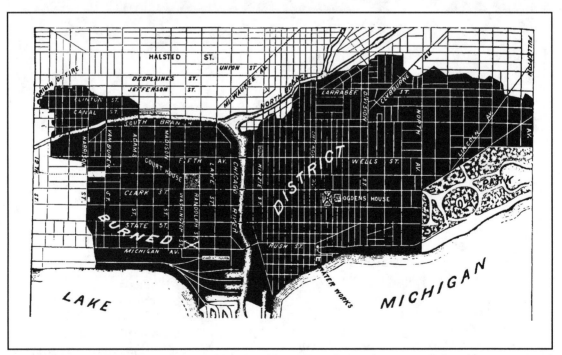

FIGURE 8-6. Map showing area destroyed by the Chicago Fire of 1871. (Courtesy of the Chicago Historical Society.)

FIGURE 8-7. Looking eastward along Randolph Street from Market Street after the Chicago Fire of 1871. Prominently visible are the gutted remains of the courthouse and city hall on the site of the present City Hall-County building. (Courtesy of the Chicago Historical Society.)

increasingly a major center of diversified manufacturing. The annexation by Chicago in 1889 of four sizable but relatively sparsely populated communities—the towns of Jefferson and Lake, the city of Lake View, and the Village of Hyde Park—increased its size from 43 square miles to 168 square miles and by 1890 the city had a population of 1,099,850 (Figure 8-8). In the preceding decade the output of many of its major industries had more than doubled.

From Fair to Fair

In 1893, just 22 years after the fire had leveled the heart of the city, Chicago again attracted worldwide attention, this time with its dazzling, classically styled World's Columbian Exposition (Figure 8-9). With its customary audacity, the city built the Exposition on an apparently impossible sandy site along the lakefront eight miles south of the river. The Exposition sparked a real estate boom on the South Side, especially in the

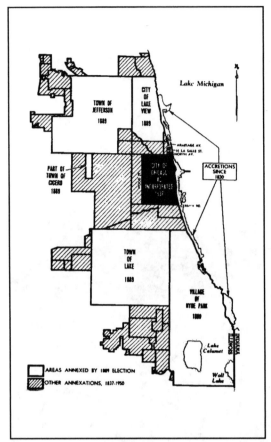

Lake Michigan

TOWN OF
JEFFERSON
1889

CITY
OF
LAKE
VIEW
1889

ARMITAGE AV.
N. LA SALLE ST.
NORTH AV.

PART OF
TOWN OF
CICERO
1889

CITY OF
CHICAGO
AS
INCORPORATED
1837

ACCRETIONS
SINCE
1830

TOWN
OF
LAKE
1889

VILLAGE
OF
HYDE PARK
1889

Lake
Calumet

INDIANA
ILLINOIS

Wolf
Lake

□ AREAS ANNEXED BY 1889 ELECTION

▨ OTHER ANNEXATIONS, 1837-1950

FIGURE 8-8. Growth of Chicago, 1837-1950. In one year, 1889, Chicago increased its size about fourfold through a series of annexations. Since 1950 Chicago has annexed about 15 additional square miles (38.9 sq km), accounted for largely by the land of O'Hare International Airport plus a few small pieces of land on the northwestern and southwestern fringes of the city. (From Chicago, Chicago Board of Education, 1951.)

...Chicago! Chicago!, queen and guttersnipe of cities, cynosure and cesspool of the world! Not if I had a hundred tongues, every one shouting a different language in a different key, could I do justice to her splendid chaos. The most beautiful and the most squalid, girdled with a twofold zone of parks and slums; where the keen air from lake and prairie is ever in the nostrils, and the stench of foul smoke is never out of the throat; the great port a thousand miles from the sea; the great mart which gathers up with one hand the corn and cattle of the West and deals out with the other the merchandise of the East; widely and generously planned with streets of twenty miles, where it is not safe to walk at night; where women ride straddlewise, and millionaires dine at mid-day on the Sabbath; the chosen seat of public spirit and municipal boodle, of cut-throat commerce and munificent patronage of art; the most American of American cities, and yet the most mongrel; the second American city of the globe, ...the first and only veritable Babel of the age; all of which twenty-five years ago next Friday was a heap of smoking ashes. Where in all the world can words be found for this miracle of paradox and incongruity?[6]

In the first three decades of the twentieth century the growth of Chicago continued unabated despite periodic depressions, the curtailment of European immigration during and after World War I, and the gangster era of the prohibition years, which created an image of lawlessness that the city's many accomplishments failed to overcome. During this period the population almost doubled, increasing from 1,698,575 in 1900 to 3,376,808 in 1930. By 1930 the entire city area, except for some small patches mainly on its fringes, had been occupied. In addition, especially since World War I, population was flowing increasingly into the suburbs (Figure 8-10).

Hyde Park-Woodlawn area around its grounds. As a legacy to the city, it left Jackson Park, the Midway, and the Museum of Science and Industry (the Fine Arts building of the Exposition).

The many facets of Chicago life in the 1890s were depicted by George W. Steevens, an English journalist in this vigorous portrayal:

Chicago's Century of Progress Exposition of 1933-34 celebrated a century of remarkable growth. The structures of the World's Fair were erected on artificially created land along the lakeshore from Roosevelt Road to 39th Street. Although the Fair was held in the depths of the Great Depression, it attracted over 39 million people and proved an unqualified financial success. More important, it showed the world how far the little muddy portage-town had come in one hundred years.

The Burnham and Subsequent Plans

Some of Chicago's finest features are the result of its pioneering efforts in city planning. In 1869 Frederick Law Olmsted developed the basic park and connecting boulevard system that has served Chicago so well for a century. In 1893 the World's Columbian Exposition gave Chicago a glimpse of the advantages of urban design that incorporated an orderly arrangement of structure and space. And in 1909 the renowned architect Daniel H. Burnham, who had presided over the construction of the World's Columbian Exposi-

FIGURE 8-9. World's Columbian Exposition, 1893. View is eastward from approximately Sixty-fifth Street. The building to the left is the Manufactures and Liberal Arts Building and on the right is the Agricultural Building. In the six months the fair was open it attracted 27 million visitors—the equivalent of almost half of the total population of the United States at that time. (Courtesy of the Chicago Historical Society.)

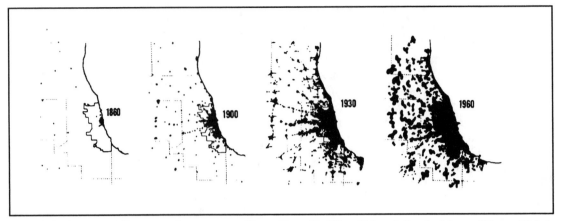

FIGURE 8-10. Metropolitan growth, 1860-1960. (From Pierre De Vise, *Chicago's People, Jobs, and Homes,* Vol. 1, Department of Geography, De Paul University, 1964.) (Reprinted by permission of Pierre de Vise.)

tion, proposed a monumental long-range *Plan of Chicago* that introduced comprehensive city planning to Chicago and the nation.

Although Burnham's plan did not adequately foresee such problems as housing and the effect of the automobile, it was bold in conception, metropolitan in scope, and comprehensive in its incorporation of the advanced concepts of the time. It stressed the "city beautiful" trend with broad avenues, parks, and civic centers. Officially adopted by the city, it has guided public improvements for many years. Although not all of Burnham's specific suggestions have been implemented, some outstanding features of present-day Chicago were either proposed or advanced by the plan, and more than a half-billion dollars has been expended in its implementation. Major recommendations and achievements of the Burnham Plan include:

1. The widening of major thoroughfares, the double-decking of Wacker Drive and part of Michigan Avenue, and most important, the creation of a regional highway system extending up to sixty miles beyond the city—a metropolitan approach that has been at least partially implemented by the expressway system.

2. The consolidation of railway terminals, largely achieved with the closing of four major stations. The Union Station was built as the plan suggested.

3. The construction of new docks and navigation facilities. In accordance with the plan, Navy Pier was built, Lake Calumet developed, the South Branch of the Chicago River straightened, and a number of bridges constructed.

4. The development of a continuous lakefront park. Twenty-five of the city's 29 miles of lakeshore are now devoted to recreational and cultural facilities. Only one of the string of contemplated off-shore islands has been completed.

5. The extension of outlying forest preserves.

An incidental but far-reaching result of the Burnham Plan was the provision for planning on a permanent basis. The plan stimulated the development of two organizations, one for the city's planning, the other for planning on a more regional scale. From these earlier agencies have evolved the city's present policy-making Chicago Plan Commission and its administrative arm, the Department of Planning Development. Serving the six counties of the northeastern part of the state is the Northeastern Illinois Planning Commission.

Subsequently, other measures were taken to ensure the orderly development of Chicago, but none of these has yet matched the effects of the Burnham Plan. In 1966 the city proposed the Comprehensive Plan of Chicago, which was later supplemented by detailed plans for each of 16 geographic areas of Chicago, including the central area. Comprehensive plans for Chicago's lakefront and downtown riverfront were also published. The recent plans have emphasized the human element—raising the quality of family life and the environment, expanding economic opportunities for the disadvantaged, and improving transportation and land use. No proposals, however, have dealt with the serious problems of segregation and white flight to the suburbs, nor have any plans helped much in the battle against crime and a deficient educational system. These plans recommended high-density communities along the lakefront, in the downtown area, and along mass transportation routes; low-density housing in the outlying parts of the city; medium density in between; and a "mix" of housing in some areas.

Major development projects for the downtown area include those over the Illinois Central land north of Randolph Street near the lakefront; the "South Loop New Town" on the railroad land south of the Loop and the adjacent "Printers Row" area; additional new construction in the area east of North Michigan Avenue and north of the river; construction on Illinois Central land south of Roosevelt Road; and scattered development in the Loop itself. These downtown area plans are included in the Chicago 21 Plan of 1973,

proposed by a non-profit organization of business and real estate people, and they are being independently implemented. While these plans for the downtown area moved ahead, improvements of the city's neighborhoods have lagged behind those of the central area. (See Figure 8-11.)

PEOPLE AND SETTLEMENT PATTERNS

Chicago's unprecedent growth resulted largely from an almost constant flow of settlers attracted by its economic opportunities. The major points of origin of the settlers changed with the passing decades. The first settlers came mainly from the eastern United States; unlike the early settlement of southern Illinois, very few of Chicago's early settlers came from the South. Then, starting around the 1840s, large numbers began to arrive first from northwestern Europe, later from eastern and southern Europe, and most recently from our South, Latin America, and Asia. In all, Chicago is an amalgam of about 80 identifiable ethnic and racial strains.

European Immigrants

By 1890, over three-fourths of Chicago's one million people were either European immigrants or the children of European immigrants, with the Germans, Scandinavians, and the Irish, in that order, being the three largest foreign-born groups. The flow of Europeans to Chicago continued unabated until the outbreak of World War I, but the geographic sources of immigration began to shift markedly about 1880. From 1880 until 1927, when national immigration quotas went into effect, the majority of the immigrants came from eastern and southern Europe—with the Poles, Italians, eastern European Jews, Czechs, Lithuanians, Russians, Greeks, Serbians, and Hungarians among the largest groups (Figure 8-12). At the peak of immigration, Chicago was the world's largest Lithuanian city, sec-

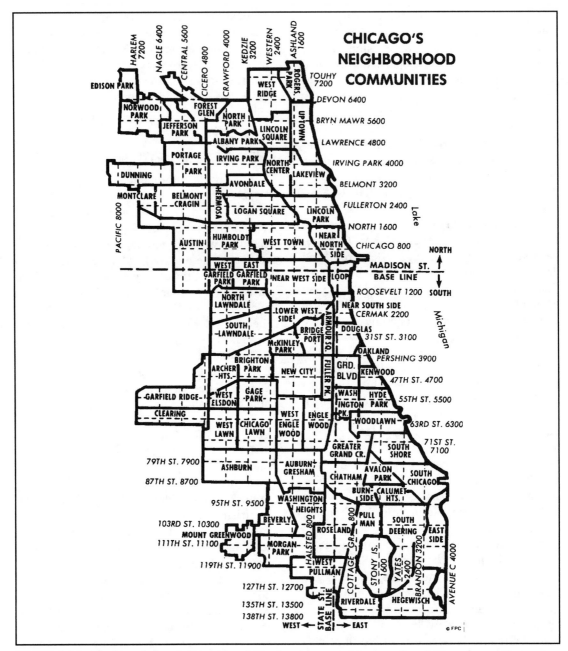

FIGURE 8-11. The boundaries of Chicago's neighborhood communities were first delineated over 60 years ago through the work of the Social Science Research Committee of the University of Chicago. Since then only minimum boundary refinements have been made, although an additional community—O'Hare—was created laragely out of land annexed in the 1950s for the airport, and recently the community of Edgewater was separated out of the northern part of Uptown, making a total of 77 communities. Over the years, most of these neighborhood communities have been changing ethnic/racial communities. (Adapted from M.S. Ratz and C.H. Wilson, *Exploring Chicago*, Follett Publishing Co., 1958.)

ond-largest Czech (Bohemian) city, and third-largest Irish, Swedish, Polish, and Jewish city.

When immigration was sharply curtailed in 1927, foreign-born whites constituted about 27 percent of Chicago's total population. This figure had fallen to 20 percent by 1940, 15 percent by 1950 and 5 percent by 1980. The number of foreign-born whites in Chicago in 1920 and in 1980 by major European country of origin is indicated in Table 8-1. These figures show that the number of foreign-born

Europeans is now comparatively small; the proportion from northwest Europe has also declined markedly.

The median age of these foreign-born Europeans is estimated at 62—foreshadowing a further decline of the group that at one time constituted a majority of Chicago's people. Their decrease also marks the decline of one of Chicago's most colorful eras—a period when much of Chicago was a microcosm of various European cultures. The city was filled with the sounds of dozens of languages, exotic dress, and myriad ethnic shops, schools, churches, synagogues, cafes, coffeehouses, and newspapers. At that time the immigrants cherished the security of their own institutions in their own neighborhoods as they worked their way upward financially and socially despite occasional hostility encountered from "native Americans," many of whom were themselves the children or grandchildren of immigrants.

FIGURE 8-12. Holy Trinity Orthodox Cathedral at 1121 North Leavitt Street was built in 1903 by the Russian community in the area. It received some financial aid from the Czar of Russia. The architect was Louis Sullivan and the stuccoed church is now an official Chicago landmark. In the tradition of the homeland, the church contains no pews. The congregants remain standing during the services. (Photograph by Irving Cutler.)

TABLE 8-1. European-born in Chicago

Native Country	1920	1980
Poland	137,611	43,338
Italy	59,215	18,593
U.S.S.R.	102,095	17,497
Germany	112,288	16,075
Ireland	56,786	8,372
United Kingdom	37,932	5,589
Austria	30,491	4,370
Czechoslovakia	50,392	3,443
Sweden	58,568	2,155

Source: U.S. Bureau of the Census

The 1990 census revealed an increase in the total foreign-born in Chicago but not in European-born. The total for all foreign-born in Chicago, which included whites, African Americans, Latinos, Asians, and other races, was 469,187 or 16.85 percent of Chicago's total population.

The Ethnic Checkerboard

The new immigrant groups usually sought housing in the congested low-rent areas around the Loop, especially on the near west side, in areas abandoned by earlier immigrant groups who had moved upward economically and outward geographically. The near west side community area has been home for a succession of groups—Germans, Irish, Czechs, Jews, and now blacks and Hispanics (Figure 8-13).

The desire of the immigrants to be close to their countrymen and to establish in their new land the institutions they had cherished in their birthplaces led to the formation of numerous ethnic neighborhoods. Some of these neighborhoods, such as the Venetian, Neapolitan, and Sicilian, were even established on the basis of ethnic subgroups. Going southward down Halsted Street earlier in this century one would pass successively through Swedish, German, Polish, Greek, Italian, Jewish, Czech, Lithuanian, and Irish neighborhoods.

On Halsted Street was Jane Addam's Hull House, which catered to immigrants who were often needy, poorly educated, and bewildered by the unfamiliar setting. Jane Addams described the conditions of the immigrant groups as follows:

Between Halsted Street and the river live about ten thousand Italians. To the south on Twelfth Street are many Germans, and side streets are given over almost entirely to Polish and Russian Jews. Still farther south, these Jewish colonies merge into a huge Bohemian colony. To the northwest are many Canadian-French and to the north are Irish and first-generation Americans. The streets are inexpressibly dirty, the number of schools inadequate, sanitary legislation unenforced, the street lighting bad, the paving miserable and altogether lacking in the alleys and smaller streets, and the stables foul beyond description. The older and richer inhabitants seem anxious to move away as rapidly as they can afford it. They make room for newly arrived immigrants who are densely ignorant of civic duties. Meanwhile, the wretched conditions persist until at least two generations of children have been born and reared in them.[7]

In time, with some acculturation and economic success, the immigrant groups and especially their offspring moved outward from their crowded islands near the downtown area, often migrating in an axial pattern (Figure 8-14). Thus, many of the Germans moved outward along Lincoln Avenue, and the Poles along Milwaukee Avenue. From their congested Pilsen colony in the Halsted-Ashland area south of 16th Street, the Czechs moved westward into South Lawndale and later into Cicero and Berwyn. Each migration outward was usually accompanied by a further loosening of Old World ties as each new generation became more assimilated, more geographically dispersed, and more active in the city's civic and economic affairs.

Recent Population Trends

The 1990 U.S. Census figures for Chicago indicated a population of 2,783,726, down substantially from the 1950 peak of 3,620,962 (Figure 8-15). This decline resulted largely from the sizable exodus of white families to the suburbs and a lower population density in the city caused by considerable redevelopment, including housing and expressway construction. (Chicago's population density declined from a peak of 17,011 per square mile in 1950 to 12,203 per square mile in 1990.)

The 1990 census data also showed that substantial changes had taken place in the

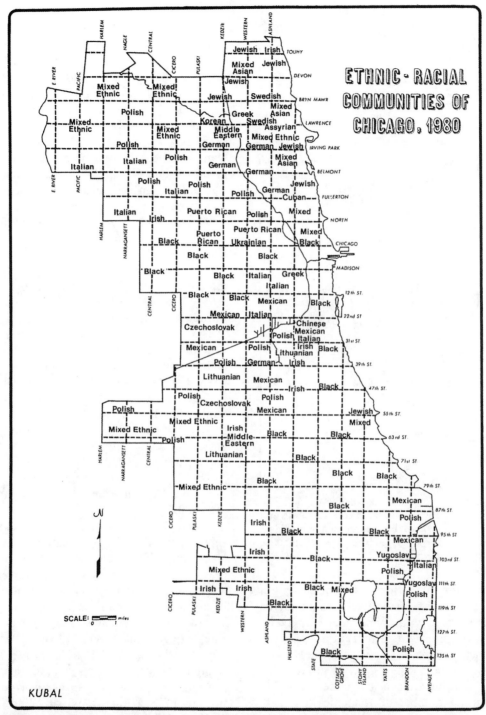

FIGURE 8-13. The approximate location of Chicago's larger ethnic and racial communities, 1980. Almost none of the communities was completely homogeneous. (Map by Joseph Kubal is based largely on U.S. Census, church, and Chicago Department of Planning data.) (Reprinted by permission of Joseph Kubal.)

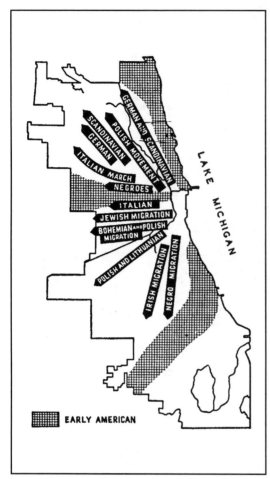

FIGURE 8-14. Outward expansion of race and nationality groups in Chicago. From *The American City and Its Church* by Samuel C. Kincheloe. Copyright © 1938 by Friendship Press, Inc., New York. Used by permission.)

composition of Chicago's population. In 1990, for the first time, blacks outnumbered whites in the city, 39 percent to 38 percent. Hispanics comprised 19 percent of the population and Asians three percent. The Hispanics have been the fastest growing group in the city, increasing from just 7 percent in 1970. The largest number of Hispanics are from Mexico, followed by those from Puerto Rico and then much smaller numbers from Cuba.

The major Hispanic groups generally live separately, with the Mexicans living mainly on the Southwest Side and the Puerto Ricans generally living on the Northwest Side (Figure 8-16). The Asian population has also grown rapidly in recent years with the arrival of sizable numbers of Filipino, Chinese, Korean, Indian, Pakistani, Thai, Vietnamese, Cambodian, Indonesian, and Arab immigrants (Figure 8-17). This changing composition of Chicago's population results from the fact that the most recent waves of migration to the city has been not as much from Europe as from the southern states, Mexico, Puerto Rico, other Hispanic areas, and most recently from Asia. The new migration was spurred by the continuing demand for labor, especially during World War II and the boom years that followed World War II.

Blacks have lived in Chicago since its earliest days. Chicago was a terminus of the underground railroad, and by the time of the Civil War there were several hundred blacks in the city. However, the percentage of blacks remained relatively small until the World I period. In 1910, blacks constituted only two percent of the population. This figure rose to seven percent in 1930, 14 percent in 1950, 32 percent in 1970, and 39 percent in 1990. The city's black population declined slightly in the 1980s.

When the number of Chicago blacks was small they were widely scattered. As the blacks grew in number they became increasingly concentrated in older, deteriorated neighborhoods of the inner city that earlier immigrant groups had abandoned. The major axis of black settlement was on the near south side, between State Street and South Parkway (Dr. Martin Luther King, Jr. Drive), with smaller black settlements on the near west side and near north side. As the black population increased during World War I, the black neighborhoods became grossly overcrowded, with all the resulting social ills.

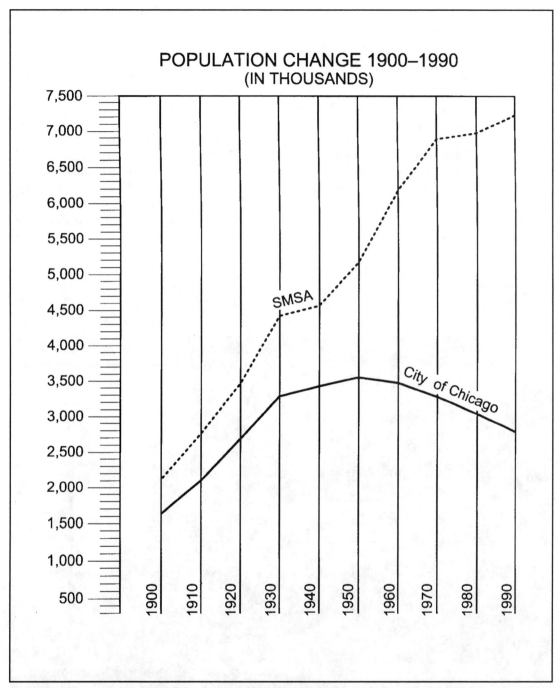

FIGURE 8-15. Chicago's population of 2,783,726 in 1990 was 837,236 less than its peak population in 1950 of 3,620,962. However, during that same period the population of the Chicago Standard Metropolitan Statistical Area (SMSA) increased 2,083,308 from 5,177,868 to 7,261,176.

The consequences were poor housing, higher rents, school segregation, loss of employment opportunities because of difficult accessibility to available jobs, and racial tension. Attempts to enlarge the ghetto were often stymied by hostility and sometimes by violence. Prejudice and restrictive real estate practices combined to confine the blacks within certain boundaries.

The large black migration to Chicago during and after World War II built up a strong housing demand. The ghetto boundaries—Cottage Grove Avenue, Stony Island Avenue, Ashland Avenue, etc.—began to give way, and on a map the black residential areas resembled an inverted *L* with the main segment pushing southward to the city limits and another segment pushing westward to the city limits (Figure 8-18). The expansion of the black ghetto has been on a block-by-block basis and has almost always been confined to the periphery of the established ghetto. Sizable parts of Chicago still have no black residents, and the city remains highly segregated but not as much as in the past.

Like the immigrant groups that preceded them, the blacks began by taking over the worst jobs and residential areas (Figure 8-19); but, unlike the earlier immigrant groups, the second and third generation blacks could not readily escape from the ghetto even when they

FIGURE 8-16. Signs of a changing neighborhood, 1982. In what once was the heart of Old Polonia at the intersection of Ashland, Milwaukee, and Division streets, a Polish travel agency still remains, but increasingly the stores cater to the growing Hispanic population of the area. (Photograph by Irving Cutler.)

FIGURE 8-17. A Korean commercial strip in the 3300 block of West Lawrence Avenue, 1980. There are numerous Korean-owned stores on Lawrence Avenue between Kedzie Avenue and Pulaski Road. (Photograph by Irving Cutler.)

could afford to—and an increasing number could. Even blacks with wealth and education were for the most part confined to the ghetto. They usually moved into adjacent neighborhoods which quickly became black as the white population fled to the city's periphery or into the suburbs. Thus, like the white settlement patterns, black residential areas in the city reflect an economic stratification. The quality of neighborhoods, homes, education level, job status, etc. generally improves as one proceeds outward from the inner city.

Although most of the suburbs have enacted open-housing laws, the percentage of suburban blacks has increased only moderately. In 1990, 76.3 percent of the blacks of the metropolitan area still lived in Chicago. Many of the suburbs still house no blacks at all, and a few have merely token integration. Less than

two percent of the west and northwest suburban population is black. Fairly sizable black populations in predominantly white suburbs are found in only a few cities such as Evanston, South Holland, Summit, Park Forest, Calumet City, Oak Park, and Dolton. Two-thirds of the suburban blacks still live in industrial satellites such as Gary, Chicago Heights, Harvey, Waukegan-North Chicago, Maywood, and Joliet; in racially changing communities such as Bellwood, Calumet Park, and University Park; and in black-ghetto suburbs such as Robbins, Phoenix, and Ford Heights.

In recent years, partly because of the impact of the civil rights movement, increasing opportunities for blacks in Chicago have been made available in education, employment, and politics. The city elected a black mayor,

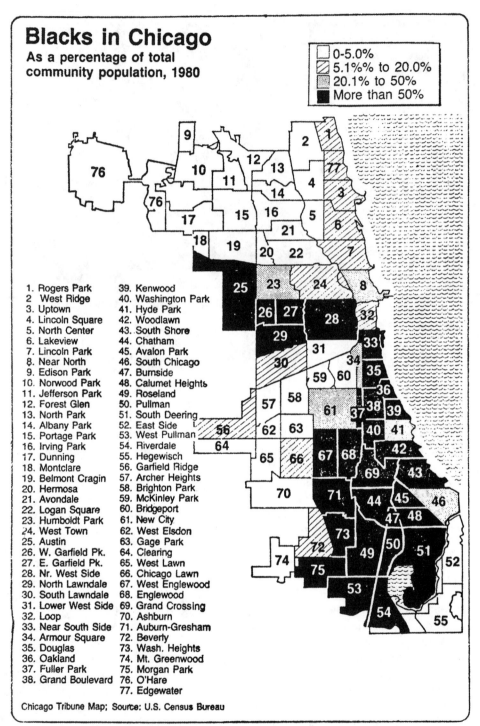

Blacks in Chicago

As a percentage of total
community population, 1980

☐	0-5.0%
▨	5.1%% to 20.0%
▦	20.1% to 50%
■	More than 50%

1. Rogers Park
2. West Ridge
3. Uptown
4. Lincoln Square
5. North Center
6. Lakeview
7. Lincoln Park
8. Near North
9. Edison Park
10. Norwood Park
11. Jefferson Park
12. Forest Glen
13. North Park
14. Albany Park
15. Portage Park
16. Irving Park
17. Dunning
18. Montclare
19. Belmont Cragin
20. Hermosa
21. Avondale
22. Logan Square
23. Humboldt Park
24. West Town
25. Austin
26. W. Garfield Pk.
27. E. Garfield Pk.
28. Nr. West Side
29. North Lawndale
30. South Lawndale
31. Lower West Side
32. Loop
33. Near South Side
34. Armour Square
35. Douglas
36. Oakland
37. Fuller Park
38. Grand Boulevard

39. Kenwood
40. Washington Park
41. Hyde Park
42. Woodlawn
43. South Shore
44. Chatham
45. Avalon Park
46. South Chicago
47. Burnside
48. Calumet Heights
49. Roseland
50. Pullman
51. South Deering
52. East Side
53. West Pullman
54. Riverdale
55. Hegewisch
56. Garfield Ridge
57. Archer Heights
58. Brighton Park
59. McKinley Park
60. Bridgeport
61. New City
62. West Elsdon
63. Gage Park
64. Clearing
65. West Lawn
66. Chicago Lawn
67. West Englewood
68. Englewood
69. Grand Crossing
70. Ashburn
71. Auburn-Gresham
72. Beverly
73. Wash. Heights
74. Mt. Greenwood
75. Morgan Park
76. O'Hare
77. Edgewater

Chicago Tribune Map; Source: U.S. Census Bureau

FIGURE 8-18. Black population in Chicago's communities, based on 1980 U.S. Census data. (©Copyrighted Chicago Tribune Company. All rights reserved. Used with permission.)

Harold Washington, in 1983. Declining birth rates and dwindling migration from the South, coupled with the continued expansion of black population into former white neighborhoods, have taken some of the pressure off some of the crowded black ghettos. By the early 1970s over a third of Chicago's non-white population had achieved "middle class" status, often because there was more than one bread winner in the family. Increasingly, such blacks were able to find improved housing, some in neighborhoods at the ghetto borders, some on a nondiscriminatory basis in such urban redevelopment or renewal areas as Sandburg Village, Lake Meadows (Figure 8-20), Prairie Shores, South Commons, and Hyde Park-Kenwood and in northside communities such as Uptown, Edgewater, and Rogers Park.

Socioeconomic Patterns

The Chicago Metropolitan Area consists of several hundred communities, of which 77 are recognized neighborhood community areas in Chicago and the remainder are suburban municipalities. The communities of the area are characterized by a great range of socioeco-

FIGURE 8-19. View eastward from the Ogden Avenue viaduct north of Division Street, 1954, showing part of the near north side area that Harvey Zorbaugh discussed in 1929 in his book entitled *The Gold Coast and the Slums.* The Gold Coast consists of the luxury high-rise apartment buildings along the lakefront, visible in the background. (Photograph by Lillian Ettinger. Courtesy of the Chicago Historical Society.)

FIGURE 8-20. Aerial view of Lake Meadows apartment community, looking northwest from south of 35th Street with the Illinois Central Gulf Railroad tracks at the lower right, 1972. The integrated community was developed under the urban renewal program by the New York Life Insurance Company on what had been one of the worst slum areas of the city. The development contains a shopping center, elementary school, public park, community center, and an office building. The high-rise apartment complex in the right background is the adjoining Prairie Shores apartment development, also integrated. More than a century ago much of this land was owned by Senator Stephen A. Douglas who lies buried in the small part just west of the tracks with a tall pillar topped by his statue capping the tomb. (Courtesy Department of Urban Renewal, City of Chicago.)

nomic diversity. This range is indicative of the diverse racial, ethnic, and educational backgrounds of the people and of the wide variation in the opportunities available to them. Per capita income in the Chicago Metropolitan Area in 1989 ranged from $4,660 annually in Ford Heights to about $70,000 annually in Kenilworth and Mettawa. In Chicago the per capita income was $12,899 in 1989. In general, the lowest income levels in Chicago are found in the overcrowded inner-city black areas on the south and west sides. Family income increases concentrically as one proceeds outward toward the more recently occupied periphery of the city—with the highest income in the peripheral Chicago communities such as Forest Glen and Beverly—and reaches its peak in the suburbs.

Both the communities of Chicago and the suburban communities vary markedly in economic status. The wealthiest suburbs are mainly to the north, especially along the lake. In addition to such physical amenities as the lake and interesting topography, wealthy North Shore suburbs have the attraction of good commuter transportation to the Loop and extensive areas free of polluting industries. The poorest suburbs are virtually all to the south of the city, with the largely black suburbs of Ford Heights, Robbins, and Phoenix having the lowest incomes.

Many of the economic elite of early Chicago lived close to the center of the city along such once fashionable streets and boulevards as Washington, Jackson, Ashland, Michigan, Wabash, Prairie, Indiana, and Calumet. With the spread of industry and immigrant groups into these areas, some of the wealthy moved farther out along north Lake Shore Drive or out into Kenwood, Hyde Park, and South Shore, and some moved into the suburbs.

In time, former high-income areas in the city deteriorated into the city's worst slums as the vacated mansions of the rich were subdivided to accommodate the poorer immigrants. However, in the post-World War II period some of these areas experienced massive redevelopment, again emerging as highly desirable locations, such as Lake Meadows, Prairie Shores, and South Commons south of the Loop and the Sandburg Village area to the north. Urban redevelopment, generally radiating outward from the downtown area, has improved inner city areas with more expensive high-rise housing. At the same time, however, the poor who lived in these areas before redevelopment have been pushed out. Once again they have crowded into what were once stable neighborhoods, creating new slums.

COMMERCE AND INDUSTRY

Chicago started as a small trading and military post. Buoyed by its excellent location in a growing area, it soon evolved into a bustling lake and river port, then into a canal terminal and railroad center, and into a major commercial and distribution center. Developing more slowly at first but generally paralleling its commercial development, Chicago's small-scale manufacturing of a century ago expanded until the city became the second-largest industrial center in the nation. Today, the Chicago area is the nation's leading producer of a variety of items ranging from snuff to steel. Moreover, about 40 percent of the nation's manufacturing and over 30 percent of the nation's wholesale and retail trade are concentrated within a 500-mile radius of the Loop.

Early Industry

The earliest industries of Chicago—milling, meat-packing, tanning, and woodworking—were closely related to the products of the surrounding fields and forests. Other industries arose in response to the need of the area's growing population for printed matter, household utensils, clothing, wagons, boat supplies, building materials, and quarry products.

Chicago early developed a lucrative symbiotic relationship with its rich hinterland. It received, processed, and distributed the products of the farm, and it produced and sent back to the farms clothing, furniture, and agricultural machinery and implements. One of Chicago's earliest major industries was the farm machinery company (the forerunner of International Harvester) established by Cyrus McCormick on the north bank of the Chicago River.

In 1856, the largest group of related firms in Chicago, 86, were in food processing, meat-packing, and industries using animal byproducts such as leather. The printing industry ranked second, with 65 firms, and next, with more than 50 firms each were the textile-garment-millinery and the building trades industries. The latter included brickyards, planing mills, lumberyards, and door and sash factories.

Certain industrial location patterns were already apparent. The area north of the river contained the fewest factories. These included the McCormick Reaper and Mower Works (Figure 8-21) and a number of breweries that clustered in the area because of its large German population. The present downtown area, south and east of the river, contained the printing industry and numerous handicraft industries, such as dressmaking, shoemaking, tailoring, and cigar manufacturing. West of the river, on the near west side, were numerous metal-using and metal-manufacturing firms.

Many of the early industries congregated along transportation routes—first along the waterways, later along the railroads. Lumber, grain, and tannery facilities were concentrated along the Chicago River. More than 500 acres of land between Halsted Street and Western Avenue along the river's south branch became the largest lumber distribution center in the world. The lumberyards, stocked from the Wisconsin, Michigan, and Canadian forests, supplied the booming home building and furniture industries of the city and the needs of the prairie farmers. Today a residual maze of tracks and lumber slips still occupy the area, but there are few lumberyards. Also, Chicago's importance as a furniture mart has declined.

In 1867, manufacturing in the Chicago area was described as follows:

At first Chicago began to make on a small scale the rough and heavy implements of

FIGURE 8-21. McCormick Reaper Works plant, 1906, on the west side along the South Branch of the Chicago River. The huge plant, as part of International Harvester, continued in operation for about another half century after this photograph was taken. (Courtesy of the Chicago Historical Society.)

husbandry. That great factory, for example, which now produces the excellent farm wagon every seven minutes of every working day, was founded twenty-three years ago by its proprietor investing all his capital in the slow construction of one wagon. At the present time, almost every article of much bulk used upon railroads, in farming, in warming houses, in building houses, or in cooking, is made in Chicago. Three thousand persons are now employed there in manufacturing coarse boots and shoes. The prairie world is mowed and reaped by machines made in Chicago, whose people are feeling their way too, into making woolen and cotton goods. Four or five miles out on the prairie, where until last May the ground had never been broken since the creation, there now stands the village of Austin, which consists of three large factory buildings, forty or fifty nice cottages for workmen and two thousand young trees. This is the seat of the Chicago Clock Factory....A few miles farther back on the prairies, at Elgin, there is the establishment of the national Watch Company, which expects soon to produce fifty watches a day....They are beginning to make pianos at Chicago, besides selling a hundred a week of those made in the East; and the great music house of Root and Cady are now engraving and printing all the music they publish. Melodons are made in Chicago on a great scale.[8]

Until the 1860s Chicago industry produced mainly for the local market and the market of the surrounding farm area. The requirements of the Civil War, however, helped to propel Chicago industry into the national market. After the Civil War, the city's industrial expansion continued unabated through an era of great technological advances and industrial consolidation. Manufacturing rapidly surpassed commerce in importance and became the dominant source of employment in Chicago.

Two of the largest industrial concentrations in Chicago in the latter half of the nineteenth century were the Union Stock Yards and the Pullman Palace Car Company. A consortium of railroad and packing interests established the former in 1865 as a replacement for the city's numerous, small, scattered stockyards. The yards were located on land that was originally outside of the city limits. Eventually they occupied about a square mile. For about a century they were the largest and busiest stockyards in the world (Figure 8-22). A peak was reached soon after World War I, when the yards employed over 30,000 people and received nearly 19,000,000 head of livestock annually. Through the years hundreds of thousands of European immigrants and later blacks from the South worked there as well. A number of future mayors worked there also.

The stockyards declined rapidly after World War II. By then they were in the congested, highly taxed geographic center of Chicago. There were labor problems, and a trend toward decentralization was taking place in the nation's meat-packing industry. Furthermore, the yard's facilities had become obsolescent; they had been designed to be served by railroads and most animals were now being shipped by truck. In 1971, after 106 years of service, the Union Stock Yards closed. On the land formerly occupied by the yards are dozens of new, small plants, virtually none of which are engaged in activities related to meat-packing.

The Pullman Palace Car Company was the foremost of the numerous railroad equipment manufacturers established in the Chicago area. It was founded by George Pullman, the inventor of the sleeping car. The Chicago area was a desirable site for the company because of its position as the railroad hub of the nation, its central location, its growing steel industry, and the availability of a vast tract of vacant land to the south of the city on the west shore of Lake Calumet. The company's operations were unique, not only in size, but in being centered in a privately constructed,

FIGURE 8-22. The Union Stock Yards, about 1905, looking west on Exchange Avenue, with sheep being driven to the slaughter pens. After more than a century of operation, the yards closed in 1971. (Courtesy of the Chicago Historical Society.)

planned, company-owned "total community" with separate residential, industrial, recreational, and commercial areas. The community, erected in the early 1880s, prospered until the prolonged and violent Pullman strike of 1894. Thereafter, under court mandate, its private character disappeared rapidly as the homes were sold to the Pullman employees. However, the basic pattern and the homes of the more than century-old, sturdily-built community are still largely intact. Although the Pullman Company's operations have long ceased there and many of the facilities are utilized by other firms, some famous landmarks remain and should continue to remain for a long time as Pullman has been declared a National Historic Landmark.

Present Industry

In this century the Chicago area's industrial growth has been so rapid that it now ranks second among U.S. metropolitan areas in the number of manufacturing employees. However, of the area's 3.25 million workers, well under a million are now engaged in the declining manufacturing base; most work in such fields as wholesale and retail trade, service, government, finance, and transportation.

Service-producing, white-color employment has gained in recent years while manufacturing blue collar employment has declined.

During this century the location, facilities, and products of Chicago's industry have changed greatly. The leading products of the pre-World War I period—meat packing, men's clothing, furniture, agricultural implements, and railway equipment—have declined in relative importance. At the same time, Chicago's industrial base has undergone widespread diversification. In addition, numerous large concerns, representing a diversity of industries, are headquartered in Chicago.

The Chicago area is a leading producer in many fields. These include telecommunications equipment, steel, metal wares, confectionery products, surgical appliances, household appliances, railroad engines and equipment, soap, paint, cosmetics, cans, industrial machinery, electrical goods, commercial printing, foods, and sports goods. To supply these industries, millions of tons of iron ore, coal, chemicals, petroleum, lumber, paper, and farm products are brought into the area annually.

Traditionally, manufacturing in the Chicago area was concentrated along the rivers and railroads, but in recent decades there has been increasing industrial development near expressways, O'Hare International Airport, and in organized industrial districts (Figure 8-23). A growing proportion of the area's industrial establishments, including numerous concerns once situated in the inner city, is now found outside of the city proper. The West and South sides of Chicago have been especially hard hit by the loss of blue-collar jobs.

Almost half of the manufacturers and well over half of the industrial jobs in the metropolitan area are now located outside the city limits of Chicago as the number of manufacturing plants in Chicago have declined from 7,330 in 1973 to 4,720 in 1993. Some suburban industries were established well before

the turn of the century, but most are of post-World War II vintage. The shift of industry to the suburbs is due to such factors as limited acreage and high land costs in Chicago as well as to the general problems that plague most large cities today—congestion, plant obsolescence, high tax and insurance rates, crime, poor schools, racial conflict, political uncertainties, labor problems, and pollution. Although the suburbs are not free from these difficulties, the problems are usually less severe. Furthermore, many suburban areas can still offer large tracts of vacant land at relatively reasonable cost—land that meets modern industry's demand for expansive one-story plants and acres of parking and landscaping. Railroad and water sites, in which Chicago excels, are not as essential to increasingly truck-oriented industry. Since 1972, the number of jobs in Chicago has declined by 11 percent while the number in the suburbs has increased 50 percent.

Chicago-area communities employing large numbers of industrial workers include Cicero, Bedford Park, Waukegan, North Chicago, Aurora, Melrose Park, Joliet, Skokie, Chicago Heights, and La Grange in Illinois and Gary, Hammond, East Chicago, and Whiting in adjacent northwestern Indiana. In the post-World War II period there was a strong movement of industry to the north, northwest, and west suburban areas, based probably on good transportation, desirable environmental factors, and minimal racial problems. More recently, lower-priced and sizable tracts of land and improving transportation in the south and southwest suburban areas have been attracting increasing numbers of industrial plants.

The industrial migration out of the city has reduced Chicago's economic base and increased unemployment among the poorly educated and the unskilled, especially the blacks and Hispanics. These groups find it difficult to take on jobs in the suburban areas far from their residences in the inner city.

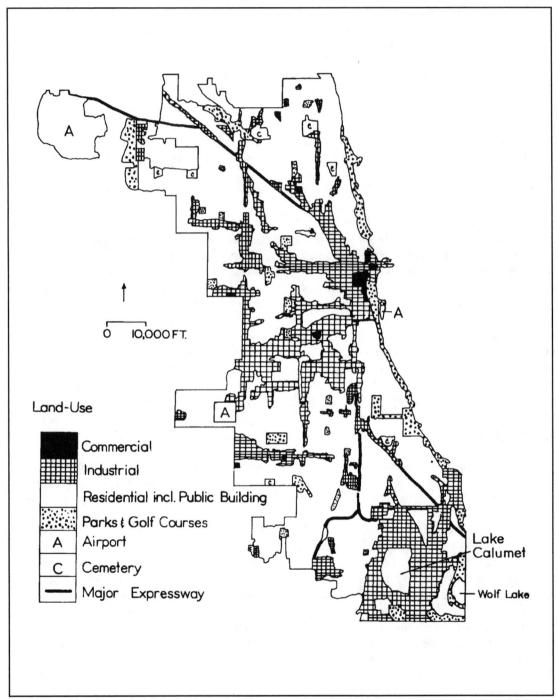

FIGURE 8-23. Land use in Chicago, 1960. Residential areas also include commercial streets. (Reprinted by permission of Ying Cheng Kiang, *Chicago*, 1968.)

Wholesale and Retail Trade

Chicago's wholesale trade generally equals its manufacturing industries in dollar volume. Chicago is a principal market for grain, machine tools, produce, fish, and flowers (Figure 8-24). Its giant Merchandise Mart and adjacent Apparel Center are leading facilities for the display of home furnishings and clothing respectively. The Chicago Board of Trade and the Chicago Mercantile Exchange are among the world's largest commodity markets. They have pioneered many new market concepts.

Because of its huge wholesale trade, its accessibility, and its accommodations, Chicago has long ranked as the nation's convention capital. More than 1,500 conventions, trade shows, and expositions are held annually. To accommodate the annual influx of well over 2 million conventioneers and buyers, Chicago offers 250 large hotels and motels, over 4,000 restaurants, and numerous exhibition facilities (including the giant and expanding McCormick Place). In recent years six large hotels have opened downtown along the river alone, with other hotels opening in

FIGURE 8-24. Randolph Street at Haymarket Square, 1892, looking westward toward Halsted Street. This area still serves as a wholesale produce market. The statue of the policeman in the right foreground commemorates the seven policemen who were killed during the Haymarket Riot of 1886. After two bombings of the statue in recent decades, it was moved into a police department building. (Courtesy of the Chicago Historical Society.)

other nearby areas. Chicago also has many museums, theaters, sporting events, night clubs, and other features of interest to visitors.

Organized Industrial Districts

Chicago has been a leader in the development of organized industrial districts and its modern version—the industrial park. The accelerating growth of such facilities has countered the previous tendency of often locating industry indiscriminately throughout residential areas, a tendency which often resulted in blight.

Chicago's Union Stock Yards and Pullman were pioneering fore-runners of organized industrial districts. One of the largest of these, the Central Manufacturing District, was begun in 1902 to develop a tract of land immediately north of the Union Stock Yards. This development proved so successful that about a dozen others, containing a total of more than 300 industrial plants, have been established by the company in the Chicago Metropolitan Area. The Clearing Industrial District was established by the railroads in 1909 in Bedford Park, just south of today's Midway Airport. It operated 10 separate industrial tracts containing over 275 plants in the southern and western fringes of the city and in some suburbs until being fragmented by the sale of industrial parks.

Today the Chicago area has over 400 industrial districts and parks. The largest is the Centex Industrial Park, begun in 1956 northwest of O'Hare Airport in an area that in recent years has experienced great industrial, commercial, and residential growth. Unlike earlier industrial districts, Centex is part of a larger complex, Elk Grove Village, which includes planned residential and commercial areas as well as industrial sections. Here more than 600 plants conform to the modern trend of well-landscaped one-story buildings with adequate parking facilities for the automobile-oriented workers. As in other districts,

the industries are mainly of a light manufacturing or service nature.

The Calumet Industrial Complex

The heavy industry of the Chicago area is confined largely to the extreme southeastern part of the city and eastward around the southern end of Lake Michigan beyond Gary to the new Burns Harbor development. In Chicago the six miles along the Calumet River from its mouth to Lake Calumet contained one of the greatest industrial complexes in the world. The river is still lined with a maze of steel plants, grain elevators, ship facilities, chemical plants, and other structures, but in many cases these facilities are closed or not fully operative.

Industry began to develop in the Calumet area more than a century ago (Figure 8-25). Here were large tracts of vacant, swampy, and sandy land, which were near plenty of fresh water, available at low cost, and strategically located near a great and growing market, but were outside of the urban areas. The lake, rivers, and railroads offered virtually unexcelled transportation.

A variety of heavy industries developed. Oil refineries concentrated around Whiting, Indiana, making it one of the largest inland refinery areas in the world. Major producers of railway equipment were scattered throughout the area. Huge soap, paint, chemical, and cement plants were constructed. Some of these plants are related to the Calumet area's main industry—steel, which is now mainly concentrated in the northwest Indiana part of the complex. The Calumet Industrial Complex is the nation's leading steel-producing area and one of the greatest steel centers in the world (Figure 8-26). It usually produces 25 to 30 percent of the steel produced in this country.

Chicago's steel industry was originally located on Goose Island in the Chicago River. Later it moved to South Chicago, where it

BIRDSEYE VIEW OF SOUTH CHICAGO—CALUMET HARBOR.

1. South Chicago Hotel.
2. Site Sinclair's Woolen Mills, & Kent, Baldwin & Co.'s Machinery Manufactory.
3. Railroad Station Buildings of Pittsburgh & Fort Wayne and Michigan Southern Railroads.
4. Location of Docks, Rolling Mills, Blast Furnaces, Elevators, Saw Mills, Etc.
5. Location of Ship Yard Dock.
6. Location of Cotton Mills.
7. Location of proposed Ship Canal to Lake Calumet.
8. Cosgrain House.
9. Office of Calumet and Chicago Canal & Dock Co.
10. South Chicago Planing Mill and Lumber Yard.
11. Lake Calumet—three miles long and navigable for vessels.
12. Congregational and Lutheran Church.
13. United States Government Engineer's Office.
14. United States Light House.

FIGURE 8-25. The Calumet River and Harbor, from an engraving originally published in 1874. (Courtesy of the Chicago Historical Society.)

grew along both the Lake Michigan shore and the Calumet River. The industry subsequently expanded into the sand dune and swamp area of adjacent Lake County, Indiana, and in the 1960s further eastward along the lake into Porter County, Indiana, where two new steel plants were built. The area's accessibility by low-cost water transportation to the Lake Superior iron ore ranges has been particularly beneficial. Today there are eight steel plants in the area with three of these plants among the largest in the nation. A number have undergone expansion and modernization. Two

large steel plants in Chicago have completely shut down in recent years due to obsolescence and a slackening of demand, partly as a consequence of foreign competition and the substitution of other materials such as plastic and aluminum.

The steel industry has been a major employer in the Chicago area, and its presence undoubtedly has aided the establishment of many other industries. However, in recent years with the closing of plants, curtailed operations, and automation, thousands of steel workers have been laid off, causing hardship

FIGURE 8-26. Aerial view looking northward from about 92nd Street and the Calumet River, 1936. The area shown is mainly the community of South Chicago. Along the lake from the Calumet River to 79th Street is the South Works of the United States Steel Corporation. Northwest of the plant, in the upper left, is the neighboring community of South Shore. (Courtesy of the Chicago Historical Society.)

FIGURE 8-27. State and Madison Street about 1905. The Louis Sullivan-designed Carson Pirie Scott store is on the right and the Mandel Brothers (later Wieboldt's) store is on the opposite side of Madison Street. Vehicular traffic includes the horse and wagon, cable car, electric streetcar, and the automobile. (Courtesy of the Chicago Historical Society.)

and distress throughout the area. The older steel mills also have brought appalling air and water pollution; dirt, grime, and congestion have blighted the areas surrounding them. Only in recent years, under public pressure, have strenuous efforts been made to improve these conditions. However, the southeast side of Chicago is the site of a number of large garbage dumps that are a continuing source of concern to the nearby residents. The area has also been suggested as the site for a third major airport for Chicago.

The Changing Central Business District

Downtown Chicago, where the city began, is now the heart of a great metropolitan area, but its structure and functions are changing rapidly. Until recent years it was relatively compact, circumscribed by physical barriers—to the east the lake, to the north and west the river, and to the south a maze of railroad facilities. Enhancing the area's importance was its position as the hub of one of the nation's greatest concentrations of rail, water-

FIGURE 8-28. (Caption on facing page.)

way, and road transportation. It also was the focal point of the highly developed internal transit system of the city.

Lake Street, just south of the river and its wholesaling activity, was the main commercial street of early Chicago. By the late 1860s, however, the principal focus of retail activity had shifted to State Street, which became the city's main retail street (Figure 8-27). Although State Street was completely gutted by the Chicago fire, it was quickly rebuilt on an even grander scale to become that "great street." The post-fire building boom helped to develop the modern skyscraper and the world-famous innovative Chicago School of Architecture which included Louis Sullivan, Dankmar Adler, John Wellborn Root, William Le Baron Jenny, Henry Hobson Richardson, and, more recently, Frank Lloyd Wright, Ludwig Miles van der Rohe, Harry Weese, Bertrand Goldberg, and Helmut Jahn. Another incidental result of the fire occurred because some of its debris was dumped into the lake along the original shoreline east of Michigan Avenue. Eventually it formed the impressive facade of downtown Chicago—Grant Park.

The commercial and manufacturing activities of the downtown area grew with the expansion of the city's population from a half million in 1880 to well over 3 million a half century later. The growth in the activities of the downtown area was accompanied by a specialization of functions. What was probably the world's most concentrated shopping district stretched for almost a mile along State Street; the wholesale produce area was situated along the main stem of the river; La Salle Street became a major financial center; Market Street was the heart of the garment district; entertainment facilities were spread along Randolph Street; and smaller enclaves contained concentrations of millinery, florist, furniture, music, and other specialty establishments. Multi-story buildings used for light manufacturing were located to the west and to a lesser extent to the north of the downtown area.

Historically, retail expansion had been slow to develop north of the river. However, the opening of the Michigan Avenue bridge in 1920 sparked a major breakout from the relatively compact downtown area. This started the development of the "Magnificent Mile" of luxury shops, hotels, and office buildings north of the Loop, a development that coincided with a major downtown building boom in the 1920s. However, from the onset of the depression until the end of World War II, virtually all major construction was brought to a standstill.

After almost a quarter century of stagnation, the completion in 1957 of the 41-story Prudential Building on the air rights over the Illinois Central tracks launched the greatest era of construction that downtown Chicago has yet experienced. The John Hancock Center soars 100 stories tall. The Sears Tower complex rose to 110 stories, to become the world's tallest building, dwarfing such

Figure 8-28. Aerial view of downtown Chicago and adjacent areas in 1975, looking northeast from approximately Harrison Street. In the lower right is the Eisenhower Expressway. The Chicago River is in the foreground before curving further north, eastward toward the lake. The five tallest buildings in the city are Sears Tower (110 stories), in the center foreground with the First National Bank Building (60 stories) behind it slightly to the right; the John Hancock Center (100 stories) is in the upper left with Water Tower Place (74 stories) just in front of it; and the Amoco Oil Building (formerly Standard Oil Building) (80 stories) is in the upper right. In the far upper right beyond the Amoco Oil Building and on the lakefront from left to right are the Central District Filtration Plant, Navy Pier, and the lock of the Chicago River Controlling Works. (Courtesy Peter J. Schulz/City of Chicago.)

erstwhile Chicago giants as the First National Bank building (60 stories), Marina City (61 stories), Lake Point Tower (70 stories), Water Tower Place (74 stories), and the Amoco building (80 stories) (Figure 8-28 and 8-29). Measured in feet, Chicago contains three of the five tallest buildings in the world.

The building boom expanded in the 1970s and 1980s especially to the north and to a lesser extent west across the river. Major multistructure skyscraper complexes arose west of the river; over the Illinois Central air rights in an 83-acre area east of Michigan Avenue from Randolph Street north to the river; and in the City-front Place development north of the river and east of Michigan Avenue. About a dozen new hotels have been built in recent years in the central area. Most of the new developments provide for well-landscaped open plazas, and a promenade area is planned for part of the riverfront. Additionally, residential and some commercial development

FIGURE 8-29. Aerial view of Chicago's lakefront and downtown area in 1974, looking northward from approximately 1900 south. Toward the center is the recreational-cultural complex of Soldier's Field, Field Museum of Natural History, the Shedd Aquarium, and the Grant Park Music Shell. At the left are the facilities of the Illinois Central Gulf Railroad and at the right is South Lake Shore Drive. The tallest buildings in the background from left to right are Sears Tower (110 stories), First National Bank building (60 stories), Amoco Oil Building (80 stories), and the John Hancock Center (100 stories). (Courtesy Peter J. Schulz/City of Chicago.)

has taken place south of the Loop on former railroad land, and there are major development plans for the area south of Roosevelt Road near the lake.

The functions of the downtown area have been changing significantly. Manufacturing and wholesale activities have declined drastically. The structures in which these activities were conducted are being replaced by office buildings and, increasingly, by tall apartment buildings on the fringe of the Loop. Some buildings, such as the John Hancock Center and Water Tower Place, combine residential, retail, and office functions (Figure 8-30). The entertainment and dining functions of the downtown area (largely nighttime activities) also have been declining, spreading into the River North, Old Town, and New Town areas to the north, and into the suburbs.

Retail growth in the downtown area has been adversely affected by the dispersal of many of the higher income families to the suburbs with their shopping malls, by proliferation of low-income, minority groups around the downtown area, by congestion and parking problems, and by the burgeoning retail activity along the Magnificent Mile of Michigan Avenue. Nevertheless, commerce in downtown has almost held its own and the area serves some 800,000 people daily who work, shop, live in the area, or attend school there. Of this number, about 500,000 are people who work in the downtown area. State Street itself, once by far the busiest shopping location in the metropolitan area, now ranks third behind Woodfield Mall in Schaumberg and Water Tower Place on Michigan Avenue. State Street has been hurt by the closing of seven major stores since the early 1980s; however, three new stores opened recently.

As the core of an ever larger metropolitan area, the downtown area is by far the front runner in many other important categories. It still has the highest concentration of daytime population within the metropolitan area, generally the highest land values, the greatest building density, and the largest array of services. The downtown area has the best accessibility for shoppers and employees, with commuter railroads and the CTA being augmented by a series of expressways that focus on downtown Chicago from most part of the city and suburbs. And the wave of redevelopment pushing outward from the downtown area should ultimately improve the environment of the surrounding area and provide better educated employees and higher income customers. The many new residential facili-

FIGURE 8-30. The Chicago Water Tower on the Magnificent Mile, Michigan Avenue, is one of the few structures to survive the disastrous Chicago fire of 1871. To the left, combining residential, retail, and office functions, is the 100-story John Hancock Center, completed in 1970. To the right is the 74-story Water Tower Place, opened in 1975, which contains an urban high-rise shopping center with many small shops and two major department stores; the 450-room Ritz-Carlton Hotel; and luxury condominiums. (Photograph by Mati Maldre, 1981.)

ties in the downtown area have already brought a substantial increase in its nighttime activity.

THE TRANSPORTATION NETWORK

Transportation has been of prime importance in the settlement and growth of the Chicago region and has provided for the spread and movement of people and goods within the area and linkages with other areas throughout the world. At first by water and wagon, then by rail and motor highway and through the air, the circulation tentacles of the growing metropolis have branched out to weave agriculture, commerce, and industry into a viable economic unit.

Chicago contains one of the greatest multilayered transportation networks in the world. In addition to being a passenger center (which includes O'Hare International Airport, the world's busiest airport), it contains an array of facilities geared to handle the millions of tons of freight used annually by its large commercial industrial base.

Water Transportation

The relative importance of the various modes of transportation has changed through the years. Water transportation dominated the early era; it has been said that Chicago was a port before it was a city. Chicago was blessed with natural waterway routes that were improved and supplemented to take advantage

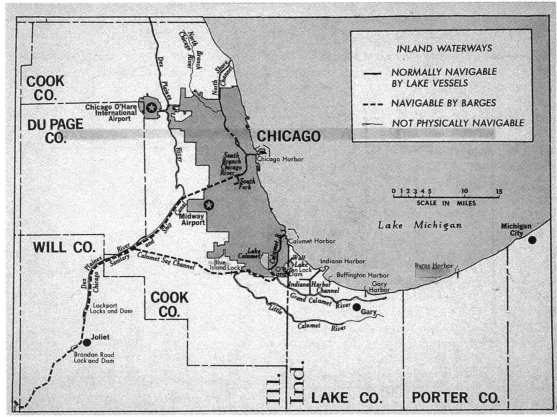

FIGURE 8-31. Harbors and Waterways of the Chicago Area. (Updated from *Mid-Chicago Economic Development Study,* Mayor's Committee for Economic and Cultural Development of Chicago, 1966.)

of its location at the junction of the major water routes (Figure 8-31). The Illinois and Michigan Canal was completed in 1848, the Chicago Sanitary and Ship Canal in 1900, and the Calumet Sag Channel in 1922. In 1959 the St. Lawrence Seaway was opened.

During the 1880s when canal and port traffic reached a peak, the arrivals and clearances of over 26,000 vessels were recorded annually for a number of seasons. For the next half century water traffic decreased, chiefly because of competition from the railroads. The once flourishing canal barges and the package freight and passenger ships on Lake Michigan largely disappeared, although on the whole the bulk industrial water traffic in iron ore, coal, and limestone for the steel mills continued to increase.

Inland waterway traffic began to revive with the completion of the Illinois Waterway in 1933. This made possible barge traffic of a nine-foot draft all the way from Chicago to the Gulf via the Illinois and Mississippi rivers as while as into a number of the Mississippi's tributaries. And although smaller ships had been operating through the old St. Lawrence Seaway for more than a century, the opening of its modern version in 1959 was an event that was to make Chicago a major world port.

The ports of the Chicago area handle river, lake, and ocean vessels and together form one of the largest ports in the country in tonnage. About a third of the tonnage consists of barge traffic on the inland waterways. Much of the tonnage is bulk cargo destined for the numerous steel mills that are now mainly in the northwest Indiana area; some is coal for the utilities, building materials, petroleum, chemicals, and grain; and a small amount is in general cargo to and from overseas ports.

The Port of Chicago accounts for about a third of the value of Great Lakes overseas trade, although the number of vessels arriving from overseas has declined markedly in re-

cent years. These overseas vessels unload autos, steel, fish, whiskey, beer, wine, olives, and furniture and load up scrap metal, machinery, farm equipment, animal and vegetable oils, hides, lumber, and a variety of food products, especially grain. Chicago's numerous waterside grain elevators store the produce of the rich hinterland and can always supply grain to complete the "topping off" of an overseas vessel.

The Port of Chicago consists of a number of harbors. The river downtown and Navy Pier, once major Chicago facilities, now handle no freight. These facilities were very limited because of neighboring traffic congestion and high land values. The South Branch of the Chicago River has some facilities for handling bulk barge cargoes and is a link in the Lakes-to-Gulf Waterway. The North Branch has a 21-foot channel to North Avenue and a 9-foot channel to the Commonwealth Edison Company plant near Addison Street; however, it has relatively little water traffic because it dead-ends in the north.

The main water traffic of the Chicago area has been concentrated in the Calumet area (Figure 8-32). The industrialized six-mile stretch of the Calumet River from Lake Michigan to Lake Calumet handles raw materials for the remaining steel plants along its banks as well as grain, chemicals, coal, and general cargo.

The harbor at Lake Calumet contains the most complete array of port services and facilities in the Chicago area. Located about a dozen miles from downtown, it is the area's major overseas port, capable of handling ocean, lake, and inland barge shipping. Grain elevators, transit sheds and warehouses, tank farms, scrap facilities, and a variety of transportation facilities enable the harbor to handle numerous ocean freighters at one time. In addition, the harbor handles barge traffic from the Illinois waterway and is a strategic point of interchange with Great Lakes over-

FIGURE 8-32. The major port of Chicago at the south end of Lake Calumet, looking northwest. At the upper left are two 6.5 million bushel grain elevators. Just beyond the lower right of the photo Lake Calumet joins the Calumet River. (Courtesy Chicago Regional Port District.)

seas traffic. The harbor's major handicaps—inadequate depth and obstructive bridge approaches—have been largely eliminated but traffic to the port continues to decline.

Two canals—the 16-mile Calumet Sag Channel and the 28-mile Chicago Sanitary and Ship Canal—are vital links in the Lakes-to-Gulf Waterway. The Calumet Sag Channel links the Calumet area waterways with the Chicago Sanitary and Ship Canal; the Chicago Sanitary and Ship Canal links the South Branch of the Chicago River with the Illinois and Mississippi rivers. The canals handle barges carrying bulk cargoes and provide scattered facilities for oil, building materials, and other products (Figure 8-33). Both were originally designed primarily to provide sewage-diversion facilities and to reverse the flow of polluted water away from Lake Michigan—Chicago's source of plentiful, low-cost

fresh water. The canals and other facilities of the Metropolitan Water Reclamation District of Greater Chicago have helped to protect both Chicago's water supply and its lakefront beaches from the type of pollution that has plagued other Great Lakes urban centers.

To the east of Chicago in Indiana are the ports of Indiana Harbor, Gary, and the new Burns Harbor Port of Indiana (in the dunes area). Their tonnage is largely bulk cargo destined for the steel mills, although they also service chemical plants and oil refineries. The Port of Indiana at Burns Harbor has been expanding rapidly and diverting freight from the Chicago ports.

The Port of Chicago has been handicapped by a nine-month shipping season, labor problems, the lack of some facilities, declining industry, inadequate promotion, the absence of a unified port authority between Illinois and Indiana, and by competitive new forms of transportation such as containerization and unit trains. The result has been a steady decline in waterway traffic in the Chicago area.

Land Transportation

The first railroad to Chicago was built in 1848, the year in which the Illinois and Michigan Canal was completed. At first the railroads were regarded as supplemental to the waterways, and many of Chicago's railroads terminated at or near the waterways of the downtown area. However, the railroads

quickly surpassed and even supplanted waterway traffic. Soon railroads led into Chicago from 27 radiating routes—and Chicago had earned the title "Player with Railroads and the Nation's Freight Handler" (Figure 8-34). Today Chicago is served by railroad companies representing about one-half of the total railroad mileage in the country, and on an average day about 35,000 freight cars—more than the combined New York-St. Louis total—are loaded and unloaded in Chicago. To facilitate distribution and interchange among its numerous industries and railroads, the Chicago

FIGURE 8-33. The Chicago Sanitary and Ship Canal at Summit, Illinois, 1961, showing oil storage tanks and an inbound tow carrying coal for the Commonwealth Edison Company in Chicago. The leading commodities handled by the area's inland waterways are coal and petroleum products. (Photograph by Larry E. Hemenway. Courtesty of the Chicago Historical Society.)

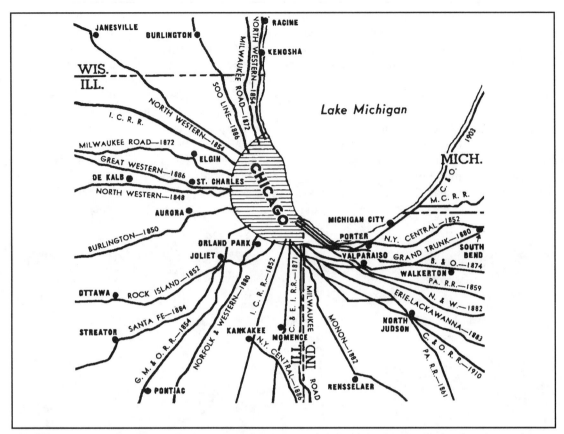

FIGURE 8-34. Historical map of Chicago's railroad network. Dates mark arrival of railroads to Chicago. (©Copyrighted Chicago Tribune Company. All rights reserved. Used with permission.)

area developed a web of a dozen intersecting belt, switching, and industrial railroads.

The railroads strongly affected Chicago's growth and national role as well as local land use and settlement patterns. First, they brought great numbers of laborers to build the roads; thereafter, they brought the permanent settlers who opened up the land. The railroads employed thousands directly, while many more worked in the manufacturing of railroad equipment. By making transportation cheaper, more reliable, and more accessible, and by offering service to virtually all parts of the country, the railroads stimulated the growth of agriculture, manufacturing, and commerce in the Chicago area. (Many railroads directly fostered industrial development.) The railroads also made possible the dispersal of population along the commuter routes into the outlying parts of the city and into the suburbs. They continue to aid the viability of Chicago's downtown area by making it readily accessible to commuters from most parts of the growing metropolitan area.

Although benefiting the city in many ways, the railroads also cut it up like a pie. This created neighborhood and traffic barriers and noise and pollution problems. In addition, the facilities of the railroads have blocked the expansion of the downtown area to the south and preempted long stretches of choice lakefront land.

Despite such technological innovations as piggybacking and the unit train, the railroads' share of intercity freight tonnage has declined in recent decades, and the once bustling intercity passenger service has virtually disappeared except for the Amtrak service. As a result, rail employment has declined sharply and many of the railroad facilities—especially the older yards, stations, and auxiliary facilities toward the inner part of the metropolitan area—are outmoded, underutilized, and available for consolidation or redevelopment, such as the Dearborn Park residential complex. More modern and less congested rail facilities have been developed in the suburban areas.

Much of the railroads' losses has been due to the competition of the automobile and the truck. Chicago has become the nation's largest trucking center, with daily scheduled service to more than 54,000 communities. Trucks now handle over one-fourth of the intercity freight tonnage.

The trucking industry has exhibited a greater locational flexibility than have other modes of transportation. Its terminal facilities have been moving generally outward, especially to the southwest, away from the former close ties with rail and water freight facilities and away from congested inner city areas with their small, poorly located facilities. The new truck terminals are usually in areas with ready access to the expressway and tollway system of over 500 miles that was superimposed in the 1950s and 1960s on the Chicago area's basic road pattern. Expressways now radiate from the center of the city in most directions (Figure 8-35 and 8-36). Connecting into and combining with the expressway system are more of the new interstate highways than are found in any other city.

Like the railroads in earlier years, the automobile and the highways are playing a major role in the location and dispersal of population and economic activities, especially in suburban areas. Settlement need no longer be aligned along railroad routes. The automobile has made huge new areas accessible for residential, commercial, and industrial uses. The expressway system, in particular, has altered area traffic patterns and affected other modes of transportation, especially mass transit.

An extensive multimodal mass transit system, consisting of commuter railroads (Figure 8-37), suburban bus lines, and the giant Chicago Transit Authority with its elevated, subway, and bus lines, has long been instrumental in establishing the urban form of the Chicago area and, by its central-area focus, in preserving the viability of the downtown area. In recent decades, the growing use of the automobile, the establishment of an expensive expressway network, the dispersal of the population into areas of lower population density, and rising costs have resulted in rapidly declining ridership and financial difficulties for the mass transit system. However, the energy and pollution crisis, renewed government interest in mass transportation because of its relative low cost and efficiency, and the creation in 1974 of the Regional Transportation Authority for the six counties of Northeastern Illinois are all examples of forces working to stabilize the steady decline of mass transportation in the area, with little success thus far.

Pipelines, although with limited visibility, are an important freight transporter accounting for about one-eighth of the freight tonnage of the area. The Chicago area has a subterranean network of 24 pipelines that carry petroleum, natural gas, and refined products.

FIGURE 8-35. Lake Shore Drive, one of the earliest prototypes of the expressway, looking south toward downtown from Lincoln Park, 1974. In the foreground is Belmont Harbor; some of Chicago's numerous beaches are visible along the lake shore. (Courtesy City of Chicago.)

EXPANSION OF THE CHICAGO METROPOLITAN AREA

The greater Chicago Metropolitan Area is a mosaic of eight counties and hundreds of fiercely independent, highly competitive, and immensely variegated communities (although often centerless), stretching approximately from the Wisconsin state line into northwestern Indiana. The area has a total population of about 8 million, with the city of Chicago containing about 35 percent of this number. And while Chicago continues to lose population, the sprawling suburban "Outer City" now has more people, more jobs, rising political power, and a far greater growth potential than the struggling, but still powerful, metropolis.

The 1990 census showed that while the city of Chicago had a net loss of 221,346 people, or 7.3 percent of its population, between 1980 and 1990, the remainder of the Chicago-Northwestern Indiana Standard Consolidated Area (which includes the Illinois counties of Du Page, Kane, Lake, McHenry, and Will, as well as Cook County outside of Chicago and the Indiana counties of Lake and Porter) had a net gain of 341,987 people or 7.2 percent.

FIGURE 8-36. The Kennedy Expressway during the afternoon rush-hour, 1975. In the median strip is the CTA Rapid Transit and on the overpass is a Chicago and North Western Railway double-deck commuter train. (Courtesy Chicago and North Western Railway.) Chicago and North Western Railway runs the trains and maintains the tracks under contract by Metra. Courtesy of Union Pacific Railroad.

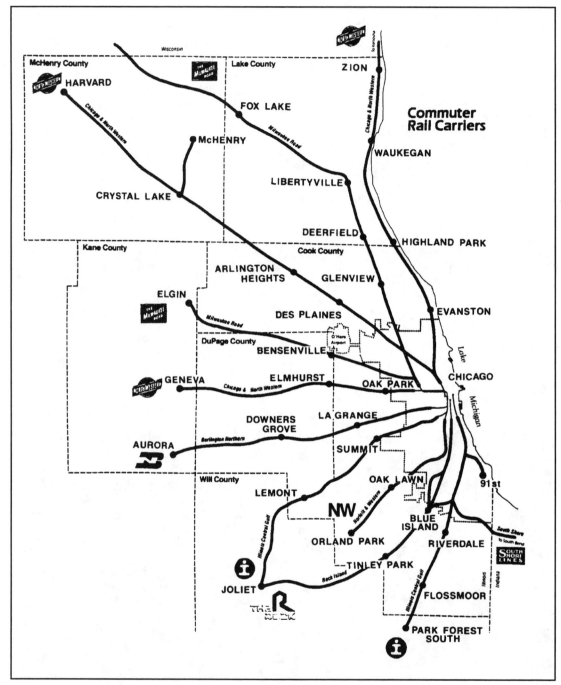

FIGURE 8-37. The commuter rail carriers of the Chicago area, 1981. Ridership on the commuter railroads increased 23 percent in the decade of the 1970s, reaching 80,800,000 riders in 1980. (Modified from a Regional Transportation Authority map.)

Similar trends were evident between 1970 and 1980, when Chicago experienced an 11 percent decrease in population, whereas the population of the remainder of the area increased about 12 percent. Much of the population loss of Chicago during the two decades was of white families with school-age children moving to the suburbs. The movement from the suburbs to the city was comparatively small; it consisted mainly of young adults without families and older adults whose children were grown.

Although a few suburbs are almost as old as Chicago, rapid suburban growth around the city is largely a phenomenon of the post-World War I period. Chicago itself is, in fact, largely an amalgamation of suburbs that in years past found it advantageous to be annexed to the city which could more readily supply needed services. Suburbs annexed by Chicago make up a sizable part of its communities: Lake View, Hyde Park, Jefferson Park, Washington Heights, West Roseland, Rogers Park, West Ridge, Norwood Park, Austin, Edison Park, Morgan Park, Clearing, Mt. Greenwood, etc. In the last half century, however, annexation of suburban areas by Chicago has been minimal, as the remaining suburbs have preferred not to become politically attached to the big city with its growing problems.

The growth of Chicago's suburbs has been closely related to the development of transportation facilities. Some early communities were established along the waterways; later, ribbons of suburbs developed along the railroads; more recently, with the advent of the motor vehicle, suburban settlement has become extremely diffuse.

Before the coming of the railroads, there were a few water-oriented communities and a number of small farm-service villages outside of Chicago. The railroads enabled people for the first time to live some distance from downtown Chicago and still commute there to work. As the railroads radiated out from downtown, they located stops every few miles. Around these stops, homes and often a few shops were built and the nucleus of a suburb developed. More than a century ago, settlements resembling widely-spaced beads on a string had been established along the main commuter railroads (Figure 8-38). Along the present Northwest Line of the Chicago and North Western Railway were the communities of Evanston, Wilmette, Winnetka, Glencoe, Highland Park, and Lake Forest. These communities usually numbered only a few hundred people because the inhabitants had to live close to the local railroad station (Figure 8-39).

The increasing use of the automobile resulted in a rapid increase in suburban growth, especially after World War I. The growth rate was substantially slowed during the depression of the 1930s and the World War II period, but thereafter it rose at an unprecedented rate. At first, the automobile chiefly allowed people to reside further from the commuter railroad stations; later, following improvements in roads, it allowed them to commute to work independently of the railroads. The result was a surge of population outward from the city into the areas between the railroad radials.

There are numerous examples of rapid suburban growth. To the north of Chicago, Skokie was a small village of 783 people in 1920 with a truck-farming economy serving the Chicago market. The general exodus to the suburbs and improved transportation—especially the opening of Edens Expressway in 1951—have resulted in an increase of its population to about 60,000. To the south of Chicago, the planned community of Park Forest, which was farmland until the late 1940s, now has a population of about 25,000. Similarly, Oak Lawn, to the southwest, increased from 3,483 in 1940 to over 56,000 today, and Addison to the west, increased from 819 in 1940 to over 32,000 today.

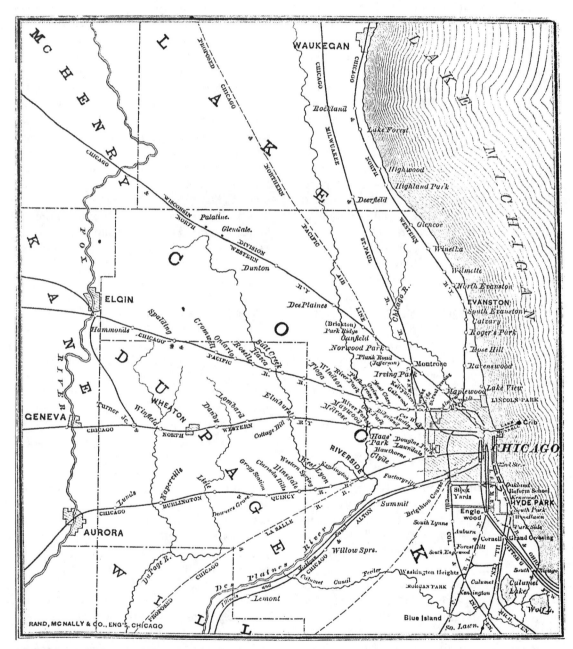

FIGURE 8-38. Chicago and vicinity, 1873. (From a suburban real estate promotion map. Courtesy of the Chicago Historical Society.)

Showing especially rapid growth in recent years, Bolingbrook has grown from 7,000 in 1970 to 41,000 in 1990 and Schaumburg has grown from 18,000 in 1970 to 69,000 in 1990.

There are many reasons for the decline in the population of Chicago and the growth in the suburban population (a pattern which is common across the nation). The central city

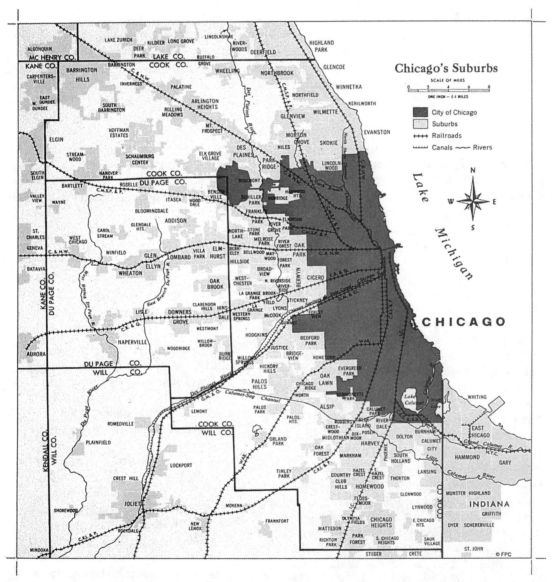

FIGURE 8-39. Originally Chicago's suburban growth was largely aligned along the railroad routes radiating out of the city. With the advent of the motor vehicle, the land between the railroad routes also became urbanized. (Reproduced by permission from M.S. Ratz and C.H. Wilson, *Exploring Chicago*, Follett Publishing Company, 1958.)

has no space for further growth; indeed, recent expressway construction and slum clearance have decreased the population density of Chicago. In addition, many inhabitants of the central city move to the suburbs to escape the city's negative aspects—racial conflict, slums, crime, high taxes, high land values, congestion, pollution, poor schools, and other problems.

The last four decades were, on the whole, periods of prosperity. Automobiles, highways, and homes were built at an unprecedented rate. The "open space, good life" attractions of the suburbs, some real and some imaginary, held strong appeal for young and growing families. Improved transportation, higher standards of living, increased leisure, industrial decentralization, government financial aid, and mass construction of homes all helped accelerate the movement to the suburbs.

In some instances, migration to the suburbs is undoubtedly a response by whites to the approach of blacks; in other cases, it is a flight of the wealthier classes from the poorer classes or of the more "Americanized" from ethnic groups. The smallness of most suburbs makes possible a more homogeneous grouping of people with their own "kind" than would be possible in the big city. Even housing may take the form of look-alike structures put up by a single developer in a single effort.

Although the suburban growth rate slowed in the 1970s and 1980s, largely urbanized stretches now radiate outward from the Loop, especially along the commuter rail lines. In some directions the built-up areas now extend about 40 miles to the vicinity of Waukegan, Barrington, the Fox River Valley cities, Joliet, University Park, Gary, and even beyond these areas in many instances. Between the railroad radials, urbanization has been filling in rapidly; there, however, the distance from Chicago of urbanized areas is usually not as great and the movement outward is often discontinuous. Patches of open space

occur because of "leap-frogging" by developers who decide to get at more distant land because it is more readily available and/or has certain amenity and cost advantages.

Suburban Characteristics

The suburbs of Chicago are far from uniform. They may differ from each other as much as do individual communities within the city. In fact, parts of Chicago are more suburban in character than some suburbs, and some suburbs are more densely populated than the average Chicago community. Chicago's suburbs are rich and poor, new and old, planned and unplanned, white collar and blue collar, industrial and dormitory, close-in and far-out, populated by young and old, successful and unsuccessful, with and without previous community tradition, homogeneous or heterogeneous in population—and everything in between. They usually differ markedly in race, religion, and ethnic backgrounds as well as in education and vocational level. In economic level, they ranged in per capita income in 1989 from under $5,000 to over $70,000.

Despite their differences, the suburbs exhibit certain general patterns and trends:

1. Their population density is lower than that of Chicago, as suburban homes generally occupy larger acreage. Chicago's population density is now about 12,000 per square mile (declining toward the periphery); in the suburbs the population density averages about 5,000 per square mile, although there is a substantial variation due to differences in zoning laws and their enforcement. A community's zoning laws control residential density and reflect the attitudes and socioeconomic status of its inhabitants, the transportation facilities, and—an increasingly

important consideration—the availability and cost of land.

2. The population density of the suburbs generally decreases as their distance from Chicago increases. Some of the older, larger, close-in suburbs, built largely in an earlier period and with good rapid transit connections to Chicago—Evanston, Oak Park, and Cicero for example—have a relatively high population density of between 9,000 and 12,000 per square mile. Some of these mature suburbs, like Chicago itself, are deteriorating around their commercial cores. In a way, such suburbs are simply extensions of the city, with population density decreasing outward from the city.

3. There is an accelerating trend toward building multiple-dwelling units in some suburban areas. In the 1960s alone some 150,000 such units were constructed. Such construction helps to circumvent high land costs, adds to local tax revenue, and better meets the needs of some families. The rapidly rising cost of single-family homes has priced them out of the range of many families.

4. Another trend is the development of planned, more self-sufficient communities embodying separate, compatible locations for residential, commercial, industrial, and recreational functions. Elk Grove Village near O'Hare Field and University Park to the south are prime examples. Other suburbs are reserving areas for industrial and commercial use in order to broaden their tax base. Some smaller planned developments are being built around recreational facilities, including artificial lakes.

5. The decentralization of industry and commerce is enabling larger numbers of suburbanites to work in the suburbs. The proportion varies somewhat with income and occupation. Thus, the suburbs with the highest proportion of rail commuters to Chicago are high-income, white-collar, residential communities such as Winnetka, Glencoe, Western Springs, Hinsdale, and Homewood. Conversely, industrial communities that are large consumers of labor, such as Whiting, Melrose Park, Northlake, Bedford Park, and the more distant satellite cities of Aurora, Elgin, Joliet, North Chicago, and Waukegan which form an arc about 40 miles from the Loop, supply few commuters.

6. Much more than Chicago, the suburban area is strongly oriented toward the automobile. The growing expressway network, in particular, has extended and opened areas for residential, commercial, and industrial development while sometimes reducing commuting time. There is very little effective public transportation for the newer commuting patterns of diffuse intersuburban movement and reverse commuting from Chicago to the suburbs as compared to the old, traditional journey from the suburb to Chicago's center. The Regional Transit Authority for the six counties of Northeastern Illinois has brought about a little of the public transportation needed to cope with the new patterns.

7. In general, the suburbs have a substantially higher median income and higher educational and professional job levels than the city of Chicago. The suburbs have a much smaller foreign-born and black population. Chicago's black population has increased rapidly in recent decades, until it now comprises about 39 percent of the city's population; during the same period, the number of blacks in the suburban area has increased slowly but is still well under 10 percent of the total suburban population. However, most of the blacks are concentrated in only about 20 of the city's 300 suburbs. Most of the Hispanics are concentrated in a few of the larger cities such as Cicero, Aurora, Elgin, Joliet, and Waukegan. Another developing socioeconomic characteristic is that the newer, more distant suburbs have a younger population than the older, closer suburbs.

8. The growth trends in the suburban areas vary widely, with suburban growth slowing down somewhat in the last decade, as is shown in Table 8-2.

9. While suburban living offers many advantages, it may also have certain drawbacks. These may include poor public transportation, long commuting distances, and inadequate facilities such as for sewage and water, especially if the suburb is new or has grown rapidly. There may also be a shortage of schools, churches, and shopping facilities. School taxes may be heavy since the suburbs, particularly the newer ones, usually contain a high proportion of young married couples with growing families. As the children grow up and leave for college, take jobs elsewhere, and settle in newer communities farther out from Chicago, the suburb may decline in population and be saddled with an over-abundance of deteriorating facilities. At the same time they may lack facilities for the elderly, including low-cost housing.

10. Although suburbia still has fairly large areas of open land, especially toward the fringes, the supply is decreasing steadily. The Northeastern Illinois Planning Commission in 1968 approved the "finger plan" to guide future development and preserve some open space in its six-county area. This plan called for the orderly development of the area along the major transportation routes, or "fingers," radiating out of Chicago. Under the plan, communities would concentrate industrial, commercial, educational, medical, and other service facilities, as well as high-density residential areas, including skyscraper complexes, along the corridors of the main transportation routes. More sparsely populated residential areas would be placed farther out, although close enough to transportation facilities. In the wedges between the corridors would be open space and recreational facilities that also would serve to absorb pollution. The plan was designed to maximize the use of land transportation facilities and to prevent chaotic, congested, and polluted urbanization. It would enhance

TABLE 8-2. County Population, 1970-1990

County	1970	1980	Change 1970-1980	% Change 1970-1980	1990	Change 1980-1990	% Change 1980-1990
Cook	5,493,766	5,253,190	-240,576	-4.4%	5,105,065	-148,123	-2.8%
Du Page	490,882	658,177	167,295	34.1	781,666	123,489	18.7
Kane	251,005	278,405	27,400	10.9	317,471	39,066	14.0
Lake	382,638	440,372	57,734	15.1	516,418	76,046	17.2
McHenry	111,555	147,724	36,169	32.4	183,241	35,517	24.0
Will	247,825	324,460	76,635	30.9	357,313	32,853	10.1
Lake (Indiana)	546,253	522,965	-23,288	-4.3	475,594	-47,371	-9.0
Porter (Indiana)	87,114	119,816	32,702	37.6	128,930	9,116	7.6
Suburban Cook	2,124,407	2,248,118	123,711	5.8	2,321,341	73,223	3.2

Source: U.S. Census

accessibility by centralizing transportation facilities in the corridor. Due to various economic and political pressures, however, the plan has not been implemented, nor does effective large-scale implementation seem likely in the foreseeable future as urbanization continues to spread almost unabated. In 1976 the plan was updated to include more emphasis on investing in existing cities such as Chicago, the suburbs, and the satellite cities (where the infrastructure is in place but often underutilized as the population declines) rather than continuing encroachment on more distant farm areas.

11. The suburban areas of Chicago can be divided broadly into north, west, and south sectors. These sectors differ in their rate of growth, economic and social status, and industrial development. Within each sector there also are noticeable, though usually less pronounced, differences.

North Suburban Growth Patterns

The north suburban area has grown more rapidly since World War II and also ranks higher economically, socially, and educationally than either the western or southern suburban areas (Figure 8-40). It contains the majority of the communities with the highest income in the Chicago area. North of Chicago there are very few lower income suburbs to supply the blue collar workers who are badly needed by local industry. The very small black population is concentrated largely in Evanston, Glencoe, Waukegan-North Chicago, and Zion, with the other northern communities having very few or no blacks.

Population has long been aligned along the three commuter railroads in the area with a more recent rapid buildup between the railroad lines. In addition to well-established transportation facilities, the northern sector has such aesthetic attractions as

1. The numerous small lakes and ponds of the inland lake area to the northwest, which developed first as a recreational area and later as the site of permanent homes
2. Rolling topography, such as that to the northwest around Barrington
3. Most important, the Lake Michigan shoreline, which is further enhanced by the picturesque North Shore Ravines from Winnetka to Waukegan.

This long-settled lakeshore area, with its attractive physical features, good rail transportation, and freedom from obnoxious industry, contains more high-income suburbs than any other sector of the Chicago area, including five of the top six—Mettawa, Kenilworth, Winnetka, Barrington Hills, and Glencoe.

The lakeshore suburbs—some in existence for more than a century—have now largely achieved maturity, and their prospects for population growth are very limited. But the suburbs just to the west of the lakeshore suburbs, such as Skokie, Glenview, Northfield, Northbrook, and Deerfield, have grown very rapidly since World War II because of the availability of land, the opening of the Edens Expressway and Tri State Tollway, and their proximity to the prestigious shoreline suburbs.

Aligned northwest in the corridor of the Chicago North Western Railway is another group of long-established suburbs—Park Ridge, Des Plaines, Mt. Prospect, Arlington Heights, Palatine, Barrington, and Crystal

FIGURE 8-40. Fountain Square in downtown Evanston about 1930. City Hall is the towered building on the left, and the Marshall Field store, one of the first suburban branches of a downtown Chicago department store, is in the center background. (Courtesy of the Chicago Historical Society.)

Lake—populated by families of a generally more modest income than that of the North Shore families. These suburbs have experienced rapid growth in recent years. The growth of northwestern communities near O'Hare Field such as Schaumburg, Elk Grove Village, Hoffman Estates, Hanover Park, and Streamwood has been especially rapid.

Light industry has expanded rapidly in the northern area, often having moved from Chicago. Much of this industry is in the growing fields of electronics, chemicals, and pharmaceuticals. Research and office facilities also are increasing rapidly. There is virtually no heavy industry in the northern sector. The major north and northwest suburban centers of employment are Evanston, North Chicago, Waukegan, Skokie, Niles, Morton Grove, Northbrook, Des Plaines, Elk Grove Village, Arlington Heights, Schaumburg, and Hoffman Estates.

Urbanization to the north has now pushed deeply, but by no means solidly, into Lake County. Much of the northern third of the county to the Wisconsin line is still largely rural, but here and there new developments are appearing. It is one of the most heterogeneous of counties, both physically and in population. Physically, Lake County ranges from the rolling topography and numerous lakes of the west to the sandy beaches along Lake Michigan in the east. It has lower income communities such as North Chicago, Round Lake Beach, Zion, and Park City and very high income communities such as Mettawa, Riverwoods, Lake Forest, and Highland Park.

West Suburban Growth Patterns

The suburbs west of Chicago are more heterogeneous than those suburbs to the north and northwest. Because of Chicago's narrow east-west width, this area includes the suburbs closest to the Loop (in some places only about six miles away), such as Cicero, Berwyn, and Oak Park. These are old, sizable, mature suburbs that have had very little room for growth in recent decades (Figure 8-41). While Berwyn and Oak Park are mainly residential, Cicero is the largest industrial employer of this western area. Oak Park, somewhat like Evanston to the north, is a community that shares characteristics of both the big city and suburbia, and also like Evanston, it has a large number of apartment buildings. Oak Park also is noted for its many homes designed by Frank Lloyd Wright.

At the far west, almost 40 miles from the Loop, is a string of satellite communities aligned along the Fox River in extreme eastern Kane County. The larger of these communities (from north to south) include Carpentersville, Elgin, St. Charles, Geneva, Batavia, and Aurora. These old residential-industrial communities were originally established largely because of their location along the river. Their sizable industrial base, especially that of such cities as Elgin and Aurora, combined with their distance from Chicago has traditionally made for self-sufficiency in employment and, thus, relatively little commuting to Chicago. Only in recent years has the frontier of urban expansion from Chicago reached into the Fox Valley and even beyond in some areas.

Between the suburbs on the periphery of Chicago and the Fox Valley suburbs is an area that has grown rapidly in the post-World War II period—western Cook County and Du Page County. Western Cook County is almost entirely urbanized. It contains a number of industrial-residential communities with very large industrial employment such as Melrose Park, La Grange, Bedford Park, Franklin Park, Bellwood, and Maywood. River Forest, Western Springs, and Riverside are attractive high income suburbs. Riverside was one of the nation's first planned communities, having been laid out in 1866 by Frederick Law Olmsted. Of the suburbs in western Cook

FIGURE 8-41. View eastward along Lake Street from Harlem Avenue in the central business district of Oak Park, in the late 1920s. Marshall Field store is on the left. (Courtesy of the Chicago Historical Society.)

County, blacks are in the majority in Maywood, Bellwood, and Broadview. However, Oak Park's black population, spilling over from the West Side of Chicago, has increased substantially in the recent years, and blacks now constitute about 18 percent of its population.

Du Page County, due west of Chicago, is rapidly urbanizing. In 1950 far more than half of its land was still devoted to agriculture, and it had a population of just 154,599. Today its agricultural land has virtually disappeared, and its population in 1990 was 781,666. Between 1970 and 1980 the population of Du Page County rose 34.1 percent, by far the largest percentage increase of any county in the Chicago Metropolitan Area. Its population increased another 18.7 percent in the 1980s.

Du Page County is largely residential, with a relatively high income status, especially in communities such as Oak Brook, Wayne, Hinsdale, and Burr Ridge. Median family income, the value of homes, and median school years completed are higher in Du Page County than in any other county of the Chicago Metropolitan Area. Du Page County has no sizable industrial communities but a number of its communities engage in some manufacturing; Naperville, West Chicago, and Addison are the largest, although relatively small industrial employers. The Fermi National Accelerator Laboratory Area is in the southwestern corner of the county. Two giant regional shopping centers, Oak Brook and Yorktown, have been developed in the eastern part of the county near the East-West Tollway. Also in the vicinity of the East-West Tollway, especially in Naperville and Oak Brook, are numerous office and research complexes of some of the largest corporations in the country. Once commuters to Chicago, the majority of Du Page residents now work within their own county.

The population is still predominantly aligned along the two major commuter railroads, although it has diffused throughout the county. Blacks constitute only about one percent of the county's population. This may be traceable to early restrictive practices, high home values, and the scarcity of industrial employment opportunities.

South Suburban Growth Patterns

The southern sector of suburbs forms a large arc stretching from Will County in the southwest, through southern Cook County in the south, into Lake and Porter Counties, Indiana, in the southeast. Although the very size of this sector allows for great diversity, in general it contains the suburbs with the lowest value of homes, the lowest levels of schooling, the greatest proportion of blacks, and the greatest amount of heavy industry. It also has a greater growth potential than that of the other sectors.

The area southwest of Chicago's border to Joliet in Will County was at one time one of the slowest growing suburban segments of the metropolitan area. It was handicapped both by its distance from the Loop and by what is probably the poorest transportation to downtown Chicago of any sector. The recent opening of several expressways plus the existence of one of the few remaining suburban areas with plentiful supplies of relatively cheap land (Will County, for example, was still almost three-fourths in farmland in 1970 but grew 44 percent in population between 1970 and 1990) has helped it become one of the fastest growing areas around Chicago. The area also has the advantages of aesthetically attractive topographic and moraine features such as in the Mount Forest-Palos area, location along the Calumet-Sag Channel and the Chicago Sanitary and Ship Canal, and good

employment opportunities at such nearby industrial centers as those located at Bedford Park (the Clearing Industrial District), Joliet, Alsip, and the East-West Tollway corridor. Because homes are relatively moderate in price, a large number of blue-collar workers have moved in, many having left the south side of Chicago after blacks moved into their neighborhoods. As yet, relatively few blacks live in the southwest suburbs. Among the fastest growing suburbs to the southwest are Orland Park, Tinley Park, Oak Forest, Hickory Hills, Romeoville, Bolingbrook, and Oak Lawn.

Due south of Chicago in southern Cook County, spilling over into Will County in an area served by the Illinois Central and Rock Island Railroads, are a large number of small and medium-sized communities. Most are older, mature suburbs, but a number developed after World War II. A few communities, such as Blue Island, Riverdale, and especially Harvey and Chicago Heights, are highly industrialized. Many communities combine industrial and residential properties and contain populations with a low-to-moderate income range and educational level. The suburbs of Robbins, Ford Heights, Phoenix, Dixmoor, Harvey, and Posen, for example, are among the poorest in the Chicago area. However, there are a few notable exceptions, especially the residential suburbs of Flossmoor and Olympia Fields. In income, these two commuter communities along the Illinois Central Railroad rank among the top suburbs in the Chicago area. The area also includes two well-planned post-World War II communities, Park Forest and University Park.

After a long period of slow growth, this area also has been developing more rapidly because of construction of new expressways, the availability of cheap land, and the mitigation of some drainage problems in the eastern section. Its growth also has been furthered by an influx of blacks continuing their southward movement from the heart of Chicago. Many were initially attracted by jobs in heavy industry, especially in the numerous steel plants nearby. Despite the fact that the south suburban area probably contains the largest percentage of blacks of any suburban sector around Chicago, most of its communities have few, if any, blacks. On the other hand, a few communities contain large percentages of black residents, with Ford Heights, Phoenix, and Robbins being almost completely black.

Across the state line in adjacent Lake County, Indiana, the area exhibits characteristics quite different from those in other sectors around Chicago. This is an area of very heavy industry, steel in particular and oil refining and chemicals to a lesser extent. The area fronts on Lake Michigan and contains part of the Indiana Dunes. Surface drainage is poor and severe air and water pollution problems have been precipitated by the heavy industry of the area. Northern Lake County contains a number of good-sized industrial cities—Hammond, Whiting, East Chicago, Gary—some large enough to be considered metropolitan areas. Gary, with a population of 116,646, down from 175,415 in 1970, and Hammond, with a population of 84,236, down from 107,790 in 1970, are the second and third largest cities of the eight-county metropolitan area. Unlike more rural southern Lake County in which growth is underway, this is an area of industrial stagnation and, in many instances, of significantly declining population. All four of the above cities lost population in the 1960-90 decades. As a whole, the area ranks low in median income, home values, and educational levels. There are very large black communities in Gary and East Chicago; in Gary most of the population is now black. A substantial Hispanic population also lives in Northern Lake County. The community of Merrillville, established in the 1960s less than 10 miles south of Gary, has

grown rapidly to a present population of 27,000, many of whom are whites who had moved from Gary.

Farther east is still sparsely populated, but growing, Porter County, Indiana. In 1990 it had 128,932 people as compared to Lake County's population of 475,594, which has been steadily declining. Porter County contains only small cities. Its growth, however, has been accelerating largely as a result of the construction in the 1960s and 1970s of two huge steel mills and Burns Harbor on the lakeshore east of Gary. The two largest cities in the county are the 1959 incorporated community of Portage with 29,000, and the county seat and university center of Valparaiso with a population of 24,000. Porter County contains Indiana Dunes State Park. In addition, after years of bitter conflict ending in the enactment of compromise federal legislation, a national dunes park was established there. In this struggle the conservationists, who wished to preserve more of the scenic dunes for the benefit of the burgeoning population of the area, were opposed by commercial and industrial interests desirous of stimulating economic growth by taking advantage of the area's excellent location for industry.

The Rural-Urban Fringe

The farm acreage around Chicago continues to decline as urbanization progresses outward from the central city. The once productive truck and dairy farms of Cook County have virtually disappeared. The same process is accelerating in the adjoining counties, although most of them still have some of their land in farms. In fact, in Kane and McHenry Counties over half of the land is still in farms.

The change of an area from rural to urban generally follows a pattern with accompanying problems. First there is an influx of nonfarm rural residents.

They come to the countryside for cheaper land and lower taxes, sunshine and open spaces, room to relax and enjoy outdoor living, a safe and healthy environment in which to raise their children, a place to grow a garden and perhaps a few chickens, a refuge during misfortune, and a home in their declining years.

The change in the rural scene often begins when a farmer sells off a front lot or two or perhaps a front tract or an acre or so. The price the farmer received was high compared with the value per acre of his farm as a whole. Because of this, other farmers are induced to sell off their frontages. More houses follow. Later, entire farms are broken up into tracts and subdivided. The change continues and as the years go by, country roads begin to look like residential streets. As the non-farm population grows, land is bought for gas stations and roadside stores and shops, then for other businesses and industrial uses.[9]

If the area is unincorporated and without adequate government regulation, new construction may be substandard, a honky-tonk atmosphere may be created, and the natural environment may be defaced.

This intrusion of non-farm rural residents often creates problems that virtually force farmers to sell out. Whereas an eighty acre farm may contain one family, an eighty acre subdivision may hold 120 families. Without adequate advance planning and zoning, the farmers may find their schools suddenly crowded, new problems of water, sewage disposal and drainage, increased traffic and higher taxes. Often there is a time lag before the subdivision assumes its full tax load, and the farmer may be assessed at land values similar to those of the subdivisions. Highway relocations may divide his farm. There may be zoning disputes.

The aggressive farmer may feel himself hemmed in as far as expansion is concerned.

He may even have to put up fencing against trespassers. He may have to use his land more intensively in order to pay the higher taxes. Necessary farm services are often curtailed as an area becomes more urbanized, and a farmer so isolated may encounter difficulty, for example, in arranging for a truck to collect his milk. The result of these problems, plus the attraction of high land prices, is that the farmer sells his land.[10]

In this way distant "exurbia" becomes ultimately suburbia, and another tier of suburbs develops around Chicago.

Chicago-Suburban Interaction

More disruptive than the more limited rural-urban conflict is that between Chicago and its surrounding suburbs, despite the fact that both have problems that neither can solve alone. Rainwater falling in one community may create flood problems for others; traffic in the heart of the central city may cause expressway backups all the way out into the suburbs; disease, pollution, and crime cannot be hemmed in by political boundaries. The major obstacle to solving such problems is that, although the Chicago Metropolitan area has a certain economic unity, it is a fragmented, overlapping, disorganized political structure—often offering inadequate, conflicting, and uncoordinated solutions at the local level. The "real Chicago" crosses state lines and encompasses more than 1,500 governmental units, ranging from counties to mosquito abatement districts. This multiplicity of government has led to an absence of coordination in road construction, the use of open space, water control, mass transit development, air pollution control, and refuse disposal. It also has resulted in increased costs because each community, regardless of size, must usually provide its own services: police, fire, garbage, school, etc. Chicago provides employment and a variety of services for suburban commuters who do not share proportionately in their cost. The suburbs, on their part, generally refuse to become enmeshed in the many problems and burdens of the central city. Political rivalry has often widened the gap between Chicago and surrounding areas; in recent decades Chicago has consistently voted Democratic, while the suburbs generally have tended to vote Republican. Racial, religious, and economic differences between Chicago and the suburbs also have been divisive.

Fortunately, there has been a growing trend toward cooperation among the numerous governmental units of the Chicago area. Chicago supplies water to about 80 other communities and the Metropolitan Water Reclamation District of Greater Chicago handles the sewage of 125 cities and villages. Comprehensive plans of Chicago have been developed in a regional context. The Northeastern Illinois Planning Commission has done general planning for the six counties in its area. The Regional Transportation Authority is designed to coordinate and improve transit facilities for the same six counties. In addition, groups of suburbs have concluded cooperative pacts dealing with such services as police, fire, water, health, street construction and maintenance, library cooperation, and refuse disposal.

MIDWEST MEGALOPOLIS

Future planning and cooperation will have to encompass a much greater area and population. By early in the twenty-first century, it is likely that the suburbs of Chicago will have virtually merged with the suburbs of Milwaukee and the other large communities to form a giant, mainly urbanized, area whose basic framework is clearly evident. Chicago is now the nucleus of this large developing urbanized sprawl—a Midwest Megalopolis—that stretches almost 300 miles between approxi-

mately Elkhart, Indiana and Green Bay, Wisconsin. This urbanizing area, encompassing about 6,000 square miles, is roughly L-shaped, around the southern and western shores of Lake Michigan through such cities as South Bend, Gary, East Chicago, Hammond, Chicago and its suburbs, and northward through Waukegan, Kenosha, and Racine, to Milwaukee and beyond. This area of developing urban coalescence already contains a population of over 10 million people. The dominance of the Chicago area is seen in the fact that the Chicago Metropolitan Area now contains two-thirds of this population.

The orderly, harmonious, and efficient development of the emerging megalopolis, with its changing scope, population, and nature of urban settlement, is one of the problems confronting the Chicago area.

NOTES

1. Adapted from Irving Cutler, *Chicago: Metropolis of the Mid-Continent*, 3rd ed. (Dubuque: Geographic Society of Chicago and Kendall/Hunt Publishing Company, 1982).

2. Wallace W. Atwood and James Goldthwait, *Physical Geography of the Evanston-Waukegan Region*, Bulletin No. 7 (Urbana: Illinois State Geological Survey, 1908), p. 4.

3. Patrick Shirreff, *A Tour Through North America; Together with a Comprehensive View of Canada and the United States* (Edinburgh: Oliver and Boyd, 1835), p. 226.

4. John Lewis Peyton, *Over the Alleghenies and Across the Prairies: Personal Recollections of the Far West One and Twenty Years Ago* (1848) (London: Simpkin, Marshall & Co., 1869), pp. 325-29.

5. Joseph Kirkland and John Moses, *History of Chicago*, vol. 1 (Chicago: Munsell and Company, 1895), p. 119.

6. George W. Steevens, *The Land of the Dollar* (New York: Dodd, Meand, and Company, 1897). p. 144.

7. Jane Addams, *Twenty Years at Hull House* (New York: Macmillan, 1910), pp. 81-82.

8. James Parton, "Chicago." *Atlantic Monthly* 19 (March 1867), pp. 325-45.

9. U.S. Department of Agriculture, *The Why and How of Rural Zoning*, Agriculture Information Bulletin No. 196 (Washington, D.C.: Government Printing Office, 1958), p. 1.

10. Irving Cutler, *The Chicago-Milwaukee Corridor: A Geographic Study of Intermetropolitan Coalescence*, Studies in Geography No. 9 (Evanston, Ill.: Northwestern University, 1965), pp. 126-27.

UNDERSTANDING ILLINOIS THROUGH THE BIER/ISGS PHYSIOGRAPHIC MAP AND EXERCISES[1]

Paul S. Anderson
Illinois State University

Albert D. Hyers
North Adams State College

Michael D. Sublett
Illinois State University

INTRODUCTION

A physiographic map, such as that of Illinois by James A. Bier accompanying this book, depicts the surface landforms in a "drawn picture" format. This map type is not the same format as a topographic map with contour lines or an aerial photograph. The physiographic landforms are drawn with both detail and generalization. Detail allows individual features, such as prominent bluffs and stream valleys, to be clearly defined. Generalization simplifies the landforms so that larger areas such as plains or complexes of hills (as in a glacial moraine) are visible but not distracting. Such maps are among the finest examples of how art and science can coexist and support each other in cartography.

On physiographic maps, the more complex the terrains to be shown on a small map, the greater the needed generalization. Therefore, on physiographic maps that show the entire USA, Illinois is viewed as near-featureless flatlands like Iowa and Kansas in comparison with mountainous California, Colorado, and West Virginia. The past great masters of physiographic maps (namely Raisz, Lobeck, Harrison, and Lentz) virtually skipped over the llinois landforms on their small scale drawings. (Note: Small scale physiographic maps of the USA and the Continents can be obtained by telephoning 1-800-242-3199).

Fortunately, Illinois has James A. Bier, a retired cartographer from the University of Illinois. Bier produced his *Landforms of Illinois* map in 1953 when military service interrupted his graduate studies of geography at the University of Illinois. He revised the map in 1980. The Illinois State Geological Survey at 615 E. Peabody Drive, Champaign, IL 61820 [tel: 217-333-4747] sells the 17 x 27 inch map for a mere twenty-five cents ($0.25) per copy, (plus $0.95 postage and handling for orders up to $6.00).

Because of the Bier map quality, enlargements and reductions of the original map are possible. Enlargements reveal the richness of detail while reductions show crisp generalization that makes the major features clearly visible. Because of the educational value of the map, the Illinois Geographical Society (IGS) is making it available as a half-scale desk map (1:2,000,000; measuring 8.5 x 14 inches) and a double-scale wall map (1:500,000; measuring 31 x 53 inches). For prices and quantities, contact the IGS Central Office at Campus Box 4400, Illinois State University, Normal, IL 61790-4400 [tel: 309-438-7649].

These maps are low in cost (less than photocopy costs), because James Bier has granted copyright permission for noncommercial purposes. One full-size copy and one half-scale copy (Figure 1-2) of the Bier map are included with this book for use with the following exercises. For teachers to utilize best the exercises and to begin using immediately this valuable cartographic resource, we suggest that they order sufficient classroom sets of the map at all three scales from the ISGS and IGS at the above addresses. For a typical classroom, one wall map plus fifteen regular size maps are the reusable instructional and reference materials, while one or two consumable desk copies are needed per student for marking and other exercises.

For reference and comparison purposes, we recommend that readers also have, in addition to this book, a reasonably recent copy of the Illinois Department of Transportation (IDOT) *Illinois Highway Map,* a free publication from the Illinois Secretary of State (1-800-252-8980).

The following pages of annotated exercises present materials for six objectives organized in four sections. For educators, these six objectives match the six essential elements and the indicated standards and pages of the National Geography Standards:[2]

Objective 1 in Section 1: Development of map use skills to maximize the value of the Bier map and exercises. [Matches Standard 1, p. 61 ff.]

Objective 2 in Section 2: Presentation (or reinforcement) of knowledge of the physical and human characteristics of places in Illinois, including the state's basic regionalization, to interpret its complexity. [Matches Standards 4 and 5, p. 69 ff.]

Objective 3 in Section 3: Understanding of continental glaciation and its impact on Illinois' surface landscapes and river patterns. [Matches Standards 7 and 8, p. 75 ff.]

Objective 4 in Section 4: Appreciation of human activities (population, settlement, etc.) in Illinois [Relates to Standards 9-13, p. 79 ff.]

Objective 5 in Section 4: Understanding of how the physical landscapes of Illinois affect the human/cultural systems of the state. [Matches Standard 15, p. 96 ff.]

Objective 6 in Section 4: Presentation of how to apply the above mentioned topics of Illinois geography to interpret the past and present and to plan for the future. [Matches Standards 17 and 18, p. 101 ff.]

These six objectives are interrelated. Use of the Bier map for objectives 2-6 will strengthen the skills in objective 1. The physical processes in objective 3 are integral components in objectives 5 and 6. These interrelationships become evident within the following discussions and exercises.

The level of the discussions and exercises is for adults, high school students, and teachers of lower grades who can adapt the materials for their grade level and local area. For many of the exercises, equivalent examples can be found in diverse sections of the state. Users are encouraged to observe closely their local area on the map to identify those equivalent features. Answers for the exercises are provided at the end of these pages.

As you do each of these activities you might want to save a copy of the marked maps for reference purposes in later exercises or class projects.

SECTION 1 AND OBJECTIVE 1: DEVELOPING MAP USE SKILLS

1.1. Marking the Bier Map

Some people frame and display the Bier map or protect it in a drawer. Others allow students to touch but not mark the maps. But maximum educational usage is benefited by marking the map, and we offer five easy suggestions to facilitate marking.

1. Provide photocopies, making sure your "original" map and photocopy machine are of high quality. Photocopying also allows you to enlarge or reduce from the original map and perhaps use the large-size (11 x 17 inch) paper. You can center your photocopies upon your local area or study region. The cost is about five cents per student or $1.50 for a class of thirty. The four-part series in the *Bulletin of the Illinois Geographical Society* has six maps at standard page size (8 1/2 x 11 inches) for easy photocopying.[1]

2. Give each student an entire map. At only $0.25 per map ($7.50 for a class of thirty) from the Illinois State Geological Survey (ISGS), this is a true bargain. Alternatively, obtain copies of the half-scale (1:2 million) map on 8 1/2 x 14 inch paper.

3. Laminate a teaching set. Clear plastic lamination offers two ma-jor advantages. (a) Lamination preserves the map from wear-and-tear and from stray marks by students. (b) Laminated copies accept color from some types of washable marking pens, thereby allowing for large see-through markings and for correction of mistakes. Also, a markable wall map for full-room viewing can be made from enlarged, large-sheet, laminated, joined photocopies.

4. Make overhead transparencies. On an overhead projector, the map can be marked either on the transparency or upon a "screen" of white paper. The paper "screen" method is especially appropriate for grade school students. Be sure to start by making tick marks (registration marks) in the four corners so you can re-align the projection onto the screen if bumped or if needed on a second day.

5. Trace onto an overlay. Sometimes a tracing is useful to highlight selected features and eliminate unnecessary detail. The best material is thin, 3-mil (0.003 inch) acetate frosted on one side, available in sheets or rolls at most art supply stores. With less transparent tracing material, you can use a light table or even a windowpane in the classroom to improve the view of the map. One advantage of tracings is that students' work can be overlain for direct comparisons. Remember to require registration marks such as the crosses of the latitude and longitude lines.

Exercise for marking on a map: (This exercise can be done for any stream on the map, so local areas can be selected.) We all know of the major rivers like the Mississippi, Ohio,

Wabash, and Illinois. They are formed by thousands of smaller streams that influence and are influenced by the landforms so well shown on the Bier map. One small drainage basin empties into the west side of the Illinois River at the town of Hennepin: 41° 20′ North latitude, 89° 20′ West longitude, or MMC 197 479 (T-148). (MMC coordinates are explained in Section 1.2.2.) Bureau Creek is the main stream. Decide if you will mark the map, a photocopy, or an overlay sheet. Blue is the traditional color for water, so mark in blue the streams of Bureau Creek and its tributaries. Be sure to include the southerly stream at MMC 183 480 (T-148), which Bier neglected to attach to Bureau Creek.

Drainage basin delimitation is also possible. To the north of the Bureau Creek basin is the Green River. The drainage divide (the separation of the waters) is somewhere in the hills between the drawn drainage patterns. **Activity:** With some color other than blue, draw a curving line to separate the two drainage basins. Where you are unsure, draw dashes. You can do this all around the Bureau Creek basin. The area of this basin can be measured on the map.

1.2. Coordinate Systems

1.2.1. Latitude and Longitude

On the full map or on extracts of the Bier map, study the appearance of the latitude and longitude lines.

(Q1-1) What symbol is found at the intersection of each whole degree?
(Q1-2) Are the lines of latitude straight?
(Q1-3) Is any longitude line straight on the Bier map? **Activity:** Connect the whole-degree intersections to draw the full extent of the latitude and longitude lines, which is called the graticule.

Geographic coordinates of latitude and longitude, with degrees and minutes, are not very user friendly, especially with curved or convergent lines.

(Q1-4) For example, what are the geographic coordinates of Springfield, Illinois? You can estimate fairly quickly, but then measure and calculate carefully the "lat-long" of the center square in the urban grid pattern for Springfield on the Bier map. Each urban grid square covers approximately 30″ of latitude and 45″ of longitude.
(Q1-5) Exactly what is shown on the map at 39° 0′ 0″ N, 87° 51′ 50″ W? This is intentionally an easy example. Clearly, geographic coordinates have both great accuracy and substantial difficulty for precise use. But other coordinate systems are available.

1.2.2. Millimetric Coordinates (MMC) and Zone Reference (ZR) Coordinates

Fortunately, we have alternative methods for easily pinpointing specific spots on the map. The methods follow the principles of "grid reference" (GR) and "area reference" (AR) systems associated with the Universal Transverse Mercator (UTM) coordinate grid. However, the alternative methods are totally independent of GR, AR, and the UTM coordinates.

One alternative method, called millimetric coordinates (MMC) only requires a pair of clearly printed axes, preferably in the bottom (X) and left (Y) margins of the map, and a millimeter ruler. (Other axes such as on the top and right margins can have supplemental uses.) **On the Bier map we use the innermost neat line (line where the landform drawing stops) for our measurements of millimetric coordinates in these exercises.** To identify a spot, measure in millimeters first to the right, and then measure up from the axes. Each measurement must have three digits, so 42 millimeters would be 042. Even without any grid lines printed on the map, everyone should be able to locate precisely any spot on the map using millimetric coordinates. As an example, the dot on the letter "i" in Bier's name on the map is at the following MMC location:

MMC 067 098 on the published full size map (scale=1:1 million)

MMC 034 049 on the half-scale 8" x 14" map (scale=1:2 million)

As another example, the location of the city of DeKalb on the regular-size and reduced-size Bier maps is at MMC 243 555 and MMC 121 277, respectively. Please note that MMC coordinates are different on even the same map when the scale is changed. In other words, millimetric coordinates are NOT transferable to different maps.

The second alternative coordinate method is called zone reference (ZR). ZR requires **printed coordinates** that increase to the right and upward, respectively, on the horizontal (X) and vertical (Y) axes on the map or on any other graphic. As an example, see the half-scale Bier map (Figure 1-2).

As with the AR (area reference) method on the UTM system, each printed ZR coordinate identifies a specific line and, also, the interval from that line to the next line to the right or upward, respectively, on the X and Y axes. Where each vertical line crosses a horizontal line, we find a very specific point location. Where a vertical interval crosses a horizontal interval, we find a square. That square is the referenced zone, and the **ZR coordinates identify the lower left corner of that square.**

For example, the coordinates ZR 24 55 refer to a square zone for which the lower left corner is the intersection of line 24 (a vertical line identified on the horizontal axis) and horizontal line 55 on the Y axis. On the Bier

map, ZR 24 55 contains the city of DeKalb. Similarly, the dot on the letter "i" in Bier's name is located within ZR 06 09. (Always use two digits.)

We note that on the standard Bier map (which was printed without ZR coordinates), the millimetric coordinates for the city of DeKalb (MMC 243 555) also can give us the zone reference (ZR 24 55). **When using centimeters as our measuring units for ZR coordinates on the original map, ZR coordinates on that map at any scale identify the same area as the millimetric MMC coordinates on the map at the original scale.** We note that ZR (zone reference) coordinates are valid at any scale but must be printed on every map except those where the ZR interval equals one centimeter and can be measured directly with a centimeter ruler. In contrast, MMC coordinates are valid only on the maps at a single scale but need not be printed on the map. MMC provides greater accuracy and only needs simple axes; ZR is faster to use but requires the printing of the ZR coordinates.

We also note that because the scales of the small desk map and large wall map are, respectively, exactly half and double the scale of the full size Bier map, the millimetric coordinates are also exactly half or double. These ratios are true for any MMC measurements on these maps.

In the remaining exercises we only provide the MMC positions on the full size Bier map because they give the ZR coordinates for all three sizes of the Bier map. For example, on any of the three maps you can easily find the number "4" located at ZR 34 23 or at MMC 340 230. Compare that ease with how you handled the same question in exercise Q1-5 using latitude and longitude! For further information about millimetric coordinates, please read the comments in the box on the next page.

In the original four-part series about the Bier map in the *Bulletin of the Illinois*

1.2.2.1. Explanation of Millimetric Coordinates

The system of millimetric coordinates (MMC) was initially developed for use with aerial photographs,[3] but it is equally helpful when several people are using exact copies of the same map. The MMC system has two requirements, two rules, four advantages, and one disadvantage.

Requirements

1. The map, image, or drawing must have clearly visible printed horizontal (X) and vertical (Y) axes. (The cut edges of the page are not adequate.) In standard usage, the MMC axes are along the bottom and left sides and all numbers are positive.
2. The user must have a sufficiently long ruler with marked millimeters.

Rules

1. For STANDARD MMC usage, measurements in millimeters expressed as three digits are made first to the right (from the Y-axis) and then up (from the X-axis). Therefore, MMC 142 005 identifies a spot 142 millimeter from the left (Y) axis and five millimeters above the X-axis.
2. If any NON-STANDARD axis is used (such as measuring down from a top axis), each non-standard measurement must be clearly identified. For example, MMC 215 T-152 uses the standard left axis and the non-standard top axis. (See Section 1.2.2.2. below.)

Advantages

1. **MMC offers ease of understanding** with great simplicity, conciseness, decimal base, international appeal, and minimal materials. The MMC system is similar to, but even less complicated, than the UTM coordinates of the Universal Transverse Mercator maps.
2. **MMC is applicable to all identical copies** of maps, figures, pictures, and photographs that are printed with clear, linear image edges (not the cut edge), even those that do not have any other coordinate system.
3. **MMC provides accuracy to a single square millimeter** anywhere on a map up to 1 x 1 meter (40 inches x 40 inches) in size, but a practical limitation for standard usage is the length of the common ruler. (12-inch = 300 millimeters and 18-inch = 460 millimeters).
4. **MMC eliminates the factors of scale, direction of orientation, and projection** that make geographic coordinates difficult to use for highly specific locations on maps.

Disadvantage

1. The MMC coordinates of a location on a map are not transferable for use on a DIFFERENT map showing the same area. Therefore, if the scale has changed, new MMC coordinates must be measured. You can easily verify that your copy matches the scale of the original map sent to the printer or to the photocopy machine if a small mark is made on the X-axis at MMC 100 000 before the copies are made.

Geographical Society, the six figures of map extracts were all single pages, making standard MMC usage easy with a regular ruler. But because of the large size of the published Bier map, we need additional instructions for non-standard usage.

1.2.2.2. Notes About Millimetric Coordinates on the Full-Size Bier Map and on Other Large Maps:

A. About Rulers

The easiest solution for MMC measurements on large maps is to have an 18 inch (460 mm) ruler for each user. These are readily available, relatively inexpensive, usually made of flexible steel, and will last a lifetime. Alternatively, tape together two cheap, flat-plastic rulers to make one that is 500 mm long (20 inches). You can also simply tape or glue an accurately measured extension of 10 or 20 centimeters onto the zero end of a standard ruler. Whatever way you do it, 380 mm (15 inches) will reach the entire width of the full-size Bier map or the half-size reduction. The long (15-18 inch) ruler is also sufficient for standard MMC usage on the north-south length of the half-size Bier map. Finally, a meter-stick or a millimeter tape measure can reach to any spot on the full-size Bier map. (A small inexpensive tape measure, such as "Executive" model by Lufkin, costs about $12 and will reach two meters/six feet. One source is Forestry Suppliers at 1-800-647-5368.)

B. About the Axes

1. Supplemental Axis

When working with maps that are larger than our ruler, we have two options. One way is to carefully draw one or more supplemental axes at appropriate spacings such as 300 mm or 400 mm from the bottom MMC axis of the Bier map. Then, the location of DeKalb (city) at MMC 243 555 is 243 mm from the left margin and either 255 mm up from the 300 mm line or 155 mm up from the 400 mm line.

The disadvantage is that each supplemental axis must be drawn carefully on each copy of the map.

2. Right and Top Axes

Many large maps have clear neat lines on all four sides. We will call them Left (L), Right (R), Bottom (B), and Top (T). The left and bottom axes are used with standard MMC designations, and we do not need to write the letters "L" and "B" with standard MMC measurements. But to measure from the Right (or Top), the measurement would be in the opposite or "negative" direction and we could use the designation "R-" and "T-". Therefore, the location of DeKalb would be MMC 243 T-074, that is, 74 mm down from the top axis. Using this variation of millimetric coordinates, each location such as DeKalb would have four (4) sets of coordinates on the exact same map: MMC nnn nnn, MMC nnn T-nnn, MMC R-nnn nnn, and MMC R-nnn T-nnn. To keep things simple, we recommend using the standard MMC measurements and a minimum of the variations.

For use with the full-size (15 inch x 25 inch) Bier map, we always provide the standard MMC coordinates. Also, if the location is in the northern half of the map, we provide for your assistance the measurement from the top axis. Example: DeKalb is at MMC 243 555 (T-074). A few exercises with MMC and the Bier map are helpful to master this map-use skill.

(Q1-6) In the next county south of DeKalb is the city of Mendota. What are the MMC coordinates of Mendota?

(Q1-7) What river flows through Yorkville at MMC 272 524 (T-105)?

(Q1-8) Name the town and give the MMC position where this river ends by flowing into a much larger river. Which direction is Yorkville's river flowing?

1.3. Scale

The Bier map has a scale. It can be given in graphic form (such as the bars segmented in miles and kilometers), as a representative fraction (such as 1:1,000,000 or 1/2,000,000 for the full size and half size maps, respectively) and as a verbal scale (such as "one inch equals sixteen miles" or "one centimeter equals ten kilometers" for the full size Bier map).

For a useful exercise in handling scale changes, make several photocopy enlargements and reductions of any part of the Illinois map in which you are most interested. Illinois residents could enlarge their local area, then calculate the scale of each enlargement or reduction and draw a graphic scale on the photocopy. You will be pleasantly surprised at how large you can make the Bier map and still maintain the excellent physiographic appearance.

Now we want to make use of scales. Illinois highway Route 47 runs in an almost straight line from Dwight (MMC 274 463 (T-166)) to Woodstock (MMC 270 600 (T-029)).

> **(Q1-9)** Determine that distance in kilometers by using two methods, one with the graphic scale and one with the representative fraction.
>
> **(Q1-10)** Multiply the kilometer measurement by 0.61 to determine the mile distance.
>
> **(Q1-11)** Check your answer to Q1-10 by measuring the mileage on the map by using both the graphic scale for miles and the representative fraction.
>
> **(Q1-12)** Was it easier working with metric measurements or with inches and miles?

Another useful measurement is the distance between the geographic coordinate grid lines.

> **(Q1-13)** How many miles and how many kilometers is the north-south distance between two consecutive whole degrees of latitude? (Note: Assign students to measure in different regions of the Bier map to check if the measurement is constant.)
>
> **(Q1-14)** Is the latitude separation equal in all parts of the Bier map?
>
> **(Q1-15)** Repeat Q1-13 and Q1-14 for longitude and explain any observed differences.

1.4. Symbols

Maps are made of point, line, and area symbols printed on paper. The Bier map has all three types. The cross-hatched or grid symbol for small towns such as Sparta (MMC 165 132) is a point symbol that shows a location without actually showing the street pattern or town size. The cross of latitude and longitude lines is also a point symbol.

> **(Q1-16)** The most easily understood symbols are the line symbols. Describe the appearance of the line symbols used to show the following on the Bier map:
>   A. Rivers,
>   B. Creeks (small streams),
>   C. Counties (straight line sections and along rivers),
>   D. State borders,
>   E. Canals,
>   F. Railroads, and
>   G. Limits of glaciation.

Feel free to explore the Bier map for other line symbols.

> **(Q1-17)** The lakes and the flat plains (white areas without any drawings) are area symbols. A cluster of area symbols is at MMC 161 162 and at MMC 135 121. Describe the symbol and tell what it represents.

(Q1-18) Glacial moraines (bumpy symbols) and plains dominate northeastern and central Illinois. Name the moraine on the north side of Mendota (MMC 216 512 (T-117)).

(Q1-19) Note that the far northwestern part of Illinois looks different (far different symbols) from the moraines and plains. Use the inset map in the upper left corner of the Bier map to name that different physiographic division.

(Q1-20) The symbol for an esker is shown on the Bier map at MMC 250 541 (T-088). We will explain these special elongated hills called eskers in Section 3. Does the esker appear to be drawn as viewed from the south, and is that viewpoint evident in the other physiographic symbols?

(Q1-21) Study the bottom and sides of the Illinois River valley in its east-west section. Starved Rock at MMC 228 486 (T-142) has beautiful tall and historic cliffs. Why does Bier not draw the cliffs but instead use a labeled arrow?

1.5. Conclusion

We can do many more map skills activities with the Bier map, but we can wait to learn and integrate those skills as we cover the other five objectives. As you continue, please remember that by keeping the map in front of you, and by actively exploring the diverse parts of the map, you will improve both your map skills and your knowledge of Illinois.

SECTION 2 AND OBJECTIVE 2: KNOWING ILLINOIS' BASIC PHYSICAL AND HUMAN CHARACTERISTICS AND REGIONS

We have already covered part of this objective with our earlier identification of specific places like DeKalb, latitudes from 37° to 42°, and hydrographic features like the Fox River. We will need this and much more knowledge of Illinois places and regions when we examine the subsequent objectives in which we focus on processes (and then, too, learn more about Illinois places). Therefore, our current objective is an important step for acquiring the basic place knowledge for daily life and for our future analyses. Geography Standards 4 and 5 recognize the importance of place knowledge and the concept of regions. This is not static information; we must understand that places and regions can change over time in both physical and human characteristics.

2.1. Illinois in the National Context

Concerning our objective to know and understand places and regions, Illinois' position within the context of North America and the Midwest region is important information. With younger students this would be major topic; our discussion assumes this basic knowledge. We also acknowledge that the Bier map only reveals small sections of the surrounding states.

(Q2-1) In terms of the artistic style and actual appearance of the images on the map, was Illinois drawn differently from the surrounding states?

(Q2-2) How does the Bier map show the state borders of Illinois?

(Q2-3) Does the Bier map show the borders between the surrounding states and, if so, how are those borders shown?

The major physiographic divisions of the United States are examples of formal regions. These can be found in most atlases, but there can be variations. One system that was used by Leighton, Ekblaw, and Horbert (1948) called the major divisions "provinces;" their regionalization of Illinois is shown in the inset map in the upper left corner of the Bier map. These landform regions are described by Arlin Fentem in Chapter 2 of this book.

> **(Q2-4)** How many and which provinces are named in the inset map, and how much of the state is in each one?

Numerous subdivisions with the names of "plain" and "country" can be seen on the inset map of physiographic divisions. They are separated by thinner lines and are discussed in Objective 3 below. A good map skill **exercise** is to transfer (with colored dashed lines) those borders onto a copy of the Bier map. This can be done later when we examine the individual areas.

2.2. Area Locations within Illinois

An example of perceptual regions is the division of Illinois into northern, central, and southern regions. There is no definitive map showing these boundary lines and the Bier map does not indicate those regions. For example, a strict definition using equal thirds of the state's length would place the cities of Peoria and Pontiac in the northern region, but most people would call them central Illinois cities.

A functional region "is organized around a node or focal point, with the surrounding areas linked to that node...."[4] For example, in terms of major airports, the airports of Chicago and St. Louis dominate most of Illinois. The midpoint between Chicago and St. Louis is near MMC 230 364, slightly north of the town of Clinton.

> **(Q2-5)** Within a radius of 50 miles (80 kilometers) of that midpoint, name the five largest urban areas, which Bier has shown with the town grid symbol having more than eight complete grid cells.

Residents of those cities can take "feeder-airline" flights from local airports or go by road to their major airport of choice, which is frequently (but not always) the closest one.

Sometimes we want to identify locations without using geographic (lat.-long.) or MMC coordinates. For example, we might simply use words to refer to a prominent city such as Springfield, and most users would easily look at the center section of the state. We also can use counties. The entire state has 102 counties. They are important administrative units, and their boundaries are on most maps. They are formal regions. The Bier map shows the straight-line boundaries plus the rivers that separate some counties. But the names are not given. **Activity:** For the entire state or for your section of it, highlight in color the county boundaries and neatly print in large letters the name of each county. Use the state highway map or Figure 1-10 in this book as a reference. Save this map for later reference.

> **(Q2-6)** How many **complete counties** are north of 41° North latitude?
> **(Q2-7)** Which one is the smallest?

Rivers and lakes are also important places. To become familiar with those in Illinois, you can follow the main ones on the Bier map. On a copy of the map, you (or students) could highlight the main streams by marking them in color (usually blue). Characteristics to note are the direction of flow, the width of the valley, the counties and towns along the streams, and their names.

2.3. Index of Local Area Topographic Maps

(This activity assumes you have access to the topographic maps of your local area. Perhaps the school library or local library has them. If not, request that they be purchased from the ISGS, which can also provide an index map.) Check the scale of the topographic maps. A map at 1:62,500 is called a 15 minute quadrangle and will show on the Bier map an area approximately 2 cm wide (longitude) by 2.6 cm high (latitude) (approximately 3/4 x 1 inch).

A "topo sheet" at 1:24,000 is a 7.5 minute "quad" and covers an area only half as long and half as wide as the 15 minute quad on the Bier map. **Activity:** Mark carefully on the Bier map (or on an enlarged photocopy) the area covered by each topographic map of your local area or of one county in Illinois. When you do this you are creating an index map that will be very useful.

Bier used these topographic maps as his basic information when he compiled the physiographic map. He and others who study maps can accurately visualize the hills and plains of the landscape by looking at the brown contour lines on the topo sheets. **Activity:** You can try your envisioning ability by comparing the Bier map with a topographic map. Can you see where the flat areas adjoin the hilly areas? When used together with selected U.S. Geological Survey (USGS) topographic maps (quadrangles), the Bier map provides a visual link between the 2nd-order large relief features (such as the Rockies and Great Plains) and 3rd-order relief features of local hills, cliffs, and stream valleys on the topographic maps.

2.4. Conclusion

In the above paragraphs we have been using the Bier map to help acquire knowledge of Illinois places and regions. Further study of this map (and others) will enrich our understanding of the basic human and physical characteristics of Illinois.

SECTION 3 AND OBJECTIVE 3: UNDERSTANDING THE PHYSICAL PROCESSES THAT SHAPED ILLINOIS

In the National Geography Standards, the two standards of physical systems focus on physical processes and the related functional units called ecosystems. Dozens, perhaps hundreds, of major physical processes have and continue to shape our planet, but at different times and places. For example, the mountain building processes and desert climate ecosystems are not well illustrated in Illinois. But Illinois does offer outstanding examples of the forces and results of one of the major physical processes.

Continental glaciation is, by far, the single most important physical process that has shaped Illinois. Continental glaciation basically covered the central United States north of the Ohio and Missouri rivers. Residents of or people studying the states of New York, Ohio, Indiana, Michigan, Missouri, Iowa, Wisconsin, Minnesota, North and South Dakota, and Montana, as well as most of Canada east of the Rockies can apply some of this Illinois information to their areas of interest.

To appreciate this subject content on physical processes, readers should know (or obtain through concurrent additional studies) at least a rudimentary understanding of physiographic regions, geomorphic processes, and geologic time, especially concerning continental glaciation. Chapter 2 in this book or a college-level introductory physical geography textbook will provide appropriate reference material.

Because the Bier map was specifically made to show the results of the physical processes, this objective and section are the largest. Readers are encouraged to read and utilize the information in the other chapters of this book to enhance their usage of the Bier map. Our discussion of this physical process and its results is regionalized according to northern, central, and southern Illinois.

3.1. Northern Illinois

3.1.1. Driftless Area

An ice sheet as much as 4,000 meters (13,000 ft) thick over Hudson Bay and 2,000 meters thick in northern Illinois covered the northern half of North America repeatedly during the Pleistocene Epoch (roughly 2,000,000 to 10,000 years ago).[5] The ice advanced and melted back at least four major times with possibly up to 23 occurrences. Evidence for the most recent major times are the clearest and are referred to as the Wisconsinan (youngest) and Illinoisan ages. The pre-Illinoisan glaciations are less understood and frequently grouped into what has been called the Kansan and Nebraskan ages. All these ice sheets either bypassed the Driftless Area (west of the dashed line in Jo Daviess County and also much of southwest Wisconsin) or perhaps inundated it only briefly before mid-Wisconsinan time. The landscape in that northwest corner of Illinois has a strong fluvial (stream-eroded) character that typifies many nonglaciated regions. Observe the general map appearance of the Driftless Area in comparison to the other areas on the Bier map.

> **(Q3-1)** Is there any other place in the state where the relief appears to be as accentuated?
>
> **(Q3-2)** What is the elevation of the tops of the hills in the Driftless Area? Is that in feet or meters? Clearly, the Driftless Area is not high and rugged, but there are ski resorts near Galena.
>
> **(Q3-3)** Also describe the appearance of the Plum, Apple, and Galena river valleys on the Bier map. Incorporate in your description such attributes as valley shape (depth and width), drainage pattern (including orientation), and interfluve (upland) appearance (shape and relief).

You can use topographic maps to study this area more closely. The *Galena, Illinois-Iowa* 15-minute quadrangle (scale 1:62,500; contour interval 20 ft.) provides a closer examination of the Driftless Area; the topographic map allows a more precise description of features visible on the landform map. For example, the interfluves are even more rugged than indicated on the Bier map.

The mighty Mississippi River has been a significant force in Illinois landscapes. One notable example is the major erosion that left escarpments and steep hills along the edge of its flat floodplain, as shown at MMC 092 614 (T-018), MMC 108 591 (T-036), and at the scenic Mississippi Palisades at MMC 128 573 (T-055). Less obvious, the ancient river course did not flow through the constrictions at MMC 113 520 (T-098) (north of Moline) nor at Rock Island (where there really is today a rock island in the river). Instead, the ancient Mississippi flowed southeastward from Fulton-Clinton (MMC 126-545 (T-083) and entered what is now the valley of the Illinois River near Hennepin at MMC 195 480 (T-147). That is one reason why the valley of the lower Illinois River is so wide. (Refer to Figure 2-13 in this book.)

> **(Q3-4)** Find the borders of Bureau County, which includes the town of Princeton at MMC 186 494 (T-134). Do you think the ancient Mississippi flowed over, through, or around the moraines in Bureau County? Or what else might have happened?

3.1.3. Recently Glaciated Northern Illinois

As recently as about 12,500 years ago the ice sheet finally retreated from the area of northeastern Illinois. Those most recent advances had erased most of the landscapes of earlier ones, leaving fresh landforms. For the following exercises, students need to know (or have access to references about) the meanings, topographic appearances, and significance of the following features: moraine, outwash plain, meltwater, esker, lake plain, terrace, beach, and shoreline.

> **(Q3-5)** Study the following moraines, describe their basic shape,

and suggest the direction *from which* the glaciers came:

1. Marseilles Morainic System: MMC 260 505 (T-121) and MMC 300 450 (T-176)
2. Bloomington Morainic System: MMC 232 555 (T-073) and MMC 260 388 (T-239)
3. Valparaiso Morainic System: MMC 300 610 (T-073) and MMC 260 388 (T-239)
4. Lake Border Morainic System: MMC 315 617 (T-012) and MMC 323 560 (T-068)

(Q3-6) Place the above-mentioned moraines in chronological order from oldest to youngest. What map evidence did you use to determine the chronology?

(Q3-7) In which counties was the ice sheet when the Kankakee Plain was developed?

Lake Chicago is the name for the Pleistocene lake that had a higher level of water than present-day Lake Michigan, but the lake basin then was often mainly filled with ice. The shore of Lake Chicago's highest level (640 ft or 195 m) was along the Tinley Moraine (MMC 333 518 (T-111)). Trace the shoreline with your finger (or mark a copy of the map).

Look for the thin hachured lines at MMC 327 545 (T-085) and MMC 346 522 (T-108)). These lines represent former beaches at different lake levels. Lake Chicago slowly diminished in size to become modern Lake Michigan. Although not apparent on the map, there also were lake stages below the level of the present lake.

(Q3-8) Eskers are drawn and marked at MMC 273 541 (T-089) and MMC 250 541 (T-089). Were DeKalb and Kane counties covered by glacial ice when the eskers formed?

3.1.4. Wheaton Morainal Country

Note: This name is misspelled as "Weaton" on the inset map of physiographic divisions.

(Q3-9) Describe the landscape in the Wheaton Morainal Country by referring only to the landform map. Your description should include relief, relative valley depth and width, drainage pattern and integration, and interfluve characteristics.

The glaciation resulted in the creation of moraines with an irregular surface that contains many small lakes and marshes. The best examples still found in Illinois are in McHenry and Lake counties between MMC 230 600 (T-029) and MMC 296 618 (T-011). But during the years since the retreat of the glaciers, many of the lakes and marshes have disappeared because of three factors:

(a) deposition has filled some
(b) farmers have drained numerous marshes
(c) natural stream formation has drained many

The influence of time on drainage development is clearly seen on the Bier map by comparing the landscape of the Wheaton Morainal Country (most recent glaciation) with that of the Rock River Hill Country and the Mt. Vernon Hill Country in southern Illinois. (See also the exercises on southern Illinois.)

(Q3-10) Explain why rivers in morainic areas generally flow parallel to the moraines.

(Q3-11) The *Grayslake Illinois-Wisconsin* 15-minute topographic map (scale 1:62,500 and contour interval 20 ft.) and others from this area nicely illustrate the 3rd order landforms of this region. If available, use this quadrangle map to describe (a) the drainage features and (b) the general appearance of the landscape.

(Q3-12) How would you describe the differences in (a) valley appearance (depth and width) and (b) drainage pattern between the Driftless Area and Wheaton Morainal Country?

(Q3-13) How can you use this information to differentiate unambiguously these two physiographic divisions? Using colored pencils, draw the borders of the Driftless Area and the Wheaton Morainal Country on a photocopy or overlay of the Bier landform map. The Driftless Area is already demarcated with a dashed line, but outlining the Wheaton Morainal Country will require some map analysis. Refer both to the inset map ("Physiographic Divisions") and observable differences in topographic character.

(Q3-14) Based on your inspection of the 1:62,500 scale quadrangles, how well does the Bier landform map illustrate the general topographic character of the two physiographic divisions?

3.2. Central Illinois

3.2.1. Wisconsinan Glacial Areas

The Wisconsinan Glacial Age was the most recent, and its glaciers left the freshest marks. The outermost (i.e., most western and southern limits) of the Wisconsinan glaciation is the morainic complex labeled Bloomington-Leroy-Shelbyville Morainic System. This moraine system is farthest west at MMC 162 500 (T-129) in Bureau County. From there it trends southeastward and then eastward through Mattoon and Charleston to Terre Haute, Indiana. The outermost edge is a major topographic boundary that appears on the inset map of Physiographic Divisions.

Exercise: Draw as a colored pencil line on the Bier map the outer edge of the Bloomington Ridged Plain.

(Q3-15) Describe the topography of the Wisconsinan glaciated area *as seen on the Bier map*. Your description should include relief, grain or arrangement of topographic elements, and drainage characteristics.

The moraines of the Wisconsinan age formed generally during the Woodfordian subage between 22,000 and 12,500 years ago; the youngest ones are in the Chicago region. Physical geographers, and others, believe that the retreating ice sheet was very active and that it fluctuated considerably during the roughly 8,000 years it took to build the full sequence of more than 30 end moraines.[6] Thirty individual moraines are not distinguishable on the Bier map because he merged several when he had to generalize for a map at this scale.

A noticeable change in moraine orientation occurs generally along a line from MMC 262 388 (T-252) near Farmer City northeastward to MMC 292 427 (T-201) between Fairbury and Gilman.

(Q3-16) Explain the reason for the change in moraine orientation.

(Q3-17-Activity) Using red and brown colored pencils (blue will become useful later for another feature), lightly color the end moraines. Use one color for moraines of the Lake Michigan Lobe and the other for the Lake Erie Lobe moraines. Pay attention to the places (towns, counties, rivers, etc.) while you color the map.

The area west of Wateska (MMC 332 430 (T-200)) and another area west of Pontiac (258 440 (T-190)) appear to be quite flat. They could conceivably consist of ground moraine or outwash plain because end moraines (the Chatsworth and Minonk) border both. Although ground moraine (or till plain) can be exceedingly low in relief, direct formation by the ice sheet or deposition by meltwater streams are not the only ways that such flat topography can form.

> **(Q3-18)** Suggest another plausible explanation for these two large, distinctive flat areas. As an activity, lightly color them blue with colored pencil.

Note the area of more subdued topography in a roughly triangular shaped area between Pekin, Beardstown, and a point (MMC 190 367 (T-260)) twenty-five kilometers east of Mason City. This area is broadest at the Bloomington-Leroy Moraine and narrows to the width of the Illinois River's floodplain near Beardstown; the northwestern border of this area is the western side of the floodplain from Beardstown to Peoria. The eastern border is the moraine.

> **(Q3-19)** Suggest a plausible origin for this area and account for the belts of uneven topography. (Remember that it borders a moraine on the east and is near a major river.)
> **(Q3-20-Activity)** Color this outwash fan light yellow.

3.2.2. Pre-Wisconsinan Areas (Illinoisan and Earlier Glacial Ages)

To the east of the Driftless Area is the Rock River Hill Country, which was only glaciated before the end of the *early* Wisconsinan time about 92,000 years ago. That hill country has a subdued glacial imprint. It is transitional in appearance between that of the nonglaciated Driftless Area and the more recently glaciated landscapes to the east. The Rock River Hill Country is similar to the other physiographic divisions that we see in west-central and south-central Illinois. The surfaces for over fifty miles west and south of the Bloomington Ridged Plain is pre-Wisconsinan in age.

> **(Q3-21)** Identify on the Bier map the following areas: Rock River Hill Country, Galesburg Plain, and Springfield Plain. Describe the general appearance in common for these areas.

Differences in drainage features can easily differentiate the Wisconsinan and older glacial areas. These drainage differences are a consequence of the very unequal amount of time that streams have been modifying the glacial deposits. The older surface has been ice free for at least 100,000 years, but ice retreated from the younger one only after about 22,000 years ago. As an illustrative example, Ruhe has an excellent map that shows drainage differences on adjacent glacial terrain of unequal age in Iowa.[7]

> **(Q3-22-Activity)** Using an overlay, trace the stream pattern in a rectangular area bounded by 39° N and 40° N latitude and by 88° W and 89° W longitude.
> **(Q3-23)** Comment on the drainage differences that are apparent on your tracing.
> **(Q3-24)** Comment on the relative amounts of postglacial erosion on the older (pre-Wisconsinan) and younger surfaces separated by the Shelbyville Moraine.

You have learned that noticeable drainage differences between the older and younger surfaces are apparent even on Bier's relatively small scale map (1:1,000,000). Examination of large scale topographic maps (1:62,500 and especially 1:24,000) reveals even more differences, similar to those illustrated by Ruhe. For example, we found statistically significant (P <.01) differences when comparing 102 first order stream basins (i.e., those without intermittent or perennial tributaries) on Illinoian deposits near MMC 105 175 (west of Havana)[8] to 25 basins on Woodfordian deposits (part of the Wisconsinan

Age) just north of Bloomington-Normal. The younger area had larger basins and longer streams but a smaller drainage density (average stream length per unit basin area) and coarser texture (stream spacing or degree of dissection). Although perhaps not directly comparable because the older basins were near the Illinois River, these results are consistent with what a major age difference would cause. This analysis also points out methods for quantifying and objectively studying such landscapes. Physical geography is strongly involved in quantitative scientific analyses.

> **(Q3-25)** Devise a way to quantify the stream differences between these areas using only the Bier map.

The analysis of the Bier map alone cannot differentiate intra-Illinoisan surfaces (the traditional Illinoisan ice sheet advanced at least three times in the map area) or even discriminate between areas of Illinoisan and pre-Illinoisan till because so much time has elapsed.

3.2.3. Other Landscapes

We have already noted the nonglaciated Driftless Area in northwestern Illinois. For central Illinois we can add the relatively small area from MMC 057 322 (T-305) to MMC 110 225 where the Illinois and Mississippi rivers meet. Glaciers affected all the rest of central Illinois but not here. Note the dashed line showing the re-entrant in the glacial border near Pittsfield (MMC 072 295). A map by Frye et. al. shows that the Illinoisan terminal moraine (which they call the Mendon moraine) defines the eastern side of the unglaciated re-entrant (near Pittsfield) and that the moraine continues northwestward to near where Bear Creek enters the Mississippi floodplain at MMC 020 355 (T-272).[9] Therefore, the glaciated area west of Pittsfield is older than Illinoisan Age. Frye et. al. also suggest that the area of unglaciated terrain

near Pittsfield may be smaller than shown on the Bier map, and that only extremely old (pre-Illinoisan) ice covered some of it.[10] The topographic evidence (greater dissection) that they cite is not observable on the Bier map because its scale is too small.

> **(Q3-26)** The Illinoisan boundary near Pittsfield separates glaciated areas with age differences at least as great as that between the Illinoisan and Wisconsinan surfaces. Why, then, is it not conspicuous?
>
> **(Q3-27)** Note the distinctive, uneven topography at MMC 084 226 and just north of Alton at MMC 122 223. Both areas have a more irregular terrain than that of the nearby loess-covered, pre-Wisconsinan glaciated landscape. Suggest a plausible explanation for these areas. [Hint: The same map symbol is also seen at MMC 122 146.]

3.3. Southern Illinois

The three major physiographic divisions in southern Illinois (see inset map) are the

(a) glaciated landscapes in the Mt. Vernon Hill Country in the Central Lowland Province
(b) unglaciated Shawnee Hills section in the Interior Low Plateau Province,
(c) floodplains of the Mississippi and Ohio rivers at the edges of the Ozark Plateaus Province and the Coastal Plain Province.

3.3.1. The Mt. Vernon Hill Country

The southernmost limits of the continental ice sheet of North America were in southern Illinois during the Illinoisan age, the second most recent of the major Pleistocene glacial ages. Specialists in glacial geomorphology can find the remnants and can accurately mark the "limit of glaciation" that Mr. Bier has drawn on his map. After the Illinoisan ice sheet receded (melted back), weathering, mass wasting, and erosion have had at least 100,000 years to modify those southern Illinois landscapes while glaciation occurred again farther north. Therefore, the glacial features of moraines and outwash plains in southern Illinois are not nearly as evident in the landscape (nor on the Bier map) as in the areas of younger deposits in central and northern Illinois.

One feature that is fairly common in the Mt. Vernon Hill Country is small marshes, as shown by the map symbol seen at MMC 162 162 and MMC 135 122. These symbols reveal extremely flat land. Be careful to avoid confusing the marsh symbols with the hachures that show relief in the hills.

This hill country also has flat-floored valleys (i.e., floodplains) of multi-branched streams bounded by low hills. Excellent examples are on the Beaucoup Creek north of Pinckneyville (MMC 193 126), the lower Kaskaskia River near MMC 143 125, and along the Little Wabash River near MMC 295 161 (east of Fairfield).

> **(Q3-28)** Find and give the millimetric coordinates of two or three other similar areas.

Some of the flat-floored valleys cannot be seen today because they are under artificially dammed lakes. Rend Lake is an excellent example. It was created in the late 1960s, so the original Bier map of 1956 showed the area as in the map extract in Figure A-1.

The lake is a major (30 mi^2 or 77 km^2) feature in southern Illinois and offers extensive recreational facilities. The greatest depth of the lake is only 35 feet (11 m), but the dam's length is 2 miles (3 km).

> **(Q3-29)** Could Rend Lake be used to generate hydroelectric power?

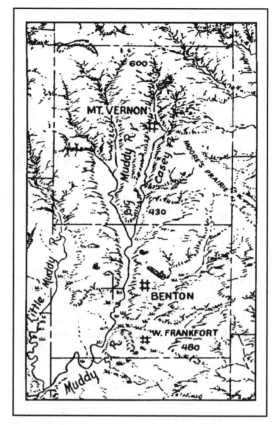

FIGURE A-1.

> **(Q3-30)** Rend Lake provides drinking water for over 300,000 people in sixty communities in seven surrounding counties. Suggest at least five of those counties.

3.3.2. Shawnee Hills

Since southern Illinois was the limit of glaciation, the ice sheet at the time was thinnest there and consequently not as powerful an agent in effecting landscape change as in areas farther north. In addition, the ice encountered higher land and thus had to move uphill.

> **(Q3-31)** Compare the spot elevations, such as 740 ft. at MMC 242

064, in the Shawnee Hills with those in the glaciated areas 40 or 50 miles (65-80 km) farther north. Are some of those *differences* in elevations more than 200 ft. (60 m)?

The Shawnee Hills have a wide range in topographic and geologic features.[11] Limestones, shales, and sandstones are plentiful. The eastern end of the Shawnee Hills, which extends into Kentucky, is intensely faulted. Specific benches and upland tracts usually correlate with resistant sandstone. The individual cliffs are not clearly shown on the Bier map.

Southern Illinois has the only National Forest in the state. Using an atlas or an Illinois highway map, transfer the border of the Shawnee National Forest onto an overlay or copy of the Bier map.

(Q3-32) Why do the areas of relief and managed forest approximately match?

3.3.3. The Mississippi and Ohio Floodplains

Activity: The large valleys of the Mississippi and Ohio rivers are highly important physical features. On a copy or on an overlay of the southern part of the Bier map, trace the following:

1. In blue, draw a single thin line in the center of the main part of the Mississippi, Ohio, Wabash, and Tennessee rivers. Do not worry about the islands because they are very flat and sometimes are covered by floods.
2. In brown (or any color other than blue), trace the outer edges of the flat valley on both sides of the rivers. Wherever you find hills in the valley (such as Fountain Bluff at MMC 184 080), draw it as an island. Do not be fooled by the symbol for marshes, such as at MMC 136 120 (NW of Chester).
3. The flat valley bottom between the two brown lines is called the floodplain.

(Q3-33) When the rivers are in full flood, do you think that the whole valley bottom will be inundated?

4. Use the scale and determine the widest, most narrow, and average width of the floodplain in each of the following areas:
   A. Mississippi River floodplain from St. Louis to Cape Girardeau (MMC 180 040).
   B. Mississippi River floodplain from Cape Girardeau to the Cairo area (MMC 211 007).
   C. The Wabash River floodplain.
   D. The Ohio River floodplain upstream from Shawneetown (MMC 303 086).
   E. The Ohio River floodplain from Shawneetown to the mouth of what Bier called Big Bay Creek (MMC 270 038). (Note: The word "Big" has been dropped from what is now simply called Bay Creek.)
   F. The Ohio River floodplain from the mouth of Bay Creek to Mound City (MMC 213 017).
   G. Note that the light marks, such as at MMC 168 003 and MMC 186 010, are probably former river terraces that do not have the same relief importance as the hills that border the valley bottom. Terraces are remnants of earlier levels of floodplains.

(Q3-34) Why is the Mississippi River floodplain south of Illinois so extremely wide? A hint is found in the inset map of physiographic divisions.

(Q3-35) How did the comparatively small Wabash River make such a wide valley? (Hint: Think back to 18,000 or so years ago.)

(Q3-36) Why is flat land so extensive just north of Shawneetown? (Hint: It is not simply because the Wabash and Ohio rivers join near there; look at the physiographic divisions shown in the inset map.)

(Q3-37) On your overlay of the floodplain you might have found two constrictions on the Mississippi River in southern Illinois. One is on the west side of Fountain Bluff (MMC 183 080). Do you think that the Mississippi River ever flowed along the east side of Fountain Bluff?

(Q3-38) What could have caused the river to start flowing on the west side and eventually erode the channel that it still uses today?

(Q3-39) The ice sheet did not go as far south as the constriction southeast of Cape Girardeau at MMC 183 030. This place is known as Thebes Gap. What could have formed this constriction?

Just like the previous example showing significantly different courses of the Mississippi River, the ancient course of the Ohio River was also different. To analyze this you need to make more tracings. They can be done on the same overlay or map.

Activity:

1. Trace in blue the current courses of the Cache ("Cash") River (MMC 220 039) and (Big) Bay Creek.
2. Trace the edges (a) of the valley bottoms along those two streams and (b) of the flat area north of Metropolis, Illinois (MMC 252 028). Be careful; there are numerous marsh symbols in the area.
3. Place another overlay on top of your tracing (or project it with an overhead projector). Assume that the Ohio River could not flow through the area of MMC 172 032.

(Q3-40) Which way did the ancient Ohio River flow? With firm dashed lines show the course of the ancient Ohio River. The shift of the river to its present course is is believed to have been caused by the same aggradation processes associated with the Mississippi River that we studied earlier (see Q3-38).

Bier shows sinkholes west of Waterloo (MMC 120 155) and by Cave in Rock (MMC 297 061). Sinkholes and caves are associated with limestone rock where ground water has dissolved subterranean passages.

Also shown are the locations of dams along the Ohio River. The Mississippi has no dams south of St. Louis, but we will discuss in the next section the numerous dams along the Illinois segment of the river.

(Q3-41) Find and count the dams along the Ohio River.

SECTION 4 AND OBJECTIVES 4, 5, AND 6: *APPRECIATION OF HUMAN/CULTURAL SYSTEMS OF ILLINOIS AS FOUND ON BIER'S PHYSIOGRAPHIC MAP*

Although the principal function of a physiographic map is to depict physical features, it is also possible to use such a map to study the human signature on the landscape and to learn of environmental and societal issues with an outlook toward uses of geography. These constitute our three final objectives, which are intermingled in this fourth section. We will look at several historical, contemporary, and future aspects of the human geography of Illinois. These aspects include the Township and Range survey system, political units, urban places, and transportation. A copy of the state highway map is necessary for some of the exercises.

4.1. Township and Range

Thomas Jefferson's genius surrounds everyone in Illinois every day of the year. Over two hundred years ago, before he became our third President, Jefferson proposed and the Continental Congress agreed to subdivide the states of the emerging nation into a grid of squares and rectangles to ease the transition from public to private ownership. With only minor modifications the United States Public Land Survey (USPLS), or simply called Township and Range, has come down to the present as Jefferson proposed it.

Surveying teams, under contract to the federal government, began their monumental task at many different places from Ohio westward. These places of beginning they called initial points. From each initial point radiated an east-west survey line (the base line) and a north-south survey line (the principal meridian). Along the base line, at six-mile intervals, they marked the intersection points of (more or less) parallel north-south lines, called range lines. Along the principal meridian, also every six miles, the surveyors laid the basis for parallel east-west lines, called town(ship) lines. Town lines and range lines extended away from their points of origin until the surveyors encountered major rivers, major lakes, or the survey lines from some other source. Although there are some exceptions, town lines and range lines formed a grid of six mile-by-six mile squares called survey townships. Further division of the survey townships, at one mile intervals, produced individual square-mile areas called sections, each with 640 acres. Local surveyors then eventually divided many of the sections into more affordable, more easily utilized half sections, quarters, eighties (eighty acres), forties, etc.

We can find evidence of Township and Range on the Bier map mainly in the county boundaries.

> **(Q4-1)** What type of line symbol does Bier use to show county boundaries? What symbol does the state use on the highway map? Which map(s) give(s) the county names?

As you look at the Bier map, almost every straight county line overlies a survey line from Township and Range.

> **(Q4-2)** What other feature did legislators like to use for delimiting counties? **(Q4-3)** A few straight county lines do not run north and south or east and west. Find two diagonal county lines in central Illinois. Why might legislators have preferred these diagonal lines on rare occasions?

Township-and-range surveys covered Illinois from three initial points, one in Indiana and two in Illinois. Surveyors worked westward from the Second Principal Meridian (starting near French Lick, in southern Indiana) until they surveyed to what is now approximately 88° West Longitude. That land equals what would become approximately one column of Illinois counties west of the Illinois-Indiana border. The northern limit in Illinois of lands surveyed from the Second PM was near Kankakee (MMC 322-467 (T-162)); in southern Illinois they almost reached the Ohio River. All of Vermilion, Edgar, Clark, Crawford, Lawrence, and Wabash counties plus parts of eleven other nearby Illinois counties have land described as being west of the Second Principal Meridian.

We can find the initial point for the Third Principal Meridian at the southwest corner of Marion County (MMC 215 170). Marion County's western border follows the Third Principal Meridian and its southern boundary is part of the base line. The Third PM stretches from the Ohio River to the Wisconsin line and is the basis of the Township and Range Survey for approximately 60 percent of Illinois's 55,000 square miles. We refer to the east-west line intersecting the Third Principal Meridian at its initial point as the Centralia Base Line, although Centralia, Illinois, did not exist when the surveyors hacked their way through the area early in the nineteenth century.

> **(Q4-4)** What is the approximate longitude of the Third Principal Meridian? What is the approximate latitude of the Centralia Base Line?

The initial point for the Fourth Principal Meridian is at MMC 102 341 (T-286). The area surveyed from the Fourth PM covers all the Illinois land west of the Illinois River and west of the Third PM where the Third PM is north of the Illinois River, starting at Peru (MMC 212 486 (T-142)).

> **(Q4-5)** If we use the name of the urban place closest to the initial point, what is the name of the base line of the Fourth PM?

The more prominent the surveyor's line, the more likely it was (and is) to appear as a county boundary. Legislators preferred principal meridians and base lines more than range lines and town lines in the same area. In turn, range and town lines took precedence over section lines.

> **(Q4-6)** Mark the county boundary segments that follow the Third Principal Meridian and those that follow the Fourth PM.

(Q4-7) On the Bier map or on an outline map with county boundaries, you can follow the Third PM from Cairo, at the junction of the Mississippi and Ohio rivers, north to Wisconsin. Roughly what percent of the entire length of the Third PM serves as county boundaries?

(Q4-8) What percent of the Fourth PM serves as county boundaries? Note: The Fourth PM passes through Jo Daviess County, in the northwest corner of Illinois; but in the calculation you should count only that portion of the Fourth PM actually in Illinois, ignoring the jump across eastern Iowa north of Rock Island County.

Here and there county boundaries take some odd jogs. The legislators mandated some of these jogs in an attempt to approximate, with short north-south and east-west segments, a diagonal line. A good example of this strategy occurred between Moultrie and Shelby counties, in the vicinity of MMC 250 284.

> **(Q4-9)** Approximately how long is each one of the steps in that segment of the Moultrie-Shelby boundary?

Another group of boundary jogs resulted because of the curvature of the earth and the convergence of the longitude lines. Surveyors sometimes made adjustments by moving range lines and north-south section lines a short distance east or west. You can see an example of such a correction at MMC 318 366 (T-263) along the boundary between Champaign and Vermilion counties.

Neither legislator strategy nor surveyor correction, however, explains the series of jogs on the southern borders of Vermilion, Edgar, and Clark counties. Here the answer lies with the aftermath of an 1809 Indian treaty. The treaty line ran northwestward from what became Indiana into what is now

Vermilion County, Illinois. In Vermilion, at MMC 346 332 (T-297), the line turned abruptly to the south-southwest and crossed what are now Edgar and Clark counties. In present-day Crawford County, just northwest of present-day Robinson, (MMC 337 230), the line turned, and on a southeastward track, extended back into Indiana. Federal surveyors soon thereafter surveyed and marked the territory within the 1809 treaty lands and other previously ceded sections of Illinois Territory to the south. Later surveys outside the 1809 Indian territory did not quite match the early surveys, hence the jogs in the survey lines and the subsequent jogs in our county boundaries.

> **(Q10-Activity)** On both the Bier and highway maps you can place a ruler or straight-edge along the three jogs on the western edge of the 1809 treaty area. Verify that they are in line.

4.2. Political Units

Partitioning of the landscape to produce political jurisdictions also is of interest to human geographers. Most prominent of all such partitions on the Bier map are, of course, the counties. The 102 Illinois counties range in size from McLean at nearly 1,200 square miles (encompassing Bloomington-Normal) to Putnam at the big bend of the Illinois River) with 160 square miles. County shapes and sizes evolved over a seventy-year period (1790-1859) as territorial governors, territorial legislators, and general assemblymen (after statehood in 1818) made political decisions concerning the state's map.[12] Early counties were often enormous, encompassing dozens of what became our modern counties. But the gigantic counties were impractical once settlement progressed beyond the frontier stage. When nearby communities vied for the right to be the county seat of their own areas, slices of the oversized counties went this way and that to help provide enough land

to form a county of reasonable dimensions. A few of these large counties were able to defend their territories against takeover of peripheral segments by the opposing factions. Champaign County (encompassing Champaign-Urbana), for example, succeeded in retaining its original, nearly perfectly rectangular 1,000 square miles despite many attempts at dismemberment. Ford County (just north of Champaign), however, presents an oddly shaped outline because by 1859, when Ford came into existence, the land it now occupies was a leftover, sparsely settled, poorly drained section of the Grand Prairie.

Each of the 102 counties has one urban place designated as the courthouse town, or county seat. County seats vary in size from Elizabethtown (MMC 289 057) with fewer than 500 people, to Chicago with nearly three million inhabitants. Bier always marked and named the county seat, but he did not distinguish seats from other communities. Therefore, the map user has no trouble picking out the seat of Schuyler County as Rushville (MMC 093 354 (T-273)) or the seat of Pike County as Pittsfield (MMC 071 297 (T-330)) because in these counties no other urban place received a Bier designation. When Bier marked two or more places with his town symbol, however, we cannot automatically recognize which serves as the seat of county justice. Sometimes we can make an educated guess on the basis of size (choose the larger) or centrality (choose the more central).

> **(Q4-11)** Which community would you say serves as the seat of Adams County, Camp Point (MMC 050 344 (T-283)) or Quincy (MMC 021 333 (T-293)?
> **(Q4-12)** Do McDonough County residents transact county business at Bushnell (MMC 098 400 (T-227)) or Macomb (MMC 084 392 (T-235))?
> **(Q4-13)** Do Kane County (MMC 270 552 (T-080)) marriages get recorded

in a courthouse in Elgin, Geneva, or Aurora?

Look immediately east of McDonough County into Fulton County (MMC 117 400 (T-228)). Lewistown, Canton, and Farmington all received from Bier the same four-line urban symbol. Canton had, in 1990, more than five times the population of either Lewistown or Farmington. Unless you know for sure where to find the Fulton County Courthouse, consult the *Illinois Highway Map.*

> **(Q4-14)** Name two ways to identify county seats on the highway map. What other information would be helpful in identifying county seats? Try to determine the seats of several other counties that have multiple urban places on the Bier map.

Neither the Bier map nor the highway map depicts civil-political units called townships. The township concept came to Illinois from the Yankee region (New England and New York). These Illinois units of local government today maintain half the roads in the state, take care of much of the property tax assessment, dispense millions of dollars in welfare, and perform various other governmental functions. Every square mile of every county in west-central Illinois and northwest Illinois is part of a distinct civil-political township. Northeastern Illinois once had townships throughout; but, in the late nineteenth century, townships essentially ceased to function in the Chicago parts of Cook County when city government there gained tremendous power. West-central and southern Illinois contain the 17 Illinois counties lacking township government. Most (15 of the 17 counties) never had townships, while the other two disadopted after trial periods. These 17 counties have remained loyal to a form of county government that came into the state from the Upper South.

> **(Q4-15)** Is there any way of telling from the Bier map which counties have never had civil-political townships?

4.3. Urban Places

Bier punctuated his physiographic maps with several hundred urban places. In the process, however, he omitted hundreds of others. His selection process raises several general questions. How did the places he chose to depict end up on the map? How did he convey the relative size and shape of his choices? Has he miscategorized any of the places? Would you have chosen a different mix, and how would you have symbolized to differentiate your choices? Has he omitted any important places?

We noted above that Bier did not discriminate between county seats and other communities. All places received the same sort of grid or mesh symbol, with names off to the side. A large majority of the places appear on the map as an interlocking set of four lines, forming a square oriented with the Township and Range grid. A handful of places received a mesh of more than one square, e.g., Peoria and Decatur.

> **(Q4-16)** Besides altering the number of squares in the mesh, how does Bier convey the relative importance of urban places?
> **(Q4-17)** How many sizes of lettering does he employ for his urban places?
> **(Q4-18)** According to the size of lettering, compile three lists (for large, medium, and small) of all urban places between the lines at MMC nnn 195 and MMC nnn 447 (MMC nnn T-180). Add the name of the place that Bier neglected to identify at MMC 024 400 (T-225).
> **(Q4-19)** Use the index to the highway map to obtain the latest population figure for each place in your

lists. Note that, contrary to the data in the 1993-94 index to the highway map, the population for Edwardsville should be 14,579 or thereabouts. Now determine the average population of the three lists and the high and low figures on each list.

(Q4-20) What appear to have been Bier's cutoff values for each list when he prepared this map in 1980?

(Q4-21) Which urban place or places should we move to a different category? Speculate on why a few places have increased so much in population.

(Q4-22) Are there any urban places not shown by Bier in this designated central region that, because of their current population, should appear on an updated version of the map? Hint: Examine the highway map for urban places depicted with an orange tint and compare that group with what you now know about the Bier map.

4.4. Transportation

The urban places discussed above do not stand in isolation. Each serves a local hinterland of farmers, shoppers, workers, etc., who use local roads to reach their destinations. Each has land links (and perhaps river and air links) with places on the Bier map. Each urban place also maintains connections (direct or indirect) with the rest of the world. In this section we will be looking at selected aspects of the transportation network: navigable water routes, early roads, railroads, superhighways, and airports.

4.4.1. Navigable Water Routes

Barges, recreational boats, and gambling palaces are among the watercraft that ply the Mississippi and Illinois rivers. Interruptions in their movement occur due to seasonal problems such as high water, low water, and ice. Dams on these rivers also pose barriers for river users. The dams, however, actually make navigation possible over certain stretches, especially for tugs pushing tows of barges heavily loaded with products like grain, petroleum, and coal. The dams back up river water to form long pools that are deep enough to permit passage. To allow vessels access to lower or higher pools, the United States Army Corps of Engineers operates locks along with each dam, locks that raise or lower vessels as needed. Locking a tow through, without splitting the tow into smaller segments, takes about thirty minutes per dam. A split tow requires three times as long at the lock.

(Q4-23) Give the MMC coordinates of the confluence of the Mississippi with its most significant right-bank tributary, the Missouri.

(Q4-24) From that confluence northward, how many dams would a tow skipper encounter on the Mississippi before reaching 41° North latitude? Circle each dam.

(Q4-25) The lock and dam at Alton (MMC 126 214) now carries the number 26R, meaning it is a replacement for number 26, once found two miles away. Using the highway map as a reference, label the lock-and-dam complexes on the Mississippi. Which number in the sequence is missing? Near what Missouri urban place would the Corps of Engineers have built that lock and dam?

Bier fails to show Lock 27, the southernmost of these lift stations on the Mississippi. Given the spatially oriented sequence of locks on the Father of Waters and the location of Lock and Dam 26R at Alton, we can infer that Lock 27 lies between Alton and Cairo, Illinois. In fact, Lock 27 is in the St. Louis area.

(Q4-26) With the official highway map in hand, find Lock and Dam 27. Place it in the proper spot on the Bier map. What is the name of the water-way that vessels must navigate in order to pass through Lock 27? What would you surmise is the name of the natural obstruction in the main part of the Mississippi northwest of Lock 27. Hint: Use the St. Louis inset map on the reverse of the main Illinois highway map.

(Q4-27) Assume you are an apprentice on the *Jennifer Ann*, a towboat pushing a tow of 15 barges of crushed limestone upstream toward the Quad Cities. The skipper asks you to determine how many hours it will take, assuming no special delays, to lock up through Lock 27 and travel up river to just beyond Lock 18. You may assume the tow will average four miles per hour on the open water. Locks 19, 26R, and 27 are large enough to permit passage of your entire tow at one time. All others require a split. Lock 27 is at mile marker 185.2; Lock 18 is at 410.4.

We find on Bier's map a total of six lock-and-dam sites on the Illinois Waterway (Illinois River and Des Plaines River). A seventh one, not shown, is at Lockport. Bier depicts one near Beardstown at MMC 104 343 (T-285), but the highway map places it just below the mouth of the La Moine River, near La Grange.

(Q4-28) What are the correct MMC coordinates of the La Grange Lock and Dam?

The preceding discussion dealt with locks on rivers, but in the centuries before people could dam major rivers, they dug canals through which draft animals pulled small barges. In Illinois, two canals were especially important. We will study them by asking

questions that utilize the Bier map. You can also refer to the canal information in earlier parts of this book.

(Q4-29) Bier uses a line-with-ticks-on-one-side as his symbol for canals. Find that symbol, name the two canals, and trace with your finger the entire lengths of the canals. In other words, find the canals on the map and study them.

(Q4-30) To what do the names Michigan, Mississippi, Illinois, and Hennepin refer?

(Q4-31) What does the Bier map tell us about the location of the locks on these two canals?

The course of the I&M Canal is obscure in the Chicago area. Basically we can say that the canal followed southwestward the lowlyng course of the Chicago River, crossed the flat lake plain, and ran parallel along the southern side of the Des Plaines River starting at MMC 324 541 (T-088).

(Q4-32) From the last mentioned point until the canal exits from Will County, use your own words to describe the course of the canal; see how much information about the canal can come from the Bier map.

You should be able to make similar meaningful statements about each major segment of the canals of Illinois. Look at the physiographic features. Think about lifestyles and historical events that impacted nineteenth century Illinois. Consider the canal purposes, the regional economic activities, and the placement of towns near the canal. As just one example, consider how the canal operators kept sufficient water in the canals while also avoiding the problems of the yearly high and low water levels of the rivers.

4.4.2. Early Roads

Land routes competed with rivers and canals as pre-railroad conduits for people and products moving to and from the Midwest. One of the most significant early nineteenth century land routes was the National Road, which stretched from Baltimore, Maryland, to Vandalia, Illinois (MMC 218 226). From 1820 to 1840 Vandalia was the capital of Illinois.

> **(Q4-33)** Trace the general path of the National Road from Terre Haute, Indiana; through county seats Marshall, Toledo, and Effingham; past St. Elmo; and into Vandalia. If federal control of the Road had not ended at Vandalia (because Illinois objected to Washington telling states where roads should go), where do you think federal authorities would have taken the National Road once they passed Vandalia?

4.4.3. Railroads

Most urban places in Illinois once had a railroad (or several) connecting them with the outside world. Many still do have rail links. Counties, civil-political townships, and communities pledged financial contributions (tax-supported or privately donated) for the right to have rails serve their space. Railroad tycoons used the competitive nature of nineteenth-century Illinoisans to raise capital for right-of-way construction, rolling stock, and operating funds.

As was the case in many parts of the country, Illinois eventually experienced a surplus of railroads. Trucks provided better freight service; automobiles lured the passenger traffic. So abandonment of rail service and trackage has become commonplace. A few abandoned rail corridors have found new life in the rails-to-trails movement, a nationwide initiative to convert old rail lines to recreational routes for hikers, bikers, etc. Thus far most of the Illinois rail-trails are in the Chicago area, but the first rail-trail owned and operated by the Illinois Department of Conservation, the Rock Island Trail State Park, stretches 28 miles from near Peoria (MMC 175 424 (T-203)) through the Bloomington Morainic System to the northwest and across the Spoon River to Toulon (MMC 152 463 (T-165)).

> **(Q4-34)** See the highway map for the approximate route of the Rock Island Trail State Park. How is it shown? Mark it on the Bier map and note the physiographic features of the area it crosses.

The route of Illinois's first railroad, the Northern Cross ran between Springfield and Meredosia, on the Illinois River at MMC 093 321 (T-306), with Jacksonville as one of several intermediate stops. The Northern Cross, originally scheduled to stretch all the way across Illinois as part of a grand scheme of transportation improvements, lasted about five years (1842-47). It preceded the first successful Illinois railroads by about a decade.

One of these railroad success stories was the Illinois Central's main line from Chicago to Centralia and beyond. We can trace that ICRR route and identify several urban places on the Bier map that owe their existence and early growth to their location beside the rail right-of-way.

> **(Q4-35)** Lay your ruler to show a straight line from Kankakee to Effingham. Your line will pass close to these Illinois Central Railroad stations, all dating from early in the second half of the nineteenth century: Gilman, Paxton, Rantoul, Champaign (then West Urbana), Tuscola, and Mattoon. Why might the cost per mile of railroad construction through this area have been fairly reasonable?

What modern American folk song celebrates this Chicago-Gulf Coast route of the Illinois Central? What other lines of transportation approximate this early route across east-central Illinois?

4.4.4. Superhighways

Expensive-to-build, high-quality superhighways link most of the larger urban places in Illinois. Many smaller places near these roads have undergone a rebirth as commuters have chosen them as residential sites instead of the crowded cities.

(Q4-36: Activity) (An Illinois highway map or a road atlas covering Illinois is necessary for this exercise.) We travel on highways and we see what is close to the highways. The Bier map does not include highways, so we would like to mark some main ones. Interstate Highway 90 from Elgin to Rockford is quite straight. Mark it on the map by paying close attention to the county boundaries. Then extend the line at both ends to Wisconsin and to the edge of Chicago. (Note: The Bier map is not sufficiently large to show the complexity inside Chicago.)

Another one that is easy is Interstate 55. Note, however, that it swings out to by-pass Dwight, Pontiac, and other towns along the old Route 66.

Some students could benefit by completing the mapping of the interstate highway network in Illinois. Others might prefer drawing more detailed highways in a smaller area, for example, in one or two counties close to home.

(Q4-37) The Interstate Highway Network evolved slowly over the last four decades. Which of the interstate highways would have been among the earliest links completed in Illinois? Which do you think are the latest additions? Why might the latter have received low priority?

4.4.5. Airports

As of January 1995, Illinois had 125 airports designated for public use. Of that number, 83 or two-thirds were in public ownership and therefore eligible for tax support. The other 42 were in private hands.

(Q4-38) Locate O'Hare Airport (MMC 316 561 (T-068)) and Midway Airport (MMC 329 540 (T-089)) and describe the physiographic feature on which the City of Chicago built these airports.

New airports come into existence infrequently; but if all goes according to plan, a new public ownership airport will begin to relieve the worsening congestion in the skies over Chicagoland soon after the turn of the millennium. The South Suburban Airport, popularly referred to as the Third Chicago Airport, is to have its first of seven projected runways in operation by the year 2001. During 1994 and 1995 the master plan and environmental impact statements have been in preparation, and other preliminary steps are occurring—including the critical task of convincing airlines to commit the resources necessary to build and maintain facilities and schedules at South Suburban. Completion of the airport would take approximately two decades, if approved.

Officials considered a final list of five potential sites. Because O'Hare and Midway Airports are in the northern and western areas of metropolitan Chicago, the five alternatives for a third major airport lay south of the city. Of the five, a pair are urban and three are rural. The urban sites (designated Lake Calumet and Gary, Indiana) are relatively close to the center of the area's population, and would displace at least 25,000 residents each, while the rural trio (designated Beecher, Peotone, and Kankakee) would each displace less than a tenth that number. Ultimately, less displacement and less cost for land acquisition had a lot to do with the decision to choose the Peotone site.

Chicago's South Suburban Airport would occupy approximately 20,000 acres (31 square miles) of farmland in Will County. Centered at MMC 334 494 (T-135), this multi-billion dollar airport would extend from Peotone to Beecher.

> **(Q4-39)** Would the new airport be on a glacial lake plain or in morainal country?
>
> **(Q4-40)** What is a potential plus and a potential minus of this (lake plain versus moraine) decision?

The Illinois Department of Transportation estimates the number of displaced residents at 1,800. Additional factors in the decision to choose this site included the proximity of two interstate highways, good wind-direction coverage, and access to the forthcoming St. Louis-Chicago high speed rail corridor. If the high speed rail route goes into service, trains would travel between South Suburban and the Chicago Loop in 17 minutes.

> **(Q4-41)** Where do you think the displaced 1,800 residents would decide to settle? Note: During World War II, fifteen miles west of Peotone, around MMC 301 493 (T-135), hundreds lost their farm homes when the fed-

eral government acquired 40,000 acres to build the Joliet Army Ammunition Plant.

CONCLUSION AND ACKNOWLEDGEMENT

The commentary, questions, and activities involving the Bier physiographic map in this appendix provide insight into the life and landscape of Illinois. Some aspects of this exercise are quite complex while others are simple or even obvious; the mix is intentional because of the wide backgrounds of users of the Bier map. We certainly have not covered all potential uses of the Bier map. We encourage you to explore your local area, making comparisons with the Bier map and with other reference maps. As you and your friends and students discover additional exercises or comments that relate to the Bier map, please send them to the authors at Illinois State University, Normal, IL 61790-4400. Also, please remember that the full size 17 x 27 inch *Landforms of Illinois* map by James A. Bier is available for only twenty-five cents per copy (plus $0.95 postage and handling for orders up to $6.00) from the Illinois State Geological Survey (ISGS) at 615 E. Peabody Drive, Champaign, IL 61820. On behalf of all who use it, the authors express their appreciation to the ISGS and James A. Bier for allowing the unrestricted noncommercial photocopying of the *Landforms of Illinois* map.

ANSWERS

(Q1-1) A cross.

(Q1-2) No, because of the Earth's curvature and the projection of the map.

(Q1-3) Yes, the central meridian, which is 89° 30' West longitude.

(Q1-4) A good estimate of Springfield's lat-long coordinates is about 39° 50' North, 89° 40' West. The central grid square is at 39° 45' 50" N, 89° 34' 21" W.

(Q1-5) The number "4" in the spot elevation of 470 feet above sea level.

(Q1-6) MMC 216 512 or MMC 216 T-117. Both answers are correct.

(Q1-7) Fox River. Do you like Bier's letter "F"?

(Q1-8) Ottawa at MMC 240 490 (T-139). Flowing to the southwest through Yorkville.

(Q1-9 to Q1-12) 136 kilometers or 83 miles.

(Q1-13 and Q1-14) The latitude answers should be approximately 70 miles or 110 kilometers in all parts of the map.

(Q1-15) The longitude distances diminish with increasing latitude.

(Q1-16) Rivers are either thin continuous black lines or, for very large rivers, black lines showing the river banks with horizontal dashed shading in the stream of the river; creeks are thin short dashed lines; counties have long dashed lines in straight sections and use the regular river symbol for borders along the rivers; state borders are thicker and occasionally double-dashed lines or they are the river symbols; canals are lines like rivers but with "tick marks" on one side of the line; railroads are not shown on the Bier map; and glaciation limits are thick dashed lines.

(Q1-17) The short vertical lines grouped together look like a clump of grass or weeds. The symbol is for a marsh or swamp.

(Q1-18) Arlington Moraine.

(Q1-19) Driftless Section.

(Q1-20) Yes. Physiographic maps usually take a view from the south with shadows appearing as if the sun would shine from the northwest.

(Q1-21) The cliffs mainly face north toward the river. Therefore, they are not so clearly seen from the southeast (refer to Q1-20 above).

(Q2-1) The drawing of Illinois is darker than the surrounding areas, but the physiographic depiction (artistic style) is the same style.

(Q2-2) Thick, occasionally dashed black lines on the straight borders, with rivers forming the other borders.

(Q2-3) Yes, because the borders are *all* rivers, and the rivers are shown.

(Q2-4) Four are shown:

(a) Ozark Plateaus Province (thin section along the Mississippi River, but extending over much of the state of Missouri)

(b) Coastal Plain Province (only one or two percent of Illinois is in this province that extends all the way to the Gulf of Mexico)

(c) Interior Low Plateau Province (This prominent but small [4-6 percent] southern Illinois area is similar to the lands extending into Kentucky)

(d) Central Lowland Province. (About 90 percent of Illinois is in this Province, the Central Lowland, which extends over much of Iowa, Indiana, and Wisconsin.)

(Q2-5) Peoria, Bloomington-Normal, Champaign-Urbana, Decatur, and Springfield.

(Q2-6) 24, if we include Kankakee County.

(Q2-7) Putnam, but in the early 1800s it was very large and extended over most of this northern Illinois area.

(Q3-1) The Shawnee Hills of southern Illinois have similar relief.

(Q3-2) 990 feet, 940 feet, 1235 feet.

(Q3-3) These valleys seem to be deeply incised into the landscape and oriented NE to SW. They have what appear to be flat bottoms—possibly floodplains—and steep sides. Their tributaries are arranged in a dendritic pattern. The wide interfluves seem to be generally low in relief but conspicuously higher than the valleys. Relief is greater near the valley walls whereas the interfluves are somewhat rounded.

(Q3-4) The moraines were not there then. The ice sheet left the moraines after it blocked the ancient Mississippi River, thus forcing the river to flow westward past Rock Island. See Figure 2-13 in this book.

(Q3-5) The ice came from the northeast in each case.

(Q3-6) The order is 2, 1, 3, 4. The moraines must decrease in age northeastward, from where the ice came.

(Q3-7) The ice was in the area of Cook and DuPage counties when it was building the Valparaiso Morainic Complex.

(Q3-8) Yes, because eskers are formed by rivers of meltwater running under stagnant (not moving) glacial ice. The sand and gravel bed of the river becomes the esker.

(Q3-9) Relief is low but the landscape is uneven. Topographic elements are clearly arranged in broad north-south concentric belts parallel to the shore of Lake Michigan. Valleys are indistinct except for those of the Des Plaines and Illinois rivers, which appear in places to have a wide floodplain. The drainage pattern is unorganized and lakes are common near the Wisconsin border. Interfluves are indistinct.

(Q3-10) They had to flow in the lower areas, between moraines. Some of these rivers were probably former ice-marginal streams and thus conveyed meltwater and related sediments.

(Q3-11) (a) Poorly drained, disordered drainage containing lakes and swamps; drainage network (pattern) is still evolving because it is so young, (b) low relief, hummocky, undulating, pitted relief landscape with juxtaposed small hills and low places. The average local relief in feet per square mile on the quadrangle is between 47 and 87 ft. (14-27 m).

(Q3-12 and Q3-13) They are so distinctly different, they must have had different histories. It would seem obvious that the drainage in the Driftless Area is much older, while that of the Wheaton Morainal Country is quite young and immature.

(Q3-14) The Bier map simplifies and generalizes to depict with accuracy the landscapes of northern Illinois.

(Q3-15) The area east and north of that moraine system has conspicuous parallel and concentric belts of irregular topography (labeled "moraines"). Zones with very low relief that seem flat on the Bier map separate the moraines. Streams there do not have obvious valleys. An underdeveloped drainage pattern prevails, and many streams flow parallel to moraines.

(Q3-16) The ice sheet had a lobate margin caused by major topographic features farther north, especially the basins of the Great Lakes. Those lake basins did not exist in preglacial time, and they became deeper and larger with each glaciation. Thus the earlier glaciations (the Nebraskan and Kansan of the traditional chronology) were less lobate in character than later ones. Ice from the Erie basin built the Iroquois Moraine and most of those to its south and southwest. The Lake Michigan Lobe made the moraines west of this major contact with the Erie Lobe.

(Q3-17) The activity of coloring the map can aid Objective 2 as well as the current Objective 3.

(Q3-18) These two areas, along with others near and northwest of Kankakee, were sites of short-lived meltwater lakes collectively called the Kankakee Flood. They formed when concentrated meltwater from the Lake Michigan, Saginaw, and Erie lobes flooded all but the higher end moraines while the ice sheet was building the Valparaiso Morainic complex near Chicago. See Schuberth for a map of Illinois' glacial lakes.[13] The large discharge from these lakes probably caused the diversion of the Mississippi River through the Thebes Gap (see Q3-33 to Q3-39).[14]

(Q3-19) This area is a large fan-shaped outwash deposit, the Bloomington Valley Train, that begins at the Bloomington-Leroy Moraine (Wanless, 1957). Its original slope (within about 50 km of the moraine) of about 0.85 m/km (4.5 ft/mi) was about 25 times steeper than the present gradient of the Illinois River. Its slope was steeper because meltwater streams had a coarse and abundant sediment load. Subsequent meltwater, such as discharge from the Kankakee Flood, has eroded it. Several terraces exist today, and Bier drew at least one of them as a line on the east side of the Illinois River from just north of Havana (at MMC 137 375 (T-252)) to Pekin; this line represents a terrace scarp. The belts of uneven topography on the terrace surfaces are areas of stabilized sand dunes. These dunes formed because seasonally dry, unvegetated sand was plentiful on the aggrading outwash fan in winter when braided meltwater streams had relatively little discharge.

(Q3-21) The landscape west and south of the moraines has essentially a fluvial (stream-created) appearance. Drainage lines are conspicuous, but areas between major streams have very low relief. The drainage pattern is primarily dendritic, although the tributaries of the LaMoine River (MMC 077 351 (T-266) south of Macomb) are parallel and join it at right angles.

(Q3-23) The drainage south of the Shelbyville Moraine has more streams and a clear dendritic pattern with many tributaries entering the main valleys. The drainage pattern on the younger surface appears relatively incomplete. The seemingly stream-free areas are smaller on the older surface at the south.

(Q3-24) Bier shows that major valleys on the older, pre-Wisconsinan surface are readily apparent, with streams well below the elevation of the upland areas. Almost none of the streams on the Wisconsinan surface have obvious valleys, thereby implying that

(Q3-25) postglacial dissection is much less than what we see on the surface beyond the Wisconsinan moraines.

(Q3-25) Try counting the number of streams per unit area within, for example, a one square inch map area outlined by a hole cut out of a piece of paper and randomly placed on the Bier map within each age region. Alternatively, and somewhat easier, try counting streams along random transects, that is, draw random straight lines of equal length in each of the two areas and count how many streams each line crosses. Calculate the average number of streams crossed per specified distance (such as 10 centimeters) for each of the two types of areas. Another possibility could involve counting the number of tributaries per X miles or Y kilometers of major streams.

(Q3-26) Postglacial erosion and mass wasting have had enough time to erase the irregular topography created directly by the older ice sheets (and still visible on the Wisconsinan surface). Therefore, the older landscape has more of a fluvial character. The degree of drainage development slows with time, especially after full drainage integration. The relief was probably never great (nor is it within the Wisconsinan area), and the sediments are fine grained and easily erodible. Also, both Illinoisan and pre-Illinoisan drift surfaces near this location have a 2.5 to 7.5 meter thick loess cover derived from the nearby Mississippi River and Illinois River valley trains.[15] The loess (fine, wind-blown silt deposits) would tend to reduce relief and mask any subtle topographic differences attributable to glaciation.

(Q3-27) These areas have sinkholes and indicate that limestone is the surface rock. A karst landscape has formed. To some extent, the topography on fresh moraines and in karst areas resemble each other.

(Q3-28) The many examples include Mary's River at MMC 163 107; upper Skillet Fork (MMC 257 167, west of Fairfield); and Casey Fork near Mt. Vernon (MMC 237 152).

(Q3-29) No, because the vertical drop is so low.

(Q3-30) Jefferson, Franklin, Jackson, Perry, and Washington counties.

(Q3-31) Yes, this is about the same as the height of a 10-story building. The ice sheet was thick, but not thick enough at the southern terminus to go over the Shawnee Hills.

(Q3-32) Rugged lands in Illinois were not desirable for crops or even for pastures, so they were left as forests by the early settlers.

(Q3-33) Most could be flooded except people have built levees to keep the river in its banks. The media coverage of the exceptionally high floods of 1993 frequently referred to these levees, which sometimes are breached.

(Q3-34) Although the Gulf of Mexico formerly extended as far north as the Cairo area, alluvial sediments now cover the deltic sediments from there to the present delta in Louisiana. During each of the Pleistocene ice ages, the Mississippi River in the map area created a wide floodplain with a braided channel pattern as it filled its valley with meltwater sediments (outwash). The outwash is 50 meters (160 ft.) thick in northern Arkansas, but the post-glacial river has cut 18 m below the top of the deposits.[16]

(Q3-35) The Wabash River was much larger when it carried meltwater. Outwash sediments deposited by the braided channel made the wide floodplain.

(Q3-36) Actually, this question refers more to the change in geology south of Shawneetown. The floodplain north of there is wide because the low relief in the Central Lowlands permitted a wide floodplain when meltwater aggradation (channel filling) was taking place. But relief in the Interior Low Plateaus Province is greater; and the Ohio River in that province could not, therefore, be as wide as in the area north of Shawneetown.

(Q3-37) Yes, because the wide flat area there is floodplain.

(Q3-38) Evidence from the map suggests that the glacier filled part of the valley with ice. The ice would have made a dam that created a lake at higher and higher elevations until it found an outlet at the location of the present stream west of Fountain Bluff. It appears that the river at that time could have eroded (down cut) long enough to attain a level equal to or even lower than the area that was blocked. According to this hypothesis, when the glacier retreated, the river stayed on the west side. Yet, the map evidence is ambiguous. Note that the Wisconsinan (the last) ice sheet did not extend that far south in Illinois; the glacial border shown on Bier's map is thus pre-Wisconsinan in age. Therefore, the flat area east of Fountain Bluff might possibly contain Wisconsinan outwash. If it does, the present course of the Mississippi River could have occurred simply because floodplain aggradation (deposition) in Wisconsinan time

may have caused the river to spill through the area of the present channel.[17]

(Q3-39) Tectonic forces, suggested by contemporary earthquakes, may at first seem a possibility. If you check an atlas, you will find that Cairo, Illinois, is less than 40 miles northeast of New Madrid, Missouri, the location of the famous New Madrid earthquakes and geologic fault. But the origin of Thebes Gap was actually due to the great volume of sediment contributed by the ice sheet's meltwater. The Mississippi River formerly flowed in the wide lowland extending southwestward from Cape Girardeau all the way to MMC 118 000. Aggradation (deposition) of glacial outwash caused the Mississippi to shift progressively eastward to the position at MMC 163 018 and later to its present route through the Commerce Hills at Thebes Gap.[18] If you did not trace in brown the earlier floodplain limits in this area of Missouri, do that tracing now and look to see how wide the historical floodplain was. If the modern Mississippi had not shifted from its earlier course, southern Illinois could have been bigger.

(Q3-40) Go "upstream" on Bay Creek and eventually join the Cache River lowlands. Earth scientists (such as geologists, geomorphologists, and physical geographers) do not believe that the ancient Ohio River initially flowed through the small gap at MMC 248 031, but instead flowed westward through the Cache River lowlands.

(Q3-41) Four are shown.

(Q4-1) Bier used a single line of alternating long and short dashes. The state highway map uses a single line, which has varied in recent years from three short dashes between longer dashes to alternating dot and short dash. County boundaries on the highway map also feature a pale yellow overprint of the black line symbol. The yellow helps the county boundaries stand out on the highway map. Bier did not name counties; the state highway map does.

(Q4-2) Rivers and lesser streams often serve as county boundaries. For instance, the Illinois River is often a county boundary. The Sangamon River and its right-bank tributary, Salt Creek, form the northern boundary of Menard County (from MMC 141 352 (T-276) to MMC 173 357 (T-271)).

(Q4-3) A diagonal separates northern Piatt County from DeWitt County (MMC 256 362 (T-267)). Another good example of a diagonal county boundary is at MMC 144 300 (T-328), between Sangamon and Morgan counties. Diagonals may have been attempts to split an area equally, to follow a drainage divide, or to leave a settlement (like Farmer City) in a particular county.

(Q4-4) 89° 9′ West Longitude. (Hint: Use the large "+" mark at MMC 228 118 to improve your accuracy to within two minutes of a degree.) 38° 28′ North latitude. The Centralia Base Line, by the way, is a westward extension of the base line of the Second PM.

(Q4-5) The Beardstown Base Line, even though Beardstown lies east of the Illinois River and is, therefore, under the jurisdiction of the Third PM.

(Q4-6) For the Third PM you should have marked seven boundary segments, beginning on the south edge with the Marion-Clinton boundary. The Fourth PM serves as a county boundary for much of its extent, but not immediately north of the initial point. Note: An additional useful exercise is to mark on a small Illinois outline map the principal meridians and base lines, and then color the areas each affects.

(Q4-7) We measured the total length of the Third PM as 390 miles and found seven county boundary segments, totalling 168 miles. The answer is approximately 43 percent. Note: Because the map scale is 1:1,000,000 (or 1 cm = 1 km), the measurements and calculations are easier if you use the metric system.

(Q4-8) Our figures for the Fourth PM are 118 total miles and 88 boundary miles (all in one stretch). The answer is 75 percent.

(Q4-9) Each step in this portion of the Moultrie-Shelby boundary equals one mile.

(Q4-10) The Illinois Department of Transportation prepares, and sells for a small fee, county highway maps. If you were to follow this treaty line across the highway maps of the affected counties, you would find other evidence of modern connections to this historic division, such as local roads running diagonally along the line. Another excellent large-scale source of map detail on this Indian treaty line is the county plat book. These county-by-county collections of maps will reveal that even property boundaries occasionally follow such treaty lines.

(Q4-11) Despite an eccentric location, Quincy has always been the seat. Around 1840, Quincy faced a major challenge to its seat status from Columbus (see the highway map), a community founded purposely right at the geometric center of Adams County. In fact the struggle between Quincy and Columbus became so intense that Quincy sponsored the creation of a new county, Marquette, which took the eastern half of Adams County and Columbus. When the people of that eastern area refused to organize as a new county, Adams eventually regained the separated portion.[19]

(Q4-12) Centrality wins here. The answer is Macomb.

(Q4-13) Although much smaller than Aurora or Elgin, Geneva is more central and is the county seat. This is an example of a functional region matching the borders of a formal region.

(Q4-14) The highway map has a small dot inside a circle to mark county seats plus bold-print type in the "Index to Cities and Villages." On some of the recent versions of the *Illinois Highway Map*, including the 1993-94 edition, Winchester (MMC 101 109) does not receive a bold-print designation in the map index. Even the most carefully prepared maps will sometimes contain errors. Other helpful information in determining seat status might include the relative age of communities because being first, even by a few years, could be a critical advantage in obtaining the courthouse.

(Q4-15) No, there is no way to identify the so-called commission counties on the Bier map. In the 17 commission counties, three people—each elected at-large—make up the county board. For the other 85 (township counties), board members usually represent a district within their county. For-

merly that district was the township. The 17 commission counties are Calhoun, Menard, Morgan, Scott, Monroe, Randolph, Perry, Edwards, Wabash, Williamson, Union, Johnson, Pope, Hardin, Massac, Pulaski, and Alexander.

(Q4-16) Bier varies the size of lettering in names of urban places.

(Q4-17) He uses three lettering sizes. For example, near MMC 182 400 (T-228), we find Peoria (large), Pekin (medium), and Morton (small).

(Q4-18) The place that Bier failed to name is Nauvoo, which in the middle of the 1840s was the largest community in Illinois. On our lists we have 3 in the large category (according to the Bier system), 9 in medium, and 75 in small—for a total of 87 places in the designated area.

(Q4-19) Population figures are, of course, always changing. Here are the answers we found using the 1993-94 highway map. The three large places averaged 100,877 with Peoria being the largest at 113,504 and Decatur the smallest at 83,900. In the medium category, the nine averaged 35,861, and ranged in size from Bloomington at 51,889 to Pekin at 32,254. Bier's 75 small places ranged from Normal at 40,023 to Hardin with 1,071. These 75 averaged 6,336 inhabitants. Note: You may want to look at medians and standard deviations when studying the data.

(Q4-20) Because of the time that has elapsed since he made the map in 1980, it is difficult to tell exactly. Perhaps he used 25,000 and 75,000.

(Q4-21) No matter what his cutoffs were, Normal must move to the list of medium size places. In fact, Normal is now larger than all but two of the places on the medium list. If we go with cutoffs of 20,000 and 50,000, Macomb and Charleston join Normal on the medium list while Bloomington and Champaign move to the list of large urban places. Part of the reason for population growth in these communities is the presence of a state university in or near each. Another factor could be called "twin city" development. If taken together, the two twin cities of Bloomington-Normal and Champaign-Urbana would have approximately 92,000 and 100,000 inhabitants, respectively, and they would merit the large category rating.

(Q4-22) Here are three that we would add: East Peoria (MMC 177 415 (T-213)), 21,378; Godfrey (MMC 125 222), 16,085; and Washington (MMC 190 422 (T-206)), 10,099. Note that all three additions are near large urban places. If we studied northern and southern Illinois, we would find additional cases near Chicago and St. Louis.

(Q4-23) The Missouri enters the Mississippi at MMC 130 210.

(Q4-24) Bier depicts eight on the Mississippi.

(Q4-25) Number 23 turned out to be unnecessary. It would have been near Louisiana, Missouri.

(Q4-26) Number 27 stands at MMC 125 195, near the southern end of the Chain of Rocks Canal, a bypass around the Chain of Rocks rapids in the main channel near the I-270 bridge. Lock and Dam 27 is really just a pair of locks in the canal; there is no dam there. There is a low-water dam across the main channel of the Mississippi River near the actual rapids.

(Q4-27) Many such journeys would involve waiting at locks for other vessels to pass, but here is how we see an undelayed trip. You would traverse 225.2 miles of river (410.4-185.2). At four miles per hour, just the traverse would require 56.3 hours. Your locking time will involve three at a half hour each and six at 90 minutes each, for a total of 10.5 hours. You should report a grand total for this stretch of 66.8 hours—nearly three calendar days.

(Q4-28) The correct coordinates are MMC 096 335 (T-298).

(Q4-29) The Illinois-Michigan Canal (at MMC 265 490 (T-139) is often called the I & M Canal. The Illinois-Mississippi Canal (at MMC 140 501 (T-127)) has the same initials, but is usually called the Hennepin Canal.

(Q4-30) Those are NOT state names. They refer to the lake, the two rivers, and the nearby town, respectively.

(Q4-31) Nothing. The locks are not shown.

(Q4-32) The I & M Canal runs southwest along the side but not in the Des Plaines River. Both the canal and river utilizes the natural cut through the Tinley, Wheaton, and West Chicago moraines. We know from our earlier map study that these cuts are large because they were made by the raging streams that flowed from Lake Chicago in the glacial times. We also see a second canal at MMC 325 258 (T-100). (The map does not tell us that this Calumet Sag Channel was finished in 1922 and is currently used with the modern river navigation.) Just north of Joliet, the I & M Canal actually entered the main channel of the Des Plaines River. South of Joliet the canal again has its own channel, this time on the north side of the river. In other words, the Canal traffic crossed through the

Des Plaines River in the Joliet area. The canal then runs through very flat land until it approaches the Minooka Moraine and the confluence of the Des Plaines and Kankakee rivers.

(Q4-33) We might guess that St. Louis was the next big stop, and a glance at the modern highway map confirms the suspicion. U.S. 40 and later I-70 followed the approximate route planned for the National Road.

(Q4-34) The symbol on the highway map is a series of pairs of green circles. An excellent source of information on such trails is *40 Great Rail-Trails in Michigan, Illinois and Indiana.*[20]

(Q4-35) Railroad building in glacial deposits with small streams was relatively inexpensive compared to construction through rugged, rocky land. *The City of New Orleans*, also the title of a folk song, is an overnight train connecting Chicago and New Orleans. U.S. 45 and I-57 follow a similar route.

(Q4-37) The earliest were I-55, I-57, I-70, I-74, I-80, I-90, and I-94. Most recent are I-39, I-155, and U.S. 36/336. The latter is an attempt to fill a gap where residents have long complained of being isolated and ignored. This area south of the Quad Cities and west of the Illinois River has received the nickname Forgottonia. Some hotheads there (half seriously) suggested that they secede. U.S. 36/366 received upgrades or completely new construction in the 1980s to connect Quincy and other poorly served parts of Forgottonia with Springfield and elsewhere. The highway map also shows a superhighway under construction from Monmouth to Macomb.

(Q4-38) On glacial lake plain.

(Q4-39) Valparaiso Morainic System.

(Q4-40) Possible plus: better natural drainage. Possible minus: more cutting and filling to level the land.

(Q4-41) Many will resettle within 20 miles, although farmers will find it difficult to acquire comparable farms without going farther.

Notes

1. This appendix is adapted from a four-part series of articles in the *Bulletin of the Illinois Geographical Society*, Jill Freund Thomas, editor. Paul S. Anderson. "Physiographic Maps of Illinois: Introduction," *Bulletin of the Illinois Geographical Society*, Vol. 34, No. 2 (Fall 1992), pp. 19-22; Paul S. Anderson and Albert D. Hyers. "Using the ISGS/Bier Physiographic Map of Illinois: Northern Examples," *Bulletin of the Illinois Geographical Society*, Vol. 35, No. 1 (Spring 1993), pp. 16-24; Paul S. Anderson and Albert D. Hyers. "Interpreting the Physical Geography of Illinois as Shown on the Bier/ISGS Physiographic Map: Northern and Southern Examples," *Bulletin of the Illinois Geographical Society*, Vol. 35, No. 2 (Fall 1993), pp. 24-34 plus 18-19 for map; Paul S. Anderson, Albert D. Hyers, and Michael D. Sublett. "Human Geography and Physical Features of Illinois as Revealed on the Bier/ISGS Physiographic Map: Focus on Central Illinois," *Bulletin of the Illinois Geographical Society*, Vol. 36, No. 1 (Spring 1994), pp. 16-38.

2. *Geography for Life: National Geography Standards-1994*. Geography Education Standards Project.

3. Paul S. Anderson. "Millimetric Coordinates (MMC) for Image Locations: Methods, Applications and Advantages," *Technical Papers*, Fall Convention of the American Congress on Surveying and Mapping and the American Society for Photogrammetry and Remote Sensing, Falls Church, Virginia, 1985, pp. 486-494, and Paul S. Anderson. "Millimetric Coordinates (MMC) for Precise Locations on Aerial Photographs and Other Images," *Photogrammetric Engineering and Remote Sensing*, Journal of the American Society of Photogrammetry, December 1985, pp. 1963-1964.

4. *Geography for Life: National Geography Standards-1994.* Geography Education Standards Project, 1994.

5. D.E. Sugden. "Reconstruction of the Morphology, Dynamics, and Thermal Characteristics of the Laurentide Ice Sheet at its Maximum," *Arctic and Alpine Research*, Vol. 9 (1977), pp. 21-47.

6. J.C. Frye, H.B. Willman, and R.F. Black. "Outline of the Glacial Geology of Illinois and Wisconsin" in Wright, H.E. and Frey, D.G., (eds.) *The Quaternary of the United States.* Princeton, NJ, Princeton University Press, 1965, pp. 45-60.

7. R.V. Ruhe. "Topographic Discontinuities of the Des Moines Lobe". *American Journal of Science*, Vol. 250 (1952), pp. 46-56.

8. A.D. Hyers. "Drainage Development on Illinoian Drift: A Comparison." Unpublished M.A. research paper, Western Michigan University, 1969.

9. J.C. Frye, H.B. Willman, and H.D. Glass. *Cretaceous Deposits and the Illinoisan Glacial Boundary in Western Illinois.* Illinois State Geological Survey, Circular 364, 1964.

10. Ibid.

11. W.D. Thornbury. *Regional Geomorphology of the United States.* New York: John Wiley and Sons, 1965.

12. Michael D. Sublett. *Paper Counties: The Illinois Experience, 1825-1867.* New York: Peter Lang, 1990.

13. C.J. Schuberth. *A View of the Past: An Introduction to Illinois Geology.* Springfield: The Illinois State Museum, 1986, p. 141.

14. M.M. Leighton, and H.B. Willman. "Loess Formations of the Mississippi Valley," *Journal of Geology*, Vol. 58 (1950), pp. 599-623.

15. H.B. Willman, and J.C. Frye. *Pleistocene Stratigraphy of Illinois.* Illinois Geological Survey Bulletin 94, 1970.

16. R.F. Flint. *Glacial and Quaternary Geology.* New York: John Wiley and Sons, 1971.

17. Thornbury, *Regional Geomorphology*, p. 225.

18. Ibid., p. 59.

19. Sublett, *Paper Counties*.

20. Roger Storm, Susan Wedzel, Karen-Lee Ryan, and Mike Ulm. *40 Great Rail-Trails in Michigan, Illinois and Indiana.* Washington, D.C.: Rails-to-Trails Conservancy, Saturn Corporation, 1994.

BIBLIOGRAPHY

Ackerman, William K. *Early Illinois Railroads.* Chicago: Fergus Printing Co., 1884.

Addams, Jane. *Twenty Years at Hull House.* New York: Macmillan, 1910.

Ahmed, G. Munir. *Manufacturing Structure and Patterns of Waukegan-North Chicago.* Research Paper No. 46. Chicago: University of Chicago Department of Geography, 1957.

Aldrich, S.R. *Illinois Field Crops and Soils.* Cooperative Extension Service Circular 901. Urbana: University of Illinois, 1965.

Alexander, John W. "Geography of Manufacturing in the Rock River Valley." *Wisconsin Commerce Papers* 1 (August 1949), 149-50.

_____. "Manufacturing in the Rock River Valley." *Annals of the Association of American Geographers*, 40 (1950), 237-58.

Allen, John W. *It Happened in Southern Illinois.* Carbondale: Southern Illinois University Press, 1968.

_____. *Legends and Lore of Southern Illinois.* Carbondale: Southern Illinois University Press, 1963.

Allman, John et al. "The Use of Standardized Values in Regionalization: The Example of a Socio-Economic Spatial Structure of Illinois, 1960." *The Professional Geographer*, XVI (May 1964), 5-11.

Altes, Jane. *East St. Louis: The End of a Decade.* RUD Report No. 3. Edwardsville, Ill.: Southern Illinois University, Regional and Urban Studies and Services, January 1970.

_____. *Population Projections for the State of Illinois and Component Regions to 2010.* Urbana: Bureau of Business and Economic Research, 1967.

Andreas, Alfred T. *History of Chicago*, 3 Vols. Chicago: A.T. Andreas, 1884-86.

Angle, Paul M., ed. *Prairie State: Impressions of Illinois, 1673-1967, by Travelers and Other Observers.* Chicago: University of Chicago Press, 1968.

Anderson, R.C. "Prairies in the Prairie State." *Transactions of the Illinois Academy of Science*, 63 (1970), 214-221.

Appleton, John B. *The Iron and Steel Industry of the Calumet District.* University of Illinois Studies in the Social Sciences, Vol. 13, No. 2, Urbana: University of Illinois, 1925.

Atlas of Illinois Resources. Springfield: Illinois Division of Industrial Planning and Development, Department of Registration and Education, 1959.

Atwood, Wallace W., and Goldthwait, James. *Physical Geography of the Evanston-Waukegan Region.* Urbana: Illinois State Geological Survey, Bulletin No. 7, 1908.

Babcock, Michael, and Folk, Hugh. "Future Employment in Illinois Industries, 1975 and 1980." *Illinois Business Review*, 30 (March 1973), 6-8.

Bach, Ira J., ed. *Chicago's Famous Buildings*, 3rd ed. Chicago: University of Chicago Press, 1980.

_____. *A Guide to Chicago's Historic Suburbs.* Chicago: Swallow Press, 1981.

Bach, Ira J. and Wolfson, Susan. *Chicago on Foot—Walking Tours of Chicago's Architecture.* Chicago: Chicago Review Press, 1987.

Baker, William. "Land Claims as Indicators of Settlement in Southwestern Illinois, Circa 1809-13." *Bulletin of the Illinois Geographical Society*, XVI (June 1974), 29-42.

Barber, S.A. "A Classification of Landforms—Illinois." Master's Thesis, Western Illinois University, 1969.

Barrows, Harlan H. *Geography of the Middle Illinois Valley.* Bulletin No. 15. Urbana: Illinois State Geological Survey, 1910.

Beimfohr, John. *Industrial Potential of Southern Illinois*. Carbondale: Southern Illinois University Press, 1952.

Belting, Natalia. *Kaskaskia Under the French Regime*. Urbana: University of Illinois Press, 1948.

Bergstrom, Robert E., ed. *The Quaternary of Illinois*. Special Publication 14. Urbana: University of Illinois College of Agriculture, 1968.

Berry, Brian J.L., ed. *Chicago: Transformation of an Urban System*. Cambridge: Ballinger Publishing Company, 1976.

_____. *Commercial Structure and Commercial Blight*. Research Paper No. 85. Chicago: University of Chicago, Department of Geography, 1963.

Bier, V.A. *Landforms of Illinois* (map). Urbana: Illinois State Geological Survey, 1956.

Blanchard, W.O. "Agricultural Provinces of Illinois." *Journal of Geography*, 23 (1922), 6-13.

Bogart, Ernest Ludlow. "The Movement of Population in Illinois, 1870-1910." *Transactions of the Illinois State Historical Society*, XXIII (1917).

Bogart, Ernest L., and Mathews, John M. *The Centennial History of Illinois*, Vol. 5, *The Modern Commonwealth 1893-1918*. Springfield: Illinois Centennial Commission, 1920.

Bogart, Ernest L., and Thompson, Charles M. *The Centennial History of Illinois*, vol. 4. *The Industrial State, 1870-1893*. Springfield: The Illinois Centennial Commission, 1920.

Boggess, Arthur C. *The Settlement of Illinois, 1776-1830*. Chicago: Chicago Historical Society, 1908.

Bogue, Margaret B. *Patterns from the Sod: Land Use and Tenure in the Grand Prairie, 1850-1900*. Vol. XXIV of *Collections of the Illinois State Historical Society*. (Land Series, Vol. 1.) Springfield: *Illinois State Historical Society, 1959*.

Bogue, Allan G. "Farming in the Prairie Peninsula, 1830-1890." *Journal of Economic History*, XXIII (March 1963), 3-29.

Bogue, Allan G. *From Prairie to Corn Belt: Farming on the Illinois and Iowa Prairies in the Nineteenth Century*. Chicago: University of Chicago Press, 1963.

Bond, John. *The East St. Louis, Illinois Waterfront: Historical Background*. Washington: U.S. Department of the Interior, Division of History, 1969.

Borchert, John. "Climate of the North American Grassland." *Annals of the Association of American Geographers*, 40 (1950), 1-44.

Bowling, Michael, and Van Es, J.C. "Changes Over a Decade in Illinois Agriculture." *Illinois Research*, 19 (Spring 1977), 14-15.

Bradbury, J.C.; Finger, G.C.; and Major, R.L. *Fluorspar in Illinois*. Circular 420. Urbana: Illinois State Geological Survey, 1968.

Breese, Gerald W. *The Daytime Population of the Central Business District of Chicago*. Chicago: University of Chicago Press, 1949.

Bretz, J. Harlen. *Geology of the Chicago Region*. Urbana: Illinois State Geological Survey, Bulletin No. 65, Part I, Geology of the Chicago Region, 1939; Part II, the Pleistocene, 1955.

Bridges, Roger, and Davis, Rodney. *Illinois: Its History and Legacy*. St. Louis: River City Publishers, 1984.

Brunn, Stanley D. "The Origin and Movement of Fresh Vegetables to Chicago." *Bulletin of the Illinois Geographical Society*, VII (June 1965), 25-30.

Brush, Daniel H. *Growing Up with Southern Illinois*. Chicago: Donnelley & Sons Co., 1944.

Buck, Solon J. *Illinois in 1818*. Springfield: The Illinois Centennial Commission, 1917.

Buder, Stanley. *Pullman: An Experiment in Industrial Order and Community Planning 1880-1930*. New York: Oxford University Press, 1967.

Buisseret, David. *Historic Illinois from the Air*. Chicago: University of Chicago Press, 1990.

Burnham, Daniel H., and Bennett, Edward H. *Plan of Chicago*. Chicago: The Commercial Club, 1909.

Burnham, Daniel H., and Kingery, Robert. *Planning the Region of Chicago*. Chicago: Chicago Regional Planning Association, 1956.

Burton, William L. *The Trembling Land: Illinois in the Age of Exploration.* W.I.U. Studies in Illinois History No. 1. Macomb: Western Illinois University, 1966.

Caldwell, Norman. *The French in the Mississippi Valley.* Urbana: University of Illinois Press, 1941.

Campbell, Edna Fay; Smith, Fanny R.; and Jones, Clarence F. *Our City—Chicago.* New York: Charles Scribner's Sons, 1930.

Carliss, Carlton J. *Trails to Rails: A Story of Transportation Progress in Illinois.* Chicago: Illinois Central Railroad, 1934.

Carlson, Theodore L. *The Illinois Military Tract: A Study in Land Occupation, Utilization and Tenure.* Vol. XXXII of *Illinois Studies in the Social Sciences.* Urbana: University of Illinois Press, 1951.

Caspall, Fred. "Parallel Drainage in West-Central Illinois." Master's Thesis, Western Illinois University, 1965.

Changnon, Stanley A., Jr. *Climatology of Severe Winter Storms in Illinois.* Bulletin 53. Urbana: Illinois State Water Survey, 1969.

Charvet, Jean-Paul. "The Recent Revolution of Farmland and Values in Illinois and in France." *Bulletin of the Illinois Geographical Society*, Vol. XXIV, No. 2, (1982), pp. 3-14.

Chicago Area Transportation Study, *Final Report*, 3 Vols. Chicago, 1959, 1960, and 1962.

Chicago Fact Book Consortium, eds. *Local Community Fact Book: Chicago Metropolitan Area.* Chicago: Chicago Review Press, 1984.

Chicago Land Use Survey. Vol. I: *Residential Chicago*, Vol. II: *Land Use in Chicago.* Chicago: Chicago Plan Commission, 1942, 1943.

Chicago Plan Commission. *Forty-four Cities in the City of Chicago.* Chicago: 1942.

_____. *Master Plan of Residential Land Use of Chicago.* Chicago, 1943.

City of Chicago Department of Development and Planning. *Chicago 21: A Plan for the Central Area Communities.* Chicago: 1973.

_____. *The Comprehensive Plan of Chicago.* Chicago: 1966.

_____. *Historic City: The Settlement of Chicago.* Chicago: 1976.

City of Chicago, Department of Public Works. *Chicago Public Works: A History.* Chicago: Rand McNally and Company, 1973.

Clayton, John (comp.). *The Illinois Fact Book and Historical Almanac: 1673-1958.* Carbondale and Edwardsville: Southern Illinois University Press, 1970.

Colby, Charles C. *Pilot Study of Southern Illinois.* Carbondale: Southern Illinois University Press, 1956.

Condit, Carl W. *Chicago 1910-29: Building, Planning, and Urban Technology.* Chicago: University of Chicago Press, 1973.

_____. *Chicago 1930-70: Building, Planning, and Urban Technology.* Chicago: University of Chicago Press, 1974.

_____. *The Chicago School of Architecture: A History of Commercial and Public Building in the Chicago Area, 1875-1925.* Chicago: University of Chicago Press, 1964.

Conoyer, John W. *Geography Through Maps: The Saint Louis Gateway.* Special Publication No. 14. Normal, Ill.: National Council for Geographic Education, 1967.

Conzen, Michael P., and Morales, Melissa J. *Settling the Upper Illinois Valley.* Chicago: Committee on Geographical Studies, University of Chicago, 1989.

Cowles, Henry C. *The Plant Societies of Chicago and Vicinity.* Chicago: Geographic Society of Chicago, Bulletin No. 2, 1901.

Cox, Henry J., and Armington, John H. *The Weather and Climate of Chicago.* Chicago: Geographic Society of Chicago, Bulletin No. 4, 1914.

Cramer, Robert E. *Manufacturing Structure of the Cicero District, Metropolitan Chicago.* Research Paper No. 27. Chicago: University of Chicago Department of Geography, 1952.

Cromie, Robert. *The Great Chicago Fire.* New York: McGraw-Hill, 1958.

Cronon, William. *Nature's Metropolis: Chicago and the Great West.* New York: W.W. Norton & Company, 1991.

Cummings, Ronald G. *Industrial Growth Trends in Illinois*. Springfield: Illinois Department of Business and Economic Development, 1967.

Cutler, Irving. *Chicago: Metropolis of the Mid-Continent*, 3rd ed. Dubuque: Geographic Society of Chicago and Kendall/Hunt Publishing Company, 1982.

_____. *The Chicago Metropolitan Area: Selected Geographic Readings*. New York: Simon and Schuster, 1970.

_____. *The Chicago-Milwaukee Corridor: A Geographic Study of Intermetropolitan Coalescence*. Northwestern University Studies in Geography No. 9. Evanston: Northwestern University Department of Geography, 1965.

Cutshall, Alden. "Illinois in Its Sesquicentennial Year." *Bulletin of the Illinois Geographical Society*, XI (June, 1969), 4-9.

Davis, James L. *The Elevated System and the Growth of Northern Chicago*. Northwestern University Studies in Geography No. 10. Evanston: Northwestern University Department of Geography, 1965.

Dedmon, Emmett. *Fabulous Chicago*, 2nd ed. New York: Atheneum, 1981.

De Meirleir, Marcel J. *Manufactural Occupance in the West Central Area of Chicago*. Research Paper No. 11. Chicago: University of Chicago Department of Geography, 1950.

De Vise, Pierre. *Chicago's People. Jobs and Homes—the Human Geography of the City and Metro Area*. 2 Vols. Chicago: De Paul University, 1964.

_____. *Chicago's Widening Color Gap*. Chicago: Inter-university Social Research Committee, 1967.

Dovring, Folke. "The Farmland Boom in Illinois." *Illinois Agricultural Economics*, 17 (July 1977), 34-38.

Draine, Edwin H. *Import Traffic of Chicago and Its Hinterland*. Research Paper No. 81. Chicago: University of Chicago Department of Geography, 1963.

Drake, St. Clair, and Cayton, Horace R. *Black Metropolis*. 2 Vols. New York: Harcourt, Brace & World, Inc., 1970.

Duddy, Edward A. *Agriculture in the Chicago Region*. Chicago: University of Chicago Press, 1929.

Duis, Perry. *Chicago: Creating New Traditions*. Chicago: Chicago Historical Society, 1976.

Duncan, Otis Dudley, and Duncan, Beverly. *The Negro Population of Chicago: A Study of Residential Succession*. Chicago: University of Chicago Press, 1957.

Federal Writers Project. *Illinois, A Descriptive and Historical Guide*, 2nd ed. Chicago: A.A. McClurg & Co., 1947.

Fehrenbacher, J.B.; Waler, G.O.; and Wascher, H.L. *Soils of Illinois*. Agricultural Experiment Station Bulletin 725. Urbana: University of Illinois College of Agriculture, 1967.

Fellman, Jerome D. *Truck Transportation Patterns of Chicago*. Research Paper No. 12. Chicago: University of Chicago Department of Geography, 1950.

Fellman, Jerome D., and Roepke, Howard G. "The Changing Industrial Structure of Illinois." *Current Economic Comment*, 18 (November 1956).

Fiedler, D.E., ed. *The Chicagoland Atlas*. Arlington Heights: Creative Sales Corp., 1980.

Fleming, Lois F., and Price, Dalias A. "The Old Order Amish Community of Arthur, Illinois, Part I." *Bulletin of the Illinois Geographical Society*, VI (June 1964), 4-13.

_____. "The Old Order Amish Community of Arthur, Illinois, Part II." *Bulletin of the Illinois Geographical Society*, VII (June 1965), 4-24.

Ford, Thomas. *History of Illinois*. Chicago: S.C. Griggs and Co., 1854.

Fowler, Melvin. *The Cahokia Atlas, A Historical Altas of Cahokia Archaeology*. Studies in Illinois Archaeology, No. 6, Illinois Historic Preservation Agency, Springfield, Ill., 1989.

Frazier, E. Franklin. *The Negro Family in Chicago*. Chicago: University of Chicago Press, 1932.

Fryxell, F.M. *The Physiography of the Region of Chicago*. Chicago: University of Chicago Press, 1927.

Garland, J.H. *The North American Midwest, A Regional Geography*. New York: John Wiley & Sons, 1955.

Gates, Paul W. "Disposal of the Public Domain in Illinois." *Journal of Economic and Business History*, III (February 1931), 216-40.

Gates, Paul W. *The Illinois Central Railroad and Its Colonization Work*. Cambridge: Harvard University Press, 1934.

Gerhard, Fred. *Illinois As It Is*. Chicago: Keen and Lee, 1857.

Goode, J. Paul. *The Geographic Background of Chicago*. Chicago: University of Chicago Press, 1926.

Guide to the Geologic Map of Illinois. Educational Series 7. Urbana: Illinois State Geological Survey, 1961.

Hansen, Harry. *The Chicago*. Rivers of America Series. New York: Farrar & Rinehart, Inc., 1942.

Harper, Robert A. *Metro East: Heavy Industry in the St. Louis Metropolitan Area*. Carbondale, Ill.: Southern Illinois University, 1958.

Harper, Robert A. *Recreational Occupance of the Moraine Lake Region of Northeastern Illinois and Southeastern Wisconsin*. Research Paper No. 14. Chicago: University of Chicago Department of Geography, 1950.

Hart, John Fraser. "The Middle West." *Annals of the Association of American Geographers*, 62 (June 1972), 258-82.

Hartnett, Harry P. *A Locational Analysis of Those Manufacturing Firms That Have Located and Relocated within Chicago, 1955-1968*. Chicago: Continental Illinois National Bank, Area Development Division, January 1972.

Hayner, Don and McNamee, Tom. *Metro Chicago Almanac*. Chicago: Bonus Books and Chicago Sun-Times, 1991.

Heise, Kenan. *The Chicagoization of America 1893-1917*. Chicago: Chicago Historical Bookworks, 1990.

Helvig, Magne. *Chicago's External Truck Movements*. Research Paper No. 90. Chicago: University of Chicago Department of Geography, 1964.

Herbst, J.H. "Twenty-Year Trends in Crop Yields: Corn, Soybeans, Wheat, and Oats." *Illinois Research*, 17 (Summer 1975), 8-9.

Hicken, Victor. *The Settlement of Illinois: 1700-1850*. W.I.U. Studies in Illinois History No. 2. Macomb: Western Illinois University, 1966.

Hillman, Arthur, and Casey, Robert J. *Tomorrow's Chicago*. Chicago: University of Chicago Press, 1953.

Hoag, Leverett P. "Location Determinants for Cash-Grain Farming in the Corn Belt." *The Professional Geographer*, 14 (May 1962), 1-7.

Holli, Melvin G., and Jones, Peter d'A. *Ethnic Chicago*. Grand Rapids: William B. Eerdmans Publishing Company, 1984.

Hopkins, M.E. "Coal in Illinois—Key to a Brighter Tomorrow?" *Illinois Business Review*, 30 (December 1973), 6-8.

Horberg, Leland. *Bedrock Topography of Illinois*. Bulletin No. 73. Urbana: Illinois State Geological Survey, 1950.

Horsley, A. Doyne. *Illinois: A Geography*. Boulder and London: Westview Press, 1986.

_____. "A Comparison of Past and Present Alternatives to Settlement in the 100 year Flood Hazard Zones of Illinois." *Bulletin of the Illinois Geographical Society*, Vol. XXIII, No. 1, (1981), pp. 20-26.

Howard, Robert P. *Illinois: A History of the Prairie State*. Grand Rapids, Mich.: William B. Eerdmans Publishing Co., 1972.

Hoyt, Homer. *One Hundred Years of Land Values in Chicago, 1830-1933*. Chicago: University of Chicago Press, 1933.

Hubbart, Henry C. *The Older Middle West: 1840-1880*. New York: Russell & Russell, 1936.

Illinois: *A Descriptive and Historical Guide*. Federal Writers Project. Chicago: A.C. McClurg and Co., 1939.

Illinois Atlas and Gazeteer. Freeport, Maine: DeLorme Mapping, 1991.

The Illinois Department of Commerce and Community Affairs Corridors of Opportunity Program and the Southwestern Illinois Corridor Council. *Southwestern Illinois 1991/1922 — Market Review and Investment Update*.

Illinois Sesquicentennial Commission. *Illinois Guide and Gazetteer.* Chicago: Rand McNally and Company, 1969.

Jakle, John A. *The American Small Town: Twentieth Century Place Images.* Hamden, Conn.: Archon Press, 1982.

Jensen, George Peter. *Historic Chicago Sites.* Chicago: Creative Enterprises, 1953.

Johnson, Charles B. *Growth of Cook County*, Vol. 1. Chicago: Board of Commissioners of Cook County, Illinois, 1960.

Karlen, Harvey. *The Governments of Chicago.* Chicago: Courier Publishing Company, 1958.

Kenyon, James B. *The Industrialization of the Skokie Area.* Research Paper No. 33. Chicago: University of Chicago Department of Geography, 1954.

Kiang, Ying Cheng. *Chicago.* Chicago: Adams Press, 1968.

Kircher, Harry B. "The Sequent Occupance of Metro East." *Bulletin of the Illinois Geographic Society*, XVI (June 1974), 3-11.

Kirkland, Joseph. *The Story of Chicago.* 3 Vols. Chicago: Dibble Publishing Company, 1892-1894.

Klove, Robert C. *The Park Ridge—Barrington Area: A Study of Residential Land Patterns and Problems in Suburban Chicago.* Chicago: University of Chicago Department of Geography, 1942.

Koepke, Robert. "Manufacturing in Metro East." *Bulletin of the Illinois Geographical Society,* XVI (June 1974), 43-55.

Koepke, Robert L., and Creed, Sara. *Directory of Manufacturers, Illinois Metro East Area.* Metro East Resources Report No. 3. Edwardsville, Ill.: Illinois Metro East Corporation, April, 1969.

Kogan, Herman, and Kogan, Rick. *Yesterday's Chicago.* Miami: E.A. Seemann Publishing, Inc., 1976.

Kogan, Herman, and Wendt, Lloyd. *Chicago: A Pictorial History.* New York: Bonanza Books, 1958.

Kollmorgen, Walter M. "Farms and Farming in the American Midwest." Chapter 7 in Saul B. Cohen (ed.) *Problems and Trends in American Geography.* New York: Basic Books, Inc., 1967.

Krausse, Gerald H. "Historic Galena: A Study of Urban Change and Development in a Midwestern Mining Town." *Bulletin of the Illinois Geographical Society*, XIII (December 1971), 3-19.

Larson, Albert J. "Prairie Temples of Justice: County Courthouses in Illinois." *Bulletin of the Illinois Geographical Society*, Vol. XXIX, No. 1, (1987), pp. 3-15.

Larson, Albert J., and Soot, Siim. "Centers of Population and the Historical Geography of Illinois." *Bulletin of the Illinois Geographical Society*, XV (December 1973), 34-52.

Lee, Judson Fiske. "Transportation: A Factor in the Development of Northern Illinois Previous to 1860." *Journal of the Illinois State Historical Society* (April 1917), 17-85.

Leighton, M.M., and Brophy, T.A. "Illinoisan Glaciation in Illinois." *Journal of Geology*, 69 (1961), 1-31.

Leighton, M.M.; Ekblaw, George E.; and Horberg, Leland. *Physiographic Divisions of Illinois.* Report of Investigations No. 129, Urbana: Illinois State Geological Survey, 1948.

Lewis, Lloyd, and Smith, Henry Justin. *Chicago the History of Its Reputation.* New York: Harcourt, Brace and Company, 1929.

Lind, Alan. *Chicago Surface Lines: An Illustrated History.* Park Forest: Transport History Press, 1974.

Lloyd, Peter, and Dickens, P. *Location in Space: A Theoretical Approach to Economic Geography.* New York: Harper & Row, 1972.

Lohmann, Karl. *Cities and Towns of Illinois—A Handbook of Community Facts.* Urbana: University of Illinois Press, 1951.

Lowe, David. *Lost Chicago.* Boston: Houghton Mifflin Company, 1975.

MacClintock, P. *Physiographic Divisions of the Area Covered by the Illinoisan Driftsheet in Southern Illinois.* Report of Investigations No. 19. Urbana: Illinois State Geological Survey, 1939.

Malhotra, Ramesh, and Halloran, Shirley. *Illinois Mineral Industry in 1974*. Illinois Mineral Note #66. Urbana: Illinois State Geological Survey, February 1977.

Mattingly, Paul F. "Population Trends in the Hamlets and Villages of Illinois: 1940-1960." *The Professional Geographer*, XV (November 1963), 17-21.

_____. "Part-Time Farming in Illinois." *Bulletin of the Illinois Geographical Society*, Vol. XXVII, No. 1 (1985), pp. 15-27.

_____. "Metropolitan Fragmentation and Governmental Competition: Some Evidence from Illinois." *Bulletin of the Illinois Geographical Society*, Vol. XXIII, No. 1, (1981), pp. 11-19.

Mausel, Paul. "General Subgroup Soil Regions of Illinois." *Bulletin of the Illinois Geographical Society*, 11 (June 1969), 70-78.

_____. "Soil Quality in Illinois—An Example of a Soils Geography Resource Analysis." *The Professional Geographer*, 23 (April 1970), 127-136.

Mayer, Harold M. *Chicago: City of Decisions*. Papers on Chicago, No. 1, Chicago: Geographic Society of Chicago, 1955.

_____. *The Port of Chicago and the St. Lawrence Seaway*. Research Paper No. 49. Chicago: University of Chicago Press, 1957.

_____. *The Railway Pattern of Metropolitan Chicago*. Chicago: University of Chicago Department of Geography, 1943.

Mayer, Harold M., and Wade, Richard C. *Chicago: Growth of a Metropolis*. Chicago: University of Chicago Press, 1969.

McClellan, Keith, "A History of Chicago's Industrial Development." *Mid-Chicago Economic Development Study*. Vol. 3, Prepared for the Mayor's Committee for Economic and Cultural Development by the Center for Urban Studies. Chicago: University of Chicago, February 1966.

McDonald, James A. *Population Projections: Economic Growth Prospects*. Springfield: Department of Business and Economic Development, 1967.

McManis, Douglas R. *The Initial Evaluation and Utilization of the Illinois Prairies, 1815-1840*. Research Paper No. 94. Chicago: University of Chicago Department of Geography, 1964.

Megee, Mary. "The American Bottoms: The Vacant Land and the Areal Image." *The Professional Geographer*, Vol. XIII (November 1961), 5-9.

Meyer, Douglas K. "Persistence and Change in Migrant Patterns in a Transitional Culture Region of the Prairie State." *Bulletin of the Illinois Geographical Society*, Vol. XXVI, No. 1, (1984), pp. 13-29.

Miller, John J. *Open Land in Metropolitan Chicago*. Chicago: Midwest Open Land Association, 1962.

Misselhorn, Roscoe. *Illinois Sketches*. Highland, Ill.; Swiss Village Book Store, 1985.

Mitchell, S. Augustus. *Illinois in 1837*. Philadelphia: Mitchell, 1837.

Moses, John. *Illinois, Historical and Statistical*. Vol. I. Chicago: Fergus Printing Co., 1889.

Nelli, Humbert S. *The Italians in Chicago, 1880-1930: A Study in Ethnic Mobility*. New York: Oxford University Press, 1970.

Northeastern Illinois Metropolitan Area Local Governmental Services Commission. *Governmental Problems in the Chicago Metropolitan Area*. Edited by Leverett S. Lyon. Chicago: University of Chicago Press, 1957.

Northeastern Illinois Planning Commission. *The Comprehensive Plan for the Development of the Northeastern Illinois Counties Area*. Chicago, 1968.

_____. *Open Space in Northeastern Illinois, Technical Report No. 2*. Chicago, 1962.

_____. *A Social Geography of Metropolitan Chicago*. Chicago, 1960.

_____. *Suburban Factbook*. Chicago, 1971.

Olson, Ernst W. *History of the Swedes of Illinois*. 2 Vols. Chicago: The Engberg-Holmberg Publishing Company, 1908.

Pacyga, Dominic A., and Sherrett, Ellen. *Chicago: City of Neighborhoods*. Chicago: Loyola University Press, 1986.

Page, John L. *Climate of Illinois*. Agricultural Experiment Station Bulletin 532. Urbana: University of Illinois, 1949.

Park, Siyoung. "Korean Residential Conentrations in Chicago." *Bulletin of the Illinois Geographical Society*, Vol. XXX, No. 2, (1988), pp. 12-22.

_____. "Rural Development and Regional Planning in Illinois." *Bulletin of the Illinois Geographical Society*, Vol. XXVI, No. 2, (1984), pp. 8-19.

Parrish, Randall. *Historical Illinois: The Romance of the Earlier Days*. Chicago: A.C. McClurg & Co., 1905.

Pease, Theodore C. *The Frontier State, 1818-1848*. Chicago: A.C. McClurg and Co., 1922.

Pease, Theodore Calvin, and Pease, Marguerite Jenison. *The Story of Illinois*. 3rd ed. Chicago: University of Chicago Press, 1965.

Peattie, Donald Culross. *A Prairie Grove*. New York: The Literary Guide of America, 1938.

Pierce, Bessie Louise. *A History of Chicago*. 3 Vols. New York: A.A. Knopf, 1937-57.

_____. *As Others See Chicago—Impressions of Visitors*, 1673-1933. Chicago: University of Chicago Press, 1933.

Piskin, Kimal, and Bergstrom, Robert E. *Glacial Drift in Illinois: Thickness and Character*. Circular 416. Urbana: Illinois State Geological Survey, 1967.

Platt, Rutherford H. *Open Land in Urban Illinois*. DeKalb: Northern Illinois University Press, 1971.

Poggi, E. Muriel. *The Prairie Province of Illinois: A Study in Human Adjustment to Natural Environment*. Vol. XIX of Illinois Studies in the Social Sciences. Urbana: University of Illinois Press, 1934.

Poles of Chicago, 1837-1937. Chicago: Polish Pageant, Inc., 1937.

Poole, Ernest. *Giants Gone—Men Who Made Chicago*. New York: McGraw-Hill Book Company, Inc., 1943.

Pooley, William V. *The Settlement of Illinois from 1830 to 1850*. Bulletin of the University of Wisconsin, No., 220 (History Series, 1). Madison: University of Wisconsin, 1908.

Price, Dalias A. "Southern Illinois Agriculture." *Bulletin of the Illinois Geographical Society,* I (April 1955), 39-41.

Price, H. Wayne. "The Barns of Illinois." *Bulletin of the Illinois Geographical Society*, Vol. XXX, No. 1, (1988), pp. 3-22.

Putnam, James W. *Illinois and Michigan Canal: A Study in Economic History*. Chicago: Chicago Historical Society, 1917.

Quaife, Milo M. *Checagou: From Indian Wigwam to Modern City, 1673-1835*. Chicago: University of Chicago Press, 1933.

_____. *Chicago and the Old Northwest, 1673-1835*. Chicago: University of Chicago Press, 1913.

_____. *Chicago's Highways Old and New: From Indian Trail to Motor Road*. Chicago: D.F. Keller & Co., 1923.

Randall, Frank A. *History of the Development of Building Construction in Chicago*. Urbana: University of Illinois Press, 1949.

Reinemann, Martin W. "The Localization and Relocation of Manufacturing within the Chicago Urban Area." (Ph.D. dissertation, Northwestern University, 1955).

Reynolds, John. *The Pioneer History of Illinois*. Chicago: Fergus Printing Co., 1887.

Ridgley, Douglas. *The Geography of Illinois*. Chicago: University of Chicago Press, 1921.

Risser, Hubert E., and Major, Robert L. *History of Illinois Mineral Industries*. Educational Series No. 10. Urbana: Illinois State Geological Survey, 1968.

Ross, R.C., and Case, H.C.M. *Types of Farming in Illinois*. Agricultural Experiment Station Bulletin 601. Urbana: University of Illinois College of Agriculture, 1956.

Rossi, Peter H. and Dentler, Robert A. *The Politics of Urban Renewal: The Chicago Findings*. New York: The Free Press of Glencoe, 1961.

Salisbury, Rollin D., and Alden, William C. *The Geography of Chicago and Its Environs*. Chicago: Geographic Society of Chicago, Bulletin No. 1, 1899.

Sauer, Carl O. *Geography of the Upper Illinois Valley and History of Development*. Bulletin No. 27. Urbana: Illinois State Geological Survey, 1916.

Scott, Roy V. *The Agrarian Movement in Illinois, 1880-1895*. Urbana: University of Illinois Press, 1962.

Sculle, Keith A. (ed). *Geography in Illinois History*. Vol. I. Illinois History Teacher. Springfield: Illinois Historic Preservation Agency, 1994.

The Sentinel's History of Chicago Jewry, 1911-1986. Chicago: Sentinel Publishing Company, 1986.

Sinclair, Upton. *The Jungle*. New York: Doubleday Page and Co., 1906.

Smith, Guy D. *Illinois Loess—Variations in Its Properties and Distribution: A Pedologic Interpretation*. Agricultural Experiment Station Bulletin 490. Urbana: University of Illinois, 1942.

Smith, Henry Justin. *Chicago's Great Century 1833-1933*. Chicago: Consolidated Publishers, 1933.

Sofranko, Andrew J. "Illinois Agriculture: The Changing Scene." *Illinois Research*, 15 (Fall 1973), 3-5.

Solomon, Ezra, and Bilbija, Zarko G. *Metropolitan Chicago: An Economic Analysis*. Glencoe: The Free Press, 1959.

Solzman, David M. *Waterway Industrial Sites, A Chicago Case Study*. Research Paper No. 107. Chicago: University of Chicago Department of Geography, 1966.

Stover, John. *History of the Illinois Central Railroad*. New York: Macmillan, 1975.

Sublett, Michael D. *Paper Counties: The Illinois Experience, 1825-1867*. New York: Peter Lang Publishing, 1990.

Sutton, Robert P., ed. *The Prairie State: A Documentary History of Illinois* (2 vols.). Grand Rapids, MI: Eerdmans, 1976.

Sutton, Robert P. *The Heartland: Pages from Illinois History*. Lake Forest, IL: Deerpath, 1982.

Taaffe, Edward J. *The Air Passenger Hinterland of Chicago*. Research Paper No. 24. Chicago: University of Chicago Department of Geography, 1952.

Townsend, Andrew J. "The Germans of Chicago." Ph.D. dissertation, University of Chicago, 1927.

Transeau, Edgar N. "The Prairie Peninsula." *Ecology*, XVI (July 1935), 423-437.

University of Chicago Center for Urban Studies. *Mid-Chicago Economic Development Study*. 3 Vols. Chicago Mayor's Committee for Economic and Cultural Development, 1966.

U.S. Bureau of the Census. *1990 Census of Population and Housing: Social, Economic and Housing Characteristics, Illinois*.

U.S. Bureau of the Census. *1990 Census of Population: General Population Characteristics, Illinois*.

U.S. Bureau of the Census. *1987 Census of Manufactures, Illinois*.

U.S. Bureau of the Census. *1987 Census of Retail Trade, Illinois*.

U.S. Bureau of the Census. *1987 Census of Agriculture, Illinois.Abbott, Edith. The Tenements of Chicago, 1908-1935*. Chicago: University of Chicago Press, 1936.

Van Arsdall, Roy N., and Elder, William A. *Economies of Size of Illinois Cash Grain and Hog Farms*. Agricultural Experiment Station Bulletin 733. Urbana: University of Illinois College of Agriculture, 1969.

Van Den Berg, Jacob, and Lowry, T.F. *Petroleum Industry in Illinois in 1975*. Illinois Petroleum No. 110. Urbana: Illinois State Geological Survey, 1976.

Vogel, Virgil J. *Indian Place Names in Illinois*. Springfield: Illinois State Historical Library, 1963.

Wade, Louise Carroll. *Chicago's Pride: The Stockyards, Packingtown, and Environs in the Nineteenth Century*. Urbana: University of Illinois Press, 1987.

Wakely, Ray E. *Growth and Decline of Towns and Cities in Southern Illinois*. Area Service Bulletin 2. Carbondale: Southern Illinois University Press, 1962.

_____. *Population Changes and Prospects in Southern Illinois*. Area Service Bulletin 1. Carbondale: Southern Illinois University Press, 1962.

Waller, Robert A. "The Illinois Waterway from Conception to Completion." *Journal of the Illinois State Historical Society* 65 (1972): 125-41.

Walters, William D., and Mansberger, Floyd. "Early Mill Location in Northern Illinois." *Bulletin of the Illinois Geographical Society* XXV, No. 2 (1983): 3-11.

Walton, Clyde F., ed. *An Illinois Reader*. DeKalb: Northern Illinois University Press, 1970.

Wanless, H.R. *Geology and Mineral Resources of the Beardstown, Glasford, Havana, and Vermont Quadrangles*. Bulletin 82, Illinois State Geological Survey, 1957.

Watters, Mary. *Illinois in the Second World War*. 2 vols. Springfield: Illinois State Historical Library, 1951-1952.

Weaver, John C. "Changing Patterns of Cropland Use in the Middle West." *Geographical Review*, 44 (1954), 560-72.

_____. "Crop Combination Regions in the Middle West." *Geographical Review,* 44 (1954), 175-200.

Weaver, John C. et al. "Livestock Units and Combination Regions in the Middle West." *Economic Geography*, 32 (1956), 237-59.

Weaver, John E. *The North American Prairie*. Lincoln: Johnsen Publishing Co., 1954.

Werner, Carol A. "Agriculture and Urban Change: The Case of Metro East." *Bulletin of the Illinois Geographical Society*, XV (June 1973), 29-40.

Wheeler, David L. "The Illinois Country, 1673-1696." *Bulletin of the Illinois Geographical Society*, VII (June 1965), 42-47.

Willman, H.B. *Summary of the Geology of the Chicago Area*. Circular 460, Illinois State Geological Survey, 1971.

Willman, H.B., and Frye, J.C. *Pleistocene Stratigraphy of Illinois*. Bulletin No. 94. Urbana: Illinois State Geological Survey, 1970.

Wilson, John W., and Changnon, Stanley A., Jr. *Illinois Tornadoes*. Circular 103. Urbana: Illinois State Water Survey, 1971.

Wirth, Louis. *The Ghetto*. Chicago: University of Chicago Press, 1928.

Woods, William I. and Holley, George R. "Upland Mississippian Settlement in the American Bottom Region." *Cahokia and the Hinterlands, Middle Mississippian Cultures of the Midwest*, ed. by Emerson, T.E. and Jewis, B.A., University of Ill. Press, Urbana and Chicago, 1991.

Yarbrough, Ronald E. "The Physiography of Metro East." *Bulletin of the Illinois Geographical Society*, XVI (June 1974), 12-28.

Zorbaugh, Harvey W. *The Gold Coast and the Slum*. Chicago: University of Chicago Press, 1929.

INDEX

Suburban patterns, 279-284
Sullivan, Louis, 261
Summit, 226, 245
Swift, Gustavus, 128, 230

Tazewell County, 180
Temperature ranges, 65
Temperatures, 63-69
Tenant farmers, 184
Tenant-operated farms, 185
Terre Haute, 121
Texas, 165
Thunderstorms, 71, 75, 78
Tiling, 49
Tinley Park, 283
Tornadoes, 75-79
Township and Range, 308-309
Tractors, 170
Trails, 126, 314
Transportation, 119-127, 214-216, 264-269, 312-316
Treaty of Paris, 111
Tribune Tower, 227
Tri-Cities, 203, 205, 212, 213
Tri-State Tornado, 78
Troy, 205
Truck crops, 179
Truck farming, 179-180
Turner's Junction, 123

Unemploment, 147
University of Chicago, 15
University of Illinois, 159
University Park, 276, 277, 283
Upland south, 6, 136
Urban centers, 13-15
Utica, 122

Valley trains, 43
Valparaiso Moraine, 222, 224
Vandalia, 105, 314
Vegetation, 83-89
Venice, 203, 205
Vermillion County, 113

Vienna, 113
Villa Ridge, 179
Vincennes, 121

Wabash River, 121, 184, 306
Ward, Aaron Montgomery, 128, 230
Warping, 26
Wars
    Black Hawk War, 7, 227
    Civil War, 10, 12, 13, 146, 231
    Korean conflict, 13
    Revolutionary War, 112
    Viet Nam conflict, 13
    War of 1812, 6, 12
    World War I, 15, 234
    World War II, 15, 147
Warsaw, 112
Washington County, 176
Water commerce, 119-123
Water Tower Place, 262, 263
Waukegan, 226, 253, 276, 277, 278, 279, 280, 281
Waves (atmospheric), 72, 73
Weathering, 89
Westerlies (winds), 71
Western Illinois Correctional Center, 159
Western Illinois University, 159
Western Springs, 277, 281
Wheat, 176
Wheaton, 123, 133
Wheaton Morainal Country, 60, 301
White County, 180
Whiting, 253, 277, 284
Will County, 136, 283
Wilmette, 273
Winnetka, 224, 273, 277, 280
Wisconsinan Till Plain, 40, 302-303
Wood River, 203, 205, 212
World's Columbian Exposition
    See Columbian Exposition
Wright, Frank Lloyd, 128, 281

Zion, 279, 281